While it seems manifest in our experience that time flows – from the past, to the present moment and into the future, there are a number of philosophical and physical objections to this. In the quest to make sense of this conundrum philosophers and physicists confront fascinating and irresistible questions such as: Can effects precede causes? Can one travel in time? Can the expansion of the universe or the process of measurement in quantum mechanics provide a direction of time?

In this book eleven eminent scholars, standing at the boundary between physics and philosophy, attempt to answer these questions in an entertaining yet rigorous way. For example, William Unruh's chapter is one of the first non-technical essays by this important cosmologist. Philip Stamp and Anthony Leggett discuss macroscopic quantum phenomena, a subject which has not been discussed much outside the specialist literature. John Earman's chapter on time travel is likely to become one of the landmarks in the literature.

The book will be enjoyed by anyone of a speculative turn of mind fascinated by the puzzle of time.

Time's Arrows Today

Time's Arrows Today

Recent physical and philosophical work on the direction of time

Edited by *Steven F. Savitt*
University of British Columbia

Published by the Press Syndicate of the University of Cambridge
The Pitt Building, Trumpington Street, Cambridge CB2 1RP
40 West 20th Street, New York, NY 10011-4211, USA
10 Stamford Road, Oakleigh, Melbourne 3166, Australia

© Cambridge University Press 1995

First published 1995

Printed in Great Britain at the University Press, Cambridge

A catalogue record for this book is available from the British Library

Library of Congress cataloguing in publication data available

ISBN 0 521 46111 1   hardback

Contents

Preface		ix
Notes on the contributors		x
Introduction		1
PART 1	**Cosmology and time's arrow**	
	1 Time, gravity, and quantum mechanics WILLIAM UNRUH, *Physics: UBC*	23
	2 Cosmology, time's arrow, and that old double standard HUW PRICE, *Philosophy: Sydney*	66
PART 2	**Quantum theory and time's arrow**	
	3 Time's arrow and the quantum measurement problem ANTHONY LEGGETT, *Physics: Urbana-Champaign*	97
	4 Time, decoherence, and 'reversible' measurements PHILIP STAMP, *Physics: UBC*	107
	5 Time flow, non-locality, and measurement in quantum mechanics STORRS McCALL, *Philosophy: McGill*	155
	6 Stochastically branching spacetime topology ROY DOUGLAS, *Mathematics: UBC*	173
PART 3	**Thermodynamics and time's arrow**	
	7 The elusive object of desire: in pursuit of the kinetic equations and the Second Law LAWRENCE SKLAR, *Philosophy: Michigan*	191
	8 Time in experience and in theoretical description of the world LAWRENCE SKLAR, *Philosophy: Michigan*	217

Contents

    9 When and why does entropy increase?   230
        MARTIN BARRETT & ELLIOTT SOBER, *Philosophy: Wisconsin*

PART 4 **Time travel and time's arrow**
    10 Closed causal chains   259
        PAUL HORWICH, *Philosophy: MIT*
    11 Recent work on time travel   268
        JOHN EARMAN, *History and Philosophy of Science: Pittsburgh*

References   311

Index   325

Preface

Most of the contributors to this volume are philosophers of science, but three are physicists (Leggett, Stamp, and Unruh) and one is a mathematician (Douglas). Their chapters are intended to be original contributions towards answers to, rather than comprehensive discussions of, some of the oddly exasperating and fascinating questions known collectively as the problem of the direction of time, but the chapters were written with an eye towards communicating their results to the scientifically literate non-specialist.

Most of the work in this volume was presented at the 'Time's Arrows Today' Conference held on the campus of the University of British Columbia in June, 1992. The exceptions are the papers by Douglas and Earman and Sklar's 'The elusive object of desire: in pursuit of the kinetic equations and the Second Law'. I am grateful to the Social Sciences and Humanities Research Council of Canada and to the Dean of the Faculty of Arts and the President of the University of British Columbia for their generous support of that conference.

In addition I wish to express my gratitude to Philip Stamp, whose advice and assistance have been invaluable, to Roy Douglas, who has long provided for me a model of rigorous thinking, and to William Unruh, without whose help my discussion of Penrose's thought experiment in the Introduction would have been much easier to write.

Steven F. Savitt
Vancouver, British Columbia

Notes on the contributors

Martin Barrett is a graduate student in philosophy at the University of Wisconsin, Madison. He has published several papers in philosophy and the design of databases for the natural sciences. Among his other interests is opera, where entropy generally rises until the Fat Lady sings.

Roy Douglas, a professor in the Department of Mathematics at the University of British Columbia, is a topologist with a life-long interest in physics. While studying both mathematics and physics as an undergraduate at Northwestern University in the late 1950s, he developed a serious desire to investigate those questions in physics which depended on a (topologically) global understanding. As a graduate student (1960–65) studying mathematics in Berkeley, he specialized in algebraic topology. From 1965 to 1985, Douglas pursued research aimed at the development of algebraic topology, but since 1985 his research has been increasingly oriented toward the application of algebraic topology to physics.

In joint work with physicist Alexander Rutherford, two interesting problems in quantum physics have been solved using algebraic topology. (1) In 1987, Douglas and Rutherford discovered an effective method of computing the Berry phase (or holonomy) associated with an arbitrary time-reversal invariant quantum mechanical system. (2) Applying similar methods to molecular spectroscopy, Douglas and Rutherford (in 1993) constructed a rigorous mathematical model of the Jahn–Teller effect. This model permits the accurate prediction of the (infra-red) spectrum of molecules.

John Earman received his Ph.D. in philosophy from Princeton University in 1968. He has taught at UCLA, The Rockefeller University, the University of Minnesota, and currently at the University of Pittsburgh, where he is Professor of History and Philosophy of Science. His research interests include the history and foundations of modern physics and confirmation theory. He is the author of *A Primer on Determinism, World Enough and Spacetime: Absolute vs. Relational Theories of Space and Time*, and *Bayes or Bust? A Critical Examination of Bayesian Confirmation Theory.*

He is editor of *Inference, Explanation, and Other Philosophical Frustrations*, and (with M. Janssen and J. Norton) of *The Attraction of Gravitation: New Studies in the History of General Relativity.* He is currently at work on a book on the philosophical implications of spacetime singularities in the general theory of relativity.

Paul Horwich is Professor of Philosophy at MIT. Born in England, he studied physics at Oxford before obtaining his doctorate in philosophy at Cornell. He is the author of *Probability and Evidence* (1982), *Asymmetries in Time* (1987), and *Truth* (1990).

Anthony J. Leggett did his first degree at Oxford in *Literae Humaniores* (philosophy and classical history) and followed this with a second undergraduate degree in physics and a doctorate in physics, under the direction of D. ter Haar. After postdoctoral work at Illinois, Kyoto, and Oxford he joined the faculty of the University of Sussex (UK) where he worked for 15 years before taking up his present position as John D. and Catherine T. Macarthur Professor and Professor of Physics at the University of Illinois at Urbana-Champaign. His main research interests are in theoretical condensed matter physics and the foundations of quantum mechanics.

Storrs McCall, born in Montreal in 1930, studied philosophy at McGill and at Oxford, where he obtained a B.Phil. and a D.Phil. He taught at McGill from 1955 to 1963, then over the next eleven years spent six at the University of Pittsburgh and five at Makerere University in Uganda, where he started courses of formal instruction in philosophy in 1965, and where he was later joined by André Gombay and Ian Hacking. In 1974–75 he worked on defining the concept of quality of life (QOL) in the Department of Environment in Ottawa, following which he returned to McGill. He is the editor of *Polish Logic 1920–39*, and is the author of *Aristotle's Modal Syllogisms* (1963) and *A Model of the Universe* (1994).

Huw Price teaches at the University of Sydney in Sydney, Australia. His interests include time asymmetry, truth and realism, and pragmatism in

Notes on the contributors

contemporary metaphysics. He is the author of *Facts and the Function of Truth* (Blackwells, 1988) and numerous articles in *Mind, The Journal of Philosophy, British Journal for the Philosophy of Science*, and other leading journals. He is currently working on a book on philosophical issues in the physics of time asymmetry.

Steven Savitt is Professor of Philosophy at the University of British Columbia. He specializes in the philosophy of space and time.

Lawrence Sklar is Professor of Philosophy at the University of Michigan, Ann Arbor and is the author of *Space, Time, and Spacetime* (1974), *Philosophy and Spacetime Physics* (1985), *The Philosophy of Physics* (1993), and *Physics and Chance: Philosophical Issues in the Foundations of Statistical Mechanics* (1993).

Elliott Sober is Hans Reichenbach Professor of Philosophy at the University of Wisconsin, Madison and the author of *The Nature of Selection, Reconstructing the Past, Parsimony, Evolution, and Inference*, and, most recently, *Philosophy of Biology*.

Philip Stamp is presently University Research Fellow and Assistant Professor at the University of British Columbia. He earned his D.Phil. at Sussex, England (1983); after various postdoctoral fellowships (Royal Society European Fellow in Grenoble, France, CSIC Personal Investigador in Madrid, and visiting scientist at the ITP in Santa Barbara) he moved to UBC in November, 1989. He is also an Associate of the Canadian Institute for Advanced Research.

His general research is in condensed matter theory. Some of this research is closely related to the subject of his chapter, such as his work on macroscopic quantum phenomena in magnets and more general work on quantum dissipation; he also has a long-standing interest in the foundations of quantum mechanics. At present his research activities are concentrated on high-temperature superconductivity.

William G. Unruh received his Ph.D. at Princeton University under John Wheeler, who helped to inspire a desire to understand the fundamental issues in physics and who originated the term 'black hole'. Unruh is now a professor of Physics at the University of British Columbia and is also the Director of the Cosmology Program of the Canadian Institute for Advanced Research.

His research interests span the range from quantum gravity, where the problem of time is one of his chief research worries, to fundamental issues in quantum mechanics. He discovered the close association between acceleration and temperature, which is related to Hawking's discovery of black hole radiation, and, with Wald, showed how a combination of the two ideas explained how black holes obeyed the Second Law of Thermodynamics.

Introduction

STEVEN F. SAVITT

1 Eddington and the running-down of the universe

The phrase 'time's arrow' seems to have entered the discussion of time in Sir Arthur Eddington's Gifford Lectures, which were published in 1928.[1] An 'arrow' of time is a physical process or phenomenon that has (or, at least, seems to have) a definite direction in time. The time reverse of such a process does not (or, at least, does not seem to) occur. Eddington thought he had found such an arrow in the increase of entropy in isolated systems. He wrote:

> The law that entropy always increases – the second law of thermodynamics – holds, I think, the supreme position among the laws of Nature.[2]

Since he held the universe to be an isolated system, he thought that its entropy, which he called its 'random element', must ineluctably increase until it reached thermodynamic equilibrium (until it is 'completely shuffled'), by which point all life, and even time's arrow itself, must have disappeared. He called this process 'the running-down of the universe'. This vision of the universe is stark, compelling, and by no means hopelessly dated. P. W. Atkins recently wrote:

> We have looked through the window on to the world provided by the Second Law, and have seen the naked purposelessness of nature. The deep structure of change is decay; the spring of change in all its forms is the corruption of the quality of energy as it spreads chaotically, irreversibly, and purposelessly in time. All change, and time's arrow, point in the direction of corruption. The experience of time is the gearing of the electrochemical processes in our brains to the purposeless drift into chaos as we sink into equilibrium and the grave.[3]

[1] Eddington (1928), 68.
[2] Eddington (1928), 74.
[3] Atkins (1986), 98.

Introduction

Eddington's view was not so grim.

> At present we can see no way in which an attack on the second law of thermodynamics could possibly succeed, and I confess that personally I have no great desire that it should succeed in averting the final running-down of the universe. I am no Phoenix worshipper. This is a topic on which science is silent, and all that one can say is prejudice. But since prejudice in favour of a never-ending cycle of rebirth of matter and worlds is often vocal, I may perhaps give voice to the opposite prejudice. I would feel more content that the universe should accomplish some great scheme of evolution and, having achieved whatever may be achieved, lapse back into chaotic changelessness, than that its purpose should be banalised by continual repetition. I am an Evolutionist, not a Multiplicationist. It seems rather stupid to keep doing the same thing over and over again.[4]

Most of this volume is given over to fairly technical discussions of issues in the foundations of physics. It is worth reminding ourselves, before we become immersed in detail, that these issues are linked in many ways to the pictures we have of the fate of the universe and our part in its story. These connections explain, in part, the perennial appeal of problems concerning the existence and nature of time's arrows.

2 G. N. Lewis and 'two-way time'

Although Eddington wrote in 1928 that he could 'see no way in which an attack on the second law of thermodynamics could possibly succeed,' the brilliant chemist G. N. Lewis had mounted just such an attack in his Silliman Lectures, published two years earlier.[5] According to Lewis, the second law is: 'Any system left to itself approaches in a single direction a definite state of equilibrium, or, in other words, its entropy increases steadily toward a maximum.'[6] Of this law, Lewis wrote:

> We ... find not even a shred of truth left in the statement that an isolated system moves toward the state of equilibrium. It will move toward it and move away from it, and in the long run as often in one direction as in the other. It is only when we start far away from the state of equilibrium, that is, when we start with some state of unusual distinction, and when we follow the system a little way along its path, that we can state that it will, as a rule, proceed toward more *nondescript* states.[7]

[4] Eddington (1928), 86.
[5] Lewis (1926), chapter 6.
[6] Lewis (1926), 143.
[7] Lewis (1926), 153.

Lewis, of course, was analysing the behaviour of isolated systems in accordance with the statistical mechanics of Boltzmann and of Gibbs. The laws of statistical mechanics are time symmetric, and hence the behaviour of isolated systems cannot distinguish future from past. On the micro-level in the long run such systems will depart from equilibrium just as often as they approach it. Appearances to the contrary on the macro-level result from the existence of far more nondescript than interesting macro-states. 'It is not true that things left to themselves approach a constant state,' wrote Lewis, 'but only that they approach a state which ordinarily appears constant to us because of the dullness of our perceptions.'[8]

Lewis employed elegant examples to help overcome what he saw as an illusion engendered by coarse-graining.

> Let us consider a box with a one-gram weight resting on its floor. Let us place this box in a bath maintained at an extremely constant temperature, we will say 65 °F, and let the whole be protected by the most perfect mechanism that we can think of to shield it from external jars. Let us, in other words, shut it off from all external influences, leaving only a small hole through which we may observe the weight. We may look into the box millions of times and always find the weight upon the floor, and we then state this to be a law of nature. But the time will come when we look in and find the weight some distance from the floor. This chance is so very small that I cannot express it in any ordinary way. We state chances as fractions, but to denote this chance I should have to put down a decimal point and zero after zero, and would spend my whole lifetime before I could write down a number not a zero. But the calculation is none the less exact.
>
> The chance becomes larger if I consider smaller weights and lesser heights from the floor. Let the height be one hundred million times as small and the weight also one hundred million times as small, and then the calculation shows that if we look in every second we shall find the weight as far off the floor as this 6.32 times in every million years. If you bet five to one on the appearance of this phenomenon, in a million years you might lose at first but would come out ahead in the long run.[9]

From this and other examples Lewis concluded that the concept of time must be split. There is a 'one-way time' that we find in consciousness, memory, and apparently irreversible phenomena, and there is a symmetric or 'two-way time' which is the time of physics. Lewis, then, sharply distinguished between a psychological arrow and the thermodynamic arrow of time, and he denied

---

[8] Lewis (1926), 147.
[9] Lewis (1926), 145–6.

that the thermodynamic arrow, the one that Eddington regarded as fundamental, was an arrow at all.[10] This clash epitomizes the philosophical problems raised by the emergence of kinetic theory in the nineteenth century and which lie at the heart of the tangle of problems known as *the problem of the direction of time*.

3 Time's arrows

Eddington had in mind, as we saw, one particular fundamental time-asymmetric process, but by 1979 Roger Penrose was concerned with seven possible 'arrows'.[11]

(1) The decay of the neutral K meson (K^0) seems to be governed by a time-asymmetric law, the only such law in particle physics.[12] The inference to this law is somewhat delicate, however, and the behaviour of K^0 seems to be unrelated to the other more salient time-asymmetric phenomena.[13]

(2) The process of measurement in quantum mechanics, along with its attendant 'collapse of the wave function', is often supposed to be a time-asymmetric phenomenon. William Unruh describes measurement in more detail in section 2 of chapter 1 below, where he, like Penrose in 1979, subscribes to a time-symmetric account. Unruh's paper deals with very broad features of time in twentieth century physical theories and the ways these broad features may need to be modified if we are to develop a successful theory of quantum gravity.

(3) The second law of thermodynamics asserts that in irreversible processes the entropy of isolated systems will increase until it reaches a maximum value. 'After that,' write the authors of a recent physics text, 'nothing else happens.'[14] An analogue of the second law can be derived in statistical mechanics.[15] The relation between the second law and its statistical mechanical analogue, and between both of these and our sense of the direction of time, is puzzling, as we have already seen in Lewis's objections to Eddington.

Barrett and Sober attempt, in a rigorous way, to apply the statistical mechanical formalism to biological examples. In so doing, they find that the

---

[10] Lewis, of course, did not know of the expansion of the universe when he wrote *The Anatomy of Science*. He did, however, react to the radiative asymmetry (to be introduced in section 3 below) by developing a time-symmetric theory of electromagnetism that was a precursor of the absorber theory. See footnote 10 of Wheeler & Feynman (1945).

[11] Penrose, R. (1979), 582.

[12] The decay violates T-invariance or, in the distinctions to be developed in section 6, it is not time reversal invariant$_1$.

[13] For further details see Horwich (1987), 54–7, or Sachs (1987), chapter 9.

[14] Olenick, Apostol, & Goodstein (1985), 328.

[15] Some elementary properties of entropy in statistical mechanics are derived in appendices A and B of chapter 9 by Barrett & Sober in this volume.

explanation of and conditions for the increase of entropy as given by Khinchin[16] generalize to the biological examples more readily than the classical explanation of entropy increase, which they explain in section 2 of their chapter.

(4) We see radiation emitted from (point) sources in spherical shells expanding in the future time direction. We do not see such shells converging to a point, even though the equations of classical electromagnetism permit such 'advanced' as well as the usual 'retarded' solutions. In section 2 of chapter 4 Philip Stamp outlines what he calls the 'orthodox standpoint' with respect to the various arrows of time, which holds that the radiative arrow is determined by the thermodynamic arrow. The radiative arrow receives no independent consideration in this volume.

(5) The fifth arrow is the direction of psychological time.

The arrow most difficult to comprehend is, ironically, that which is most immediate to our experiences, namely the feeling of relentless forward temporal progression, according to which potentialities seem to be transformed into actualities.[17]

This 'arrow' is rather a grab-bag, for under this heading Penrose mentions our feeling (a) that the future is mutable but the past is fixed, (b) that we know more about the past than the future, and (c) that causation acts toward the future only.

The feeling of 'relentless forward temporal progression' is discussed below in section 4 on 'Becoming'. The future-directedness of causation seems incompatible with the possibility of time travel into the past, a possibility that is defended by Horwich and Earman in the final two chapters in this book.

(6) The expansion of the universe has been invoked to explain the thermodynamic arrow.[18] Would the thermodynamic arrow, then, reverse in a contracting universe? Would our sense of time reverse as well? These questions, amongst others, are taken up by Huw Price in chapter 2.

(7) According to the general theory of relativity (GTR), the gravitational collapse of a sufficiently massive star results in a *black hole*. After the collapse (and neglecting quantum-mechanical effects), 'the hole settles down and remains unchanging until the end of time.'[19] The time reverse of this process, which is in principle permitted by the equations of GTR, would be a singularity, known as a *white hole*, that sits for some indeterminate amount of time from the beginning of the universe and then erupts in a shower of ordinary matter. Black holes may well exist; we have no evidence that white holes do. The 'orthodox standpoint', as presented by Stamp, holds that we do not yet know

---

[16] Khinchin (1949).
[17] Penrose, R. (1979), 591.
[18] For instance, in Layzer (1975).
[19] Penrose, R. (1979), 600–1.

enough about the nature and existence of such singularities to evaluate this arrow and its relation(s) to the other arrows.

The existence of *any* of these arrows, if they do exist, is puzzling, because all basic theories of physics seem to be time symmetric or time reversal invariant.[20] If these theories do govern the fundamental physical processes in our universe, then it is unclear how processes which are directed in time or are not time symmetric arise. A thorough examination of a number of recent attempts to explain the origin of the thermodynamic arrow, for instance, will be found in Lawrence Sklar's chapter, 'The elusive object of desire'. This chapter is necessarily condensed,[21] but in general he finds that each attempt fails because either it introduces a time-asymmetric postulate which, overtly or covertly, begs the question at issue or because it fails to recognize that its rationale for the increase of entropy of a system in the future time direction would also serve as a rationale for increase of entropy in the past time direction (the 'parity of reasoning' problem). Sklar sees time symmetry at the micro-level, time asymmetry at the macro-level, and no fully compelling connection between the two.

In chapter 2 Huw Price looks at modern cosmology from a standpoint that is similar to Sklar's. Price calls failure to meet the parity of reasoning problem applying a 'double standard', and he shows the double standard at work via a detailed examination of recent cosmological theories of Davies, Penrose, and Hawking.

Difficult questions can be raised about the nature and existence of other arrows as well. Does the wave function really 'collapse', for instance, and what is the nature of this collapse? Must causes precede their effects? Other kinds of problems concern the order of dependence or explanatory priority of pairs of the arrows. Is one of them fundamental, are there clusters of dependencies, or are they unrelated to one other? Does the expansion of the universe, for example, explain the thermodynamic asymmetry (or perhaps *vice versa*)? These problems, lumped under the label *the problem of the direction of time*, straddle the border between physics and philosophy.[22]

The authors in this volume tackle more precise or more narrowly defined versions of these or of closely related questions that may crop up in various

[20] Time reversal invariance will be discussed in section 6 below.
[21] He examines these issues in more detail in Sklar (1993).
[22] In the physics literature extended discussions are to be found in Davies (1974), Sachs (1987), and Zeh (1989). Recent important philosophical works on this subject include Reichenbach (1956), Earman (1974), Horwich (1987), and Sklar (1993).

branches of physics (and, in one case, biology); but before turning to their contributions, a brief examination of a *metaphysical* 'arrow of time' may prove helpful in dispelling some confusions and providing some distinctions that are necessary background for the chapters to follow.

4 Becoming

The problem of the direction of time, as raised by Eddington and Lewis, might seem to have an obvious solution. It seems manifest in our experience that time flows – from the past, to the present moment, and into the future. Newton wrote that absolute time

> of itself, and from its own nature, flows equably without relation to anything external, and by another name is called duration.[23]

According to P. T. Landsberg 'the time variable is rather like a straight line on which a point marked "The Now" moves uniformly and inexorably.'[24] Santayana, somewhat more poetically, wrote, 'The essence of nowness runs like fire along the fuse of time.'[25] The arrow *of* time, one could say, points in the direction in which time flows, moves, or runs. The other arrows are arrows *in* time.

Natural as the view of flowing time is, there are a number of philosophical and physical objections to it. The first type of philosophical objection is rooted in the difficulty of giving a literal explanation of the notion of temporal flow. Here is the metaphysician Richard Taylor:[26]

> Of course there is a temptation to say that the present moves in some sense, since the expression 'the present' never designates the same time twice over; a moment no sooner emerges from the future and becomes present than it lapses forever into an ever receding past. But this kind of statement, gravely asserted, says only that the word 'now' never, i.e., at no time, designates more than one time, viz., the time of its utterance. To which we can add that the word 'here' likewise nowhere, i.e., at no place, designates more than one place, viz., the place of its utterance.

Taylor's challenge is to find a literal meaning for the assertion that The Now moves that distinguishes The Now from The Here, the temporal from the

---

[23] Newton (1686), 6.
[24] Landsberg (1982), 2.
[25] Quoted in Williams (1951).
[26] Taylor (1955), 388–9.

spatial, a challenge that has proved surprisingly difficult to meet. An object moves if it occupies different spatial positions at different times. Does The Now move by occupying different temporal positions at different times? If one thinks of The Now's motion in this way (i.e., by analogy with change of spatial position), then The Now's motion, which is supposed to *be* the passage of time, must take place *in* time, so we are covertly postulating a second temporal dimension in which The Now moves. An infinite hierarchy of temporal dimensions beckons. One might try to avoid the spatial analogy by claiming, as did C. D. Broad, that the movement of The Now, which he called 'absolute becoming', is conceptually basic or primitive and hence cannot be explained or conceived in other terms without loss or distortion.[27] Is this more than to point to an unsolved problem?

One might hope to deal with the moving Now in the manner of Austin Dobson.

> Time goes, you say? Ah no!
> Alas, Time stays, *we* go ... [28]

Dobson is contrasting our passage *through* time with the passage *of* time, but there is less to this contrast than meets the eye. Recall Landsberg's description of dynamic time: 'the time variable is rather like a straight line on which a point marked "The Now" moves uniformly and inexorably.' Suppose, as in a classical spacetime, that all events can be put into exhaustive and mutually exclusive equivalence classes under the relation 'x is simultaneous with y' and that these equivalence classes can be linearly ordered by the relation 'x is later than y' (or, equivalently, 'x is earlier than y'). If we imagine that The Now is a tiny, intense, *static* light and that sets of simultaneous events flow past The Now in the direction from future to past, being illuminated for an instant, we have Dobson's 'Time stays, *we* go ... ' The relativity of motion, however, tells us that this view of time is equivalent to Landsberg's original picture with the light gliding along the static sets of events.

A second philosophical objection is J. M. E. McTaggart's famous argument that the dynamic concept of time is self-contradictory.[29] A moving present, he said, and its cognate notions of past and future (which he called A-determinations, in contrast to the unchanging B-determinations 'x is later than y', 'x is earlier than y', and 'x is simultaneous with y') involve a

---

[27] Broad (1938), 281.
[28] Dobson (1905), 163.
[29] McTaggart (1908).

contradiction because, on the one hand, every event can either be present or have but one definite degree of pastness or futurity while, on the other hand, every event must be past, present, and future. This argument might sound like an obvious sophism, but thoughtful philosophers like Michael Dummett, D. H. Mellor, and Paul Horwich[30] have contended that it is either a serious or a conclusive objection to the possibility of a moving Now.

These abstract arguments run counter to everyday experience, which seems to involve at the very deepest level and in the most direct manner the passage of time or, as Eddington called it, 'Becoming'.

> [I]f there is any experience in which this mystery of mental recognition can be interpreted as *insight* rather than as *image-building*, it would be the experience of 'becoming'; because in this case the elaborate nerve mechanism does not intervene. That which consciousness is reading off when it feels the passing moments lies just outside its door.[31]

Recognition of Becoming leads, however, to an uncomfortable dilemma that is delineated by Lawrence Sklar (following Eddington) in chapter 8 of this volume, 'Time in experience and in theoretical description of the world'. Many distinguished thinkers, especially Boltzmann and Reichenbach, held that all the various time-asymmetric features of the world were 'reducible to' or somehow 'grounded in' the entropic asymmetry. But, as Sklar points out, straightforward identification of experienced succession with increased entropy seems wildly implausible.

> We know from perception what it is for one state of a system to be temporally after some other state. And we know what it is for one state to have a more dispersed order structure than another state. We also know that these two relations are not the same.

On the other hand, once we sunder the two, we have no reason to suppose that the 'time' of the physical world is anything like or has any feature like experienced succession. Since the time of Galileo various ostensible sensory qualities or properties of things, like warmth and colour, have been held by reflective scientists and philosophers to be *secondary qualities*, mind-dependent reflections of whatever real properties or structures in the world give rise to them in our consciousness. If succession joins the ranks of the secondary

---

[30] Dummett (1960), Mellor (1981), chapter 6, and Horwich (1987), chapter 2. I argue in Savitt (1991) that not all the premises in Horwich's version of McTaggart's argument can be simultaneously true.

[31] Eddington (1928), 89.

qualities as the subjective reflection of entropy increase, then, Sklar argues, we would thereby be deprived of the last insight we have into the real or intrinsic qualities of the world around us.

> Put most crudely, the problem is that at this point the veil of perception has become totally opaque, and we no longer have any grasp at all upon the nature of the physical world itself. We are left with merely the 'instrumental' understanding of theory in that posits about nature bring with them predicted structural constraints upon the known world of experience.

The unpalatability of either horn of this dilemma might lead one to re-think the privileged epistemological status accorded to our knowledge of becoming by both Eddington and Sklar.

5 Physical arguments against Becoming

Some physicists, like Lewis, are sceptical of Becoming as a physical concept. David Park, for example, wrote:

> No formula of mathematical physics involving time implies that time passes, and indeed I have not been able to think of a way in which the idea could be expressed, even if one wanted to, without introducing hypotheses incapable of verification.[32]

In the same vein P. C. W. Davies wrote that 'present day physics makes no provision whatever for a flowing time, or for a moving present moment.'[33] From time to time a stronger claim, that some portion of physics is actually *incompatible* with a dynamic conception of time, is made. Two important arguments of this type appear, conveniently, in one source, Kurt Gödel's 'A remark about the relationship between relativity theory and idealistic philosophy'.[34]

The first argument does not originate with Gödel, and he does not endorse it. A moving Now is supposed to have some special metaphysical significance, to be connected with existence or with the objective lapse of time. In the special theory of relativity (STR), however, the principle of the relativity of simultaneity assures us that there are a nondenumerable infinity of now's, and the standard symmetries assure us that no one of them can have special significance. Becoming does not fit easily into Minkowski spacetime.

[32] Park (1972), 112.
[33] Davies (1977), 3.
[34] Gödel (1949b).

Gödel was concerned, however, that in the general theory of relativity (GTR) it seemed possible to single out certain privileged frames and hence to re-establish something like an 'objective lapse of time'. But GTR, as he had recently discovered, also permitted the existence of a spacetime that contained closed timelike curves (CTCs). Gödel argued that in the model spacetime that he discovered (a) there could not be an objectively lapsing time, and (b) there could therefore be no objective lapse of time in our world either. Gödel's paper is terse, and the exact formulation of his argument is controversial.[35]

Storrs McCall has long tried to show that the argument from STR above lacks force by exhibiting what he calls 'a model of the universe' that is consistent with STR and also has a dynamic time. It is important to note that dynamic time has two metaphysical variants. If one focuses on the present moment, conceiving of 'The Now' as moving smoothly along an ordered series of events from past to future, one could hold that only the present exists – the future not yet having become, the past having become but also having vanished. As an alternative to this *presentism*, one might hold that although future events have indeed not yet come to be out of the myriads of unrealized possibilities, the past as well as the present are fully real. This *probabilist* picture or tree model of time (future possible paths being represented as branches on a tree-structure) most likely traces back to Aristotle,[36] and is the one that McCall attempts to reconcile with the multitude of now's to be found in STR.

Roy Douglas, motivated by the belief that the indeterministic foundation of quantum mechanics indicates an incomplete theory, looks at branching temporal structures from the standpoint of a topologist. In a nearly self-contained presentation, Douglas constructs (though he does not necessarily advocate) a mathematical model of a branching, deterministic spacetime which is specified in terms of local properties only. Roger Penrose concluded an earlier discussion of branching temporal structures by remarking, 'I must ... return firmly to sanity by saying to myself three times: "spacetime is a *Hausdorff* differentiable manifold; spacetime is a *Hausdorff* ... " '[37] Douglas's conclusion is, however:

> The (global) Hausdorff property certainly simplifies the mathematics of our models; unfortunately, it is also an entirely inappropriate restriction for models of spacetime, precisely because ... it is a strictly global constraint ... [38]

[35] See Yourgrau (1991) and Savitt (1994) for somewhat different views of it.
[36] *De Interpretatione*, chapter 9.
[37] Penrose, R. (1979), 595.
[38] Douglas, R. (this volume), chapter 6, section 2.

He supports this claim with a detailed analysis of the global/local distinction.

Although Gödel's argument clearly aimed to undermine the notion of an 'objectively elapsing time', most commentators on it were struck by the fact that it raised the possibility of a certain kind of time travel. In fact, his model raised the novel possibility of travelling into one's own past by travelling around a CTC. But what sort of 'possibility' is this? John Earman explores in detail the idea that the global causal structure of spacetime (e.g., the existence of CTCs) may constrain the notion of physical possibility beyond the local constraints of GTR. These constraints, in turn, serve to illuminate the concept of physical law.

The possibility of travelling into one's own past, whether in a spacetime with CTCs or in some other manner, raises the possibility of the existence of closed causal loops. It is frequently alleged that purely conceptual arguments suffice to rule out the existence of closed causal loops, since it is thought that they must permit arrangements with contradictory outcomes. Paul Horwich rejects this conclusion and tries to show the precise senses and circumstances in which closed causal loops are possible.

6 Time reversal invariance

A question that it is natural to ask is: when does a physical theory pick out a preferred direction in time? In fact, the more usual way of approaching this question is by asking: when is it that a theory does *not* pick out a preferred direction in time? When is it, that is, that a theory is *time reversal invariant*?

Suppose that some set of laws, L, are at least a significant component of a scientific theory, T. If these laws involve a time parameter, t, then one can define a *time reversal transform* as the mapping $t: t \rightarrow -t$. If the laws, L, are differential equations, then one can say that T is *time reversal invariant$_1$* if and only if every solution of L is mapped to a (not necessarily distinct) solution of L under the time reversal transform, t. One can find this characterization of time reversal invariance explicitly in P. C. W. Davies' *The Physics of Time Asymmetry*.[39]

Immediately following his official characterization of time reversal invariance, Davies remarks

> The references to 'time reversal' are purely mathematical statements, and have nothing to do with a return to the past. It is to be identified physically with process or velocity reversal.[40]

[39] Davies (1974), 22–7. See also Zeh (1989), 2.
[40] Davies (1974), 23.

It is, in fact, in terms of this latter notion that Davies discusses three specific examples which cast some *prima facie* doubt on the time reversal invariance of classical physical theories. His second example, for instance, is the familiar one of the damped motion of a particle moving through a viscous medium or across a surface that exerts friction on it. If the viscous drag or frictional force is assumed to be proportional to $\dot{r}$, where $r(t)$ is the position of the particle at time t and $\dot{r}$ is the total derivative of $r(t)$, then the equation describing the motion of the particle is

$$m\ddot{r} = f(r) - \alpha\dot{r}, \qquad (1)$$

where α is a positive constant and m is the particle's mass. If no additional force is acting on the particle, that is, if $f(r) = 0$, then the solutions of (1) that describe the motion of the particle have the form

$$|\dot{r}| \propto e^{[-(\alpha/m)]t}, \qquad (2)$$

where $|\dot{r}| \to 0$ as $t \to \infty$. Under the time reversal, t, (2) becomes

$$|\dot{r}| \propto e^{[(\alpha/m)t]}. \qquad (3)$$

Solutions of the form (3), according to Davies, 'are clearly unphysical as they correspond to a body being spontaneously accelerated to an infinite velocity as a result of its contact with a viscous medium or frictional surface.'[41] Does Davies then conclude from this example that the failure of 'physicality' under the time reversal transform implies that classical mechanics is not time reversal invariant? By no means. He argues, rather, that one has not yet fully factored in the 'process or velocity reversal' mentioned above.

> In the second example of viscous or frictional damping, the motion of the body is slowed by the communication of kinetic energy to the medium atoms in the form of *heat*. It follows that if the motions of the individual atoms are also reversed then, because of the invariance of the laws of physics governing the atomic interactions, each collision will be reversed, causing a cooperative transfer of momentum to the large body, which would then become exponentially accelerated.[42]

Let us pin down the concept of invariance that underlies the above argument. Suppose that the laws, L, of theory T concern some set of properties (or parameters) of a system, S, such as the positions or momenta at some time t of the set of particles making up S. A *state* of a system, S_j (relative to

---

[41] Davies (1974), 25.
[42] Davies (1974), 26.

theory T), is some specification of values for all parameters of the components of the system.⁴³ A sequence of states of a system is *dynamically possible* (relative to theory T) if the sequence of states $S_i \to S_f$ (indicating that S_i is before S_f) is consistent with the laws of T (is 'a permissible solution of the equations of motion' of T). Finally, let $(S_j)^R$ be the *time-reversed state* of the state S_j. How the time-reversed state is to be specified will, in general, depend upon the theory that is under consideration.⁴⁴ We can now say that a theory T is *time reversal invariant₂* under the following circumstances: a sequence of states $S_i \to S_f$ is dynamically possible (relative to the laws of T, of course) if and only if $(S_f)^R \to (S_i)^R$ is dynamically possible (relative to T). Davies's argument above successfully, I believe, defends the thesis that classical mechanics is time reversal invariant₂ from a purported counter-example.

Unfortunately, but importantly, there is yet a third concept of time reversal invariance that is sometimes used and sometimes conflated with the time reversal invariance₂. Let us say that a theory, T, is *time reversal invariant₃* if and only if should S_i evolve to S_f, according to T, then $(S_f)^R$ must evolve to $(S_i)^R$.⁴⁵ Time reversal invariance₃ captures the essential idea of what John Earman introduced, by way of contrast with what he thought of as genuine time reversal invariance, as the 'reversal of motion transformation – when one asks whether the laws governing a system of particles are "time reversible", one is simply asking whether the laws imply that if the three velocities of the particles were reversed, the particles would retrace their trajectories but with opposite three velocities.'⁴⁶ This third concept of time reversal invariance is different from and stronger than the second. No indeterministic theory, T, can be time reversal invariant₃, since even in circumstances in which $(S_f)^R$ might evolve to $(S_i)^R$ (and so T might be time reversal invariant₂), the laws of T will in general permit $(S_f)^R$ to evolve in other ways as well.

Although I might seem to be multiplying distinctions beyond necessity, teasing out (at least) these three senses of time reversal invariance can help to clarify some recent arguments. Three authors in this volume (Leggett,

---

⁴³ See section 2.1 of Hughes (1989) for further specification of the state of a classical system.

⁴⁴ It should, however, be true in general that $((S_j)^R)^R = S_j$. As Davies's examples show, specification of the time-reversed state can be a matter of some subtlety.

⁴⁵ In an unpublished manuscript, *Explaining Time's Arrow: a How-To Manual*, Craig Callender (Department of Philosophy, Rutgers University) suggests some reasonable names for some of the different sorts of time reversal invariance that I distinguish above. He calls my *time reversal invariance₁* 'Formal TRI', *time reversal invariance₂* 'Motive TRI', and *time reversal invariance₃* 'Actual History TRI'.

⁴⁶ Earman (1974), 25–6.

Stamp, and Unruh) defend the thesis that quantum mechanics (QM) is time reversal invariant. Leggett has shown that one can experimentally push QM surprisingly far into the 'classical' realm without any 'collapse' or 'reduction' of the state-vector, and in chapter 3 in this volume he incorporates the view that a macroscopic system cannot be in a superposition of states into a broader position that he calls *macrorealism* and then shows that macrorealism can in principle entail an experimental result that is in conflict with orthodox QM.

Stamp develops some of the philosophical implications of current research in the area of macroscopic quantum phenomena by using the idea of a 'quantum computer' as a model of a reversible measurement. This line of thought seems in conflict with a simple thought experiment by means of which Roger Penrose, in a recent, prominent book, *The Emperor's New Mind*,[47] argues that QM is time asymmetric – that it is, in some sense, not time reversal invariant. As an example of the utility of the above distinctions, I want to show that Penrose's argument establishes only the unsurprising conclusion that QM is not time reversal invariant$_3$ and does not contradict the view of Leggett, Stamp, and Unruh.

Penrose claims that a simple experiment shows that the measurement process, R, in QM is 'time-irreversible'. He imagines a source of photons, a lamp L, located at A and aimed precisely at a photon detector P, which is located at C.[48] In between is a half-silvered mirror, M, which is turned at a 45° angle to the path from L to P, permitting half the photons to travel from L to P and reflecting half to a point D on the laboratory wall. Emission and reception of photons is recorded at L and P, and by implication one can determine when photons emitted at L wind up at D.

The crux of Penrose's argument seems to lie in the comparison of a pair of questions regarding conditional probabilities:

(Q) Given that L registers, what is the probability that P registers? and the *time-reverse* question

(QR) Given that P registers, what is the probability that L registers?

I shall assume that the experiment envisaged by Penrose involves a run of photons emitted at L and (shortly thereafter) either detected at P or not detected at P and assumed to have travelled to D. I understand the question

---

[47] Penrose (1989), 354–9. The relevant section is entitled 'Time-asymmetry in state-vector reduction'.

[48] My discussion is keyed to figure 6 in chapter 1, where Unruh discusses Penrose's example.

(Q) to ask, given that L registers the emission of a photon, what is the probability that P registers its receipt? One-half, calculates Penrose. Of (QR) he then observes that

> the *correct* experimental answer to this question is not 'one-half' at all, but 'one'. If the photo-cell indeed registers, then it is virtually certain that the photon came from the *lamp* and not from the laboratory wall! In the case of our time-reversed question, the quantum-mechanical calculation has given us *completely the wrong answer*![49]

In light of this remark, it is clear that (QR) is intended to ask something like, given that P registers the receipt of a photon, what is the probability that the photon was emitted by the lamp (and not from anywhere else, like the wall for instance)? It is also clear that the answers provided by Penrose to (Q) and (QR) so understood are correct. What is not clear at all is under what understanding of time reversal invariance the fact that (Q) and (QR) have different answers would lead one to conclude that QM is not time reversal invariant.

Penrose in an earlier publication pointed out that the time evolution of the state of a quantum system according to the Schrödinger equation is time reversal invariant$_1$.[50] He there added that, although the process of measurement might seem to be time asymmetric, he explicitly endorsed a theory of measurement that is time reversal invariant$_1$.[51] Nothing in the present thought experiment *directly* addresses either of those two former positions.

I further claim that QM is time reversal invariant$_2$. As long as the half-silvered mirror is characterized in statistical terms, it is not actually inconsistent with the laws of QM (that is, it is dynamically possible according to the laws of QM) that *all* photons emitted at P wind up back at L. Of course, such a perverse reverse run will be extraordinarily rare, perhaps as rare as the exponential acceleration of the particle by friction in Davies's example if a sufficient number of photons are involved. But the *possibility* of such a run is all that is required for the preservation of time reversal invariance$_2$.

Penrose's thought experiment does illustrate the fact that QM is not time reversal invariant$_3$. The failure of this type of time reversal invariance, however, seems to be fully accounted for in this example by the indeterministic action of the half-silvered mirror and (in the absence of additional,

---

[49] Penrose (1989), 358.
[50] Penrose, R. (1979), 583.
[51] See Aharonov, Bergmann, & Lebowitz (1964).

controversial assumptions about metaphysical preconditions for indeterminism) tells us little of interest about temporal asymmetry.

There is yet another sense of time reversal invariance (let us refer to it as *time reversal invariance₄*) that Penrose may have in mind. As Lawrence Sklar has pointed out, '[I]n the quantum-theoretic context, it is transition probabilities between reversed states which must equal the probabilities of the unreversed states taken in opposite temporal order for the laws to be time-reversal invariant.'[52] Since the transition probabilities (presumably represented in this case by (Q) and (QR)) are unequal, it seems that one should conclude that the relevant laws of QM are not time reversal invariant₄.[53]

Whatever the importance of this last claim might be, I believe that Penrose's argument is not sufficient to establish it. Two matters must be settled before the status of his claim is clear. First of all, what does Penrose consider to be the time reverse of his proposed thought experiment? One possible reading of what Penrose has in mind is that the time reverse of the experiment he describes is that sequence of events that would be shown by playing a videotape of the original experiment backward. Another suggestion is, however, that a time reverse of a given experiment is one in which the 'initial' state is the time reverse of the final state of the original experiment, and then the time reversed experiment evolves from its 'initial' state according to the laws of the theory whose time reversal invariance is in question. For an indeterministic theory like QM, a time reverse of a given sequence of events in the second sense need not coincide with the time reverse in the first sense.

Second, one can ask how the transition probabilities are to be calculated. One might challenge Penrose's assertion that the answer to (Q) is 'one-half' while the answer to (QR) is 'one' by questioning whether, even if it is true that in the course of the time reverse (on the tape-played-backward understanding) of Penrose's proposed experiment all the photons that register at P will register at L, the transition probabilities are to be established by counting the results of that one run. Let us, for convenience, call this approach the *narrow frequentist* way of determining the transition probabilities. In opposition to the narrow frequentist approach, one might suggest that the transition probabilities are to be determined by averaging over the set of *all* physically possible runs having the same 'initial' conditions. In this latter way of determining the

---

[52] Sklar (1974), 368.
[53] This might constitute an indirect challenge to the claim that the laws of QM are *in toto* time reversal invariant₁.

transition probabilities, which seems unavoidable if one uses the second sense of the time reverse of a sequence of events, both (Q) and (QR) receive the same answer, 'one-half'. The question of how to determine the transition probabilities is independent of, though it may become confused with, the question of choosing between the two senses described above of the time reverse of the original experiment.

The success of Penrose's argument seems to depend upon choosing the videotape-played-backward version of the time reverse of his proposed thought experiment and the narrow frequentist way of determining the answers to (QR). I see no positive reason to look at the matter in this fashion, and I see two reasons against it. First of all, the answer to (Q) is determined by broad theoretical considerations, rather than by counting the results of one run. How could one justify determining the answer to (QR) differently without begging questions about temporal asymmetry?

Second, consider an experimental set-up much like the one Penrose proposed, except that L shoots out at varying intervals little metal pellets or ball bearings aimed directly at P. Suppose also that in place of the half-silvered mirror there is an aperture controlled by a device that opens and closes it such that (1) the aperture is open exactly half the time during any experimental run and (2) there is no correlation or connection between the device that controls the aperture and the firing mechanism for the ball bearings. It seems that here we have described a classical experimental set-up about which one can ask exact analogs of the questions (Q) and (QR) and give precisely the same answers that Penrose gave in the QM case above, using the videotape-played-backward time reverse or the original experiment and the narrow frequentist method of determining probabilities. If the fact that (Q) and (QR) receive different answers in Penrose's experimental set-up implies that QM is not time reversal invariant, then the fact that the analogues of (Q) and (QR) receive different answers in the variant of Penrose's experiment that I propose should *mutatis mutandis* imply that classical mechanics (CM) is not time reversal invariant in that same sense. I think that Penrose would find this result uninteresting for CM, but to be consistent he should find it uninteresting for QM as well.

There are, I have shown, several assertions that might be intended by the claim that QM is time asymmetric. Despite Penrose's argument, I believe that no stronger claim has been justified than the claim that QM is not time reversal invariant$_3$.

7 Conclusion

Penrose's seemingly simple thought experiment has called forth a host of distinctions, illustrating a point made by John Earman some years ago.

> [V]ery little progress has been made on the fundamental issues involved in 'the problem of the direction of time.' By itself, this would not be especially surprising since the issues are deep and difficult ones. What is curious, however, is that despite all the spilled ink, the controversy, and the emotion, little progress has been made towards clarifying the issues. Indeed, it seems not a very great exaggeration to say that the main problem with 'the problem of the direction of time' is to figure out exactly what the problem is supposed to be![54]

I hope that the papers contained in this volume will help to clarify and perhaps even to resolve some of the problems of the direction of time.

[54] Earman (1974), 15.

PART 1 **Cosmology and time's arrow**

1 Time, gravity, and quantum mechanics

WILLIAM UNRUH

The role that time plays in Einstein's theory of gravity and in quantum mechanics is described, and the difficulties that these conflicting roles present for a quantum theory of gravity are discussed.

1 Gravity and time

The relation of any fundamental theory to time is crucial as was evident from the earliest days of physics. If we go back to Newton's *Principia*, in which he established a general theoretical structure for the field of physics, we find an odd series of sentences in the first few pages. He tells us that it is unnecessary to define time as it is obvious to all, but then proceeds to do just that. And his definition is, certainly to modern eyes, rather strange. To quote[1]

> I do not define time, space, place, and motion as being well known to all. Only I observe, that the common people conceive these quantities under no other notions but from the relation they bear to sensible objects. And thence arise certain prejudices, for the removing of which it will be convenient to distinguish them into absolute and relative, true and apparent, mathematical and common.
> I. Absolute, true and mathematical time, of itself, and from its own nature, flows equably without relation to anything external, and by another name is called duration: relative, apparent, and common time, is some sensible and external (whether accurate or unequable) measure of duration by means of motion, which is commonly used instead of true time; such as an hour, a day, a month, a year.

Reading this definition today, it is hard to see what the fuss is about. Time for us common folk is exactly Newton's true time. Taught about time since we were small, we know that there is something, insensible but present in the universe,

[1] Newton (1686), 7.

called time, something that is separate from the other objects in the universe and the same everywhere. Newton would certainly not need to include his definition today, nor would he ascribe to common people that which he did.

It is precisely because we have so thoroughly absorbed Newton's lesson that we all have immense difficulty in coming to terms with the revolutions in physics of the twentieth century and that we in physics now have such difficulty in producing a unified theory of quantum mechanics and gravity.

It is the purpose of this chapter to sketch the changes in the notion of time in twentieth-century physics. I will show how in General Relativity the idea of time is radically different from that of Newton and how gravity itself becomes an aspect of time. I will then examine the role time plays in quantum physics. Finally, I will sketch ways in which the lack of a proper understanding of time seems to be one of the chief impediments to developing a quantum theory of gravity.

The change began with Special Relativity, the first theory in which time lost some part of its absolute and invariant character. Time became, at least in some small sense, mutable. It is precisely this conflict between a mutable notion of time and the absolute and unitary notion of time inherited from Newton that has caused consternation and confusion. This confusion came about not because of any innate violation of the sense of time that we are born with. Time for children is flexible and changeable, and certainly need not be the same here as it is there. Throughout our early years we were taught the lessons of Newton. Time was something out there, something that our watches measured, and something that really was the same everywhere. We learnt while very young that our excuse to the teacher that our time was different from the teacher's time was not acceptable.

The conclusions of Special Relativity came into direct conflict with these early lessons. The 'twins paradox' is the epitome of this confusion, because there is, of course, no paradox at all except in the conflict between the notion of time as expressed in this theory and the notion of time as expressed by Newton. It is because we have so thoroughly absorbed Newton's definition of time that we become confused when time in Special Relativity behaves differently. In Special Relativity time, at least time as measured by any physical process, became not the measure of that unitary non-physical phenomenon of universal time, but a measure of distance within the new construct of 'space-time'. No-one expresses any surprise, or considers it a paradox, that the difference in the odometer readings on my car from the start of a trip to its end is not the same as the difference on your odometer for our trips from Vancouver to Toronto, especially if I went via Texas, while you went via the Trans-Canada Highway. Distances are simply not functions only of the end points of the journey but depend on the complete history of the journey. Within Special Relativity the same is true of time. Times are no longer dependent only on the beginning and end of our journey, but are history-dependent and depend on the details of the journey themselves in exactly the same way that distances do. Mathematically this is expressed by having the time in Special Relativity be given by an extended notion of distance in an extended space called spacetime. Just as the spatial distance between two points depends on the details of the path joining the two points, so the temporal distance joining two points at two separate instants depends on the details of the path and the speed along that path joining the two points at the two instants.

Even though time as a measure of the duration of a process became mutable, Special Relativity was still a theory which retained some of the Newtonian picture. Just as space, for Newton, was another of those non-material but real invariant externals, so in Special Relativity spacetime is also a real non-material invariant external. Instead of having two such concepts, i.e., space and time, Special Relativity has them unified into one single concept which retains most of the features of space.

This changed in General Relativity, Einstein's theory of gravity. Within Special Relativity, the immutability, the sameness, the independence of space and time from the other attributes of the universe, was kept inviolate. Although time, as measured by a watch, was path-dependent, it was the same everywhere, and was independent of the nature of matter around it. In General Relativity this aloofness vanished.

One often hears that what General Relativity did was to make time depend on gravity. A gravitational field *causes* time to run differently from one place to the next, the so-called 'gravitational red shift'. We hear about experiments like the one done by Pound and Rebka[2] in which the oscillation frequency of iron nuclei at the top and the bottom of the Harvard tower were found to differ, or about Vessot's[3] 'disposal' procedure of one of his hydrogen masers in which such a maser was shot up 10 000 km above the earth by rocket before dropping into the Atlantic. During that journey, he noted that the maser ran more quickly at the top of its trajectory than at the

---

[2] Pound & Rebka (1960).
[3] Vessot, Levine, Mattison, Bloomberg, Hoffmann, Nystrom, *et al.* (1980).

bottom, in agreement with General Relativity to better than one part in five thousand. The lesson of these experiments would appear to be that gravity alters the way clocks run. Such a dependence of time on gravity would have been strange enough for the Newtonian view, but General Relativity is actually much more radical than that. A more accurate way of summarizing the lessons of General Relativity is that gravity does not *cause* time to run differently in different places (e.g., faster far from the earth than near it). Gravity *is* the unequable flow of time from place to place. It is not that there are two separate phenomena, namely gravity and time and that the one, gravity, affects the other. Rather the theory states that the phenomena we usually ascribe to gravity are actually caused by time's flowing unequably from place to place.

This is strange. Most people find it very difficult even to imagine how such a statement could be true. The two concepts, time and gravity, are so different that there would seem to be no way that they could possibly have anything to do with each other, never mind being identical. That gravity could affect time, or rather could affect the rate at which clocks run, is acceptable, but that gravity is in any sense the same as time seems naively unimaginable. To give a hint about how General Relativity accomplishes this identification, I will use an analogy. The temptation with any analogy is to try to extend it, to think about the subject (in this case time and gravity) by means of the analogy and to ascribe to the theory (General Relativity) all aspects of the analogy, when in fact only some of the aspects are valid. For this reason I will emphasize the features that carry the message I want to convey, and I will point out a few of the features that may be misleading.

In this analogy I will use the idea from Special Relativity that some of the aspects of time are unified with those of space and that the true structure of spacetime is in many ways the same as our usual notions of space. I will therefore use a spatial analogy to examine certain features of spacetime in the vicinity of the earth. In order to be able to create a visual model, I will neglect two of the ordinary dimensions of space and will be concerned only with the physical spatial dimension of up and down along a single line through the center of the earth chauvinistically chosen to go through my home city of Vancouver. In this model, time will be represented by an additional spatial dimension, so that my full spacetime model will be given by a two-dimensional spatial surface. I will argue that I can then explain the most common manifestation of gravity that when I throw something up into the air, it slows down and

1 Time, gravity, and quantum mechanics

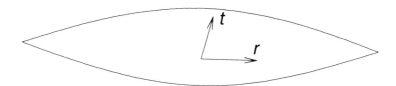

Fig. 1. The lens shape used to create the model of spacetime near the earth

finally stops its ascent and then comes back down to the surface of the earth (i.e., that which goes up must come down). Usually one ascribes this behaviour to the presence of a force called gravity which acts on the object, pulling it down toward the center of the earth. The crucial point is that one can alternatively explain this essential attribute of gravity by assuming that time flows unequably from place to place, without calling into play any 'force of gravity' at all.

In order to develop the analogy, we must first interpret the phrase 'time flows unequably' in terms of our model. We can assume, for the purposes of our discussion, that the physical phenomena near the earth are the same from one time to the next. That is, if we are to construct the two-dimensional spacetime (up–down, and time) near the earth out of pieces representing the spacetime at different instants, those pieces should all look the same. I will use pieces that look like lenses (see figure 1). The direction across the lens I will take as time (t), and the direction along the lens (r) I will take as space. These lenses have the feature that the physically measurable time, which Special Relativity teaches is just the distance across the lens, varies from place to place along the lens. I have thus interpreted the phrase 'the flow of time is unequable' as the statement that the distance (time) across each lens is unequal from place to place. I will now glue a large number of these lenses together along their long sides, giving us the two-dimensional shape in figure 2, for which I will use the technical term 'beach ball'. I will take this beach ball to be a model of the spacetime near the earth, a spacetime made out of pieces on which time flows unequably from place to place.

Let us now locate the surfaces of the earth in this model. The earth in the up–down direction has two surfaces – one here in Vancouver and the other one near Isle Crozet in the south Indian Ocean. Since the distance through the earth from Vancouver to this island is constant, the distance between the two strips on the surface of the beach ball must be constant from one time to the next to model this known fact about the earth accurately. Furthermore, since we expect the

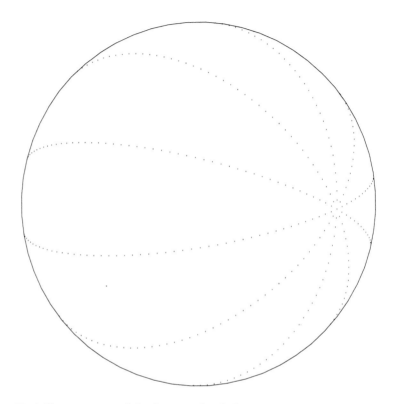

Fig. 2. The spacetime made by gluing together the lenses

system to be symmetric, we expect that 'up' here at Vancouver, and 'up' at Isle Crozet should behave in exactly the same way. Thus, the strips should be placed symmetrically on the beach ball. I thus have figure 3, with the two black strips representing the two surfaces of the earth and the region between the strips representing the interior of the earth.

I stated that I would use this model to explain why, when something is thrown into the air, it returns to the earth. To do so, we must first decide how bodies move when plotted in this spacetime. I go back to the laws of motion first stated by Newton, in particular his first law of motion.

> Every body continues in its state of rest, or of uniform motion in a right line, unless it is compelled to change that state by forces impressed upon it.

As I stated above, I will dispense with the idea of a gravitational force. I want to describe gravity not as a force but as the unequable flow of time from place to place. A body thrown up into the air is thus not acted upon by any forces. By the

1 Time, gravity, and quantum mechanics

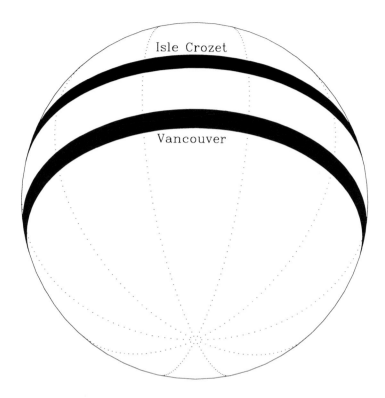

Fig. 3. The two surfaces of the earth at Vancouver and at Isle Crozet drawn in on the spacetime model. Between the two strips is the interior of the earth

above law the body will continue its motion in a 'right', or as we would now say, a straight line. But what does 'straight' mean in this context of plotting the path of the body on the surface of this beach ball? I go back to one of the oldest definitions of a straight line, namely that a line is straight if it is the shortest distance between any two points along the line.

The beach ball is the surface of a sphere. On the surface of a sphere the shortest distance between two points is given by a great circle. Thus, applying this generalization of Newton's law, the free motion of a body on the two-dimensional spacetime modeled by the beach ball will be some great circle around the ball. If we plot the vertical motion of an object thrown into the air at Vancouver on this model of the spacetime, that plot will have the object following a great circle (straight line) on the surface of the beach ball. Consider the great circle given in figure 4. Starting at point A, it describes the behaviour of a particle leaving the surface of the earth. Initially, as time

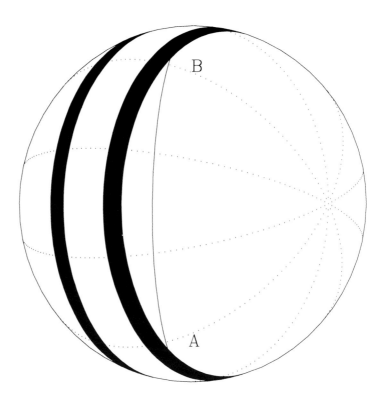

Fig. 4. A great circle leaving the earth's surface at A and returning at B

increases, the particle goes progressively further from the surface of the earth. As gravity is not a force, the particle continues to travel along the straight line (great circle). Eventually, the distance from the earth stops increasing and begins to decrease. The straight line, the great circle, re-intersects the band representing the surface of the earth at Vancouver at point B. That is, the particle has returned to the surface of the earth, just as a real body thrown away from the surface of the earth will eventually return thereunto.

Although one cannot construct as simple a model of the situation, one can show that the same concept of the unequable flow of time can describe the behaviour of the moon as it orbits the earth. The moon also is following a straight line through the spacetime surrounding the earth, a spacetime constructed so that the flow of time far from the earth differs from its flow near the earth. The line is certainly not straight in space, but it is straight if plotted in spacetime, straight in the sense of always following a path which either maximizes or minimizes the distance between any two points along the path.

1 Time, gravity, and quantum mechanics

With the above simple two-dimensional model one can also explain another aspect of gravity, namely the pressure we feel on the soles of our feet as we stand. Usually we ascribe this pressure to the response of the earth to the force of gravity pulling us down. As Einstein pointed out in 1908, there is another situation in which we feel the same pressure, namely in an elevator accelerating upwards. In that case the pressure is not due to the resistance of the floor of the elevator to some force pulling us down; rather, it is the force exerted on us by the elevator in accelerating us upwards. Thus another possible explanation for the force we feel under our feet is that the surface of the earth is accelerating upwards. Of course the immediate reaction is that this seems silly – if the earth at Vancouver was accelerating upwards and that at Isle Crozet was also accelerating upwards (since people there also feel the same force on the soles of their feet when they stand), the earth must surely be getting larger. The distance between two objects accelerating away from each other must surely be changing. In the presence of an unequable flow of time this conclusion does not necessarily follow, as I can again demonstrate with the beach ball. Both sides of the earth can be accelerating upwards even though the distance between them does not change.[4]

In our beach ball model, the diameter of the earth (the distance between the two black lines) is clearly constant at all times. Let me carefully cut out one of the black strips, the one representing the surface of the earth at Vancouver say, as in figure 5. I will lay the strip out flat, as I have done with the peeled portion of the strip in figure 5. The resulting graph is just the same as the graph of an accelerating object in flat spacetime. Local (within the vicinity of the strip) to the surface of the earth, the spacetime is the same as that around an accelerating particle, and one can therefore state that the surface of the earth at Vancouver is accelerating upwards. It is not following a straight line, it is following a curved line, and by Newton's first law must therefore have a force (the one we feel on the soles of our feet) which causes that acceleration. It is the

---

[4] It is often said that Einstein's theory describes gravity by means of a curved spacetime. Note that the model of the beach ball is in fact curved. The primary attribute of gravity, that it explains why things that go up come down again, does not require a curved spacetime. It is the fitting together of the various aspects of gravity (e.g., that the earth can have a surface that is accelerating upwards at all points while still retaining the same size) which does require a curved spacetime. Furthermore, 'curvature' is a more invariant concept than is 'unequal flow of time', which, as I will mention later, has difficulties associated with the definition of what one means by time. Unfortunately, 'curvature of spacetime' does not carry very much meaning to the average person and tends to lead to the belief that General Relativity is impossible to understand for any but a select few.

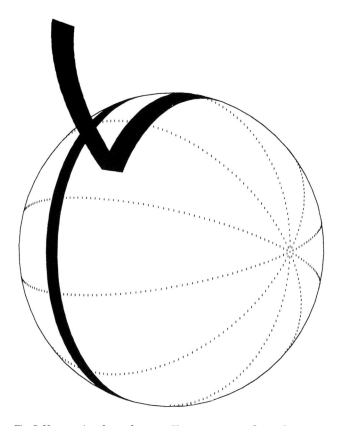

Fig. 5. Unwrapping the surface near Vancouver onto a flat surface

acceleration upward of the surface of the earth which leads to the sensation of a force on the soles of our feet.

I must at this point insert a few warnings. As I stated earlier, an analogy will often contain features which are not applicable to the system being modeled. This holds true here as well. In the lenses I used to make up the spacetime near the earth, the time decreases when we get further from the earth (i.e., the distance across the lens, and thus the time as measured by a clock, is less far away than nearby). However, the Vessot experiment showed that clocks far from the earth run faster not slower. Had we constructed our lenses so as to have them wider far from the earth than near the earth objects would have been repelled rather than attracted to the surface of the earth as represented on the beach ball. However the beach ball analogy also assumes that the notion of distance in time is identical to the notion of distance in space.

In particular it assumes that distances in time and space obey the usual Pythagorian formula, that the square of the distance along the hypotenuse of a triangle is the sum of the squares along the other two sides. However in General and Special Relativity spacetime is not strictly analogous to space. In particular, the Pythagorian formula for time contains a minus sign – the square of the distance in spacetime looks like $\Delta x^2 - \Delta t^2$ – the square of the distance along the hypotenuse equals the difference in the squares of the other two sides, if one of the sides is in the time direction. These two effects, the difference in sign and the difference in rate of flow between our model and reality, cancel so that both the analogy and reality have the feature that thrown objects return to the earth.

Another feature, that time reoccurs on the beach ball – i.e., by going forward in time you get back to where you started – is not shared by the structure of time near the earth. It is however interesting that once we have allowed time to become dynamic, once we have allowed time to flow unequably, situations like spacetimes with closed time-like curves (time travel) are at least a theoretical possibility. The past five years have seen an explosion of interest in such situations in General Relativity.[5]

A third feature of the model, that at a sufficient spatial distance from the earth the flow of time goes to zero (each lens comes to a point at the poles of the beach ball) is inaccurate for the earth and is actually also unnecessary in the model. It would have been better to use lenses whose sides became parallel far from the earth, with only a slight thickening near and through the earth. Such a model would also have allowed a demonstration of escape velocity. On the surface obtained by gluing a number of identical copies of this better lens shape together, straight lines which have a sufficient slope at the earth's surface do not return, as they always do on the beach ball, but continue to spatial infinity. This added realism would, however, have made it more difficult to see what the straight lines are in the model.

I have discussed the role of time in General Relativity at such length and in such an elementary fashion, in order to emphasize the radical nature of the change wrought by Einstein in this theory, and to emphasize that the revolution in the nature of time is not simply some abstruse and technical change, as is often claimed for General Relativity. Rather the change is simple and fundamental. General Relativity is not simply another rather complicated

---

[5] For a review of some of the recent work on spacetimes with closed time-like curves, see chapter 11 in this volume by John Earman.

field theory, but changes the very nature of physical explanation (what is gravity?) in a way totally unexpected and still largely unappreciated by the general populace. We shall see furthermore in section 3 that this change of the role of time in General Relativity has in fact led to a reintroduction of precisely the opposite of Newton's definition of time, namely the conception he ascribed to the common people (and which is today a conception utterly foreign to the common people).

2 Quantum mechanics and time

Time is also an important player in the theory of quantum mechanics, but it is in many ways a different player here than in General Relativity. It does not itself become a participant in the action, but compensates for this by assuming a much more important role in the interpretation of the theory.

I will identify and discuss, at varying lengths, four places in quantum mechanics where time plays a crucial role. Two of these are in the area of what is usually called the dynamics of the theory, and the other two are in the area of the interpretation of the theory.

(I) A theory of physics operates by elevating certain features of the physical world to the status of fundamental features. These features are modeled by certain mathematical structures, and the theory delivers its predictions and explanations in terms of those structures. One of the key features in physics is that these fundamental structures are taken to be the same at all times.

At each instant of time there is a set of fundamental dynamical variables, the fundamental quantities in terms of which the physical nature of the world is described. In quantum mechanics these are represented mathematically as linear operators on some Hilbert space. For example, if our world of discourse consists of some single particle, the fundamental variables are position and momentum. These are represented by Hermitian operators, and all physical attributes are assumed to be represented by functions of these operators. Furthermore, as long as one's world of discourse remains that single particle, these same variables will maintain their fundamental role. On the other hand, if our world of discourse is a quantum field theory (e.g., electromagnetism) these physical attributes will be the value of the field at each spatial point, together with the conjugate momentum for that particular field value.

1 Time, gravity, and quantum mechanics

It is one of the peculiar assumptions in physics that for any physical situation, the fundamental set of physical attributes in terms of which we describe that situation are the same at all times. The universe of discourse does not change from time to time. This is of course very different from most other fields of human endeavour, in which any attempt to pretend that the essential entities in terms of which the world is constructed are always the same would be silly. Institutions fail and disappear, department heads retire. In physics, on the other hand, one believes that the physical world can be described at all times by the same physical structures, and that the changes in the world are simply the changing relations between these fundamental entities or changes in the values of their properties.

One can regard this as either an admission by physicists that they will limit their interest only to those situations in which such a universality and invariance of the fundamental attributes is accurate or as the claim that all situations, no matter how complex and how changeable, can be described by such a single unchanging set of fundamental attributes. The almost universal belief among physicists is that the latter is the case, that at heart there are some universal fundamental structures that can be used to describe any physical process in the world.

Let me state this assumption in a slightly more technical vein. To do so I will review the basic mathematical structure of quantum mechanics. I would refer the reader who is not familiar with quantum mechanics to various books on the subject or to the article by Shimony.[6] In quantum mechanics, the basic entities used to model the physical world are linear operators on a Hilbert space. A Hilbert space is a collection of things, called vectors, which you can add together to get another vector or which you can multiply by a complex number to get another vector. I will denote these vectors by $|name\rangle$ where '*name*' is simply some symbol to name a particular vector. Thus linearity is expressed by the statement that if $|a\rangle$ and $|b\rangle$ are both vectors, then $\alpha|a\rangle + \beta|b\rangle$ is also a vector for arbitrary complex numbers α and β. These vectors also have associated with them a concept of 'inner product' (or 'dot product') designated by $\langle a|b\rangle$ which is a complex number associated with any two vectors $|a\rangle$ and $|b\rangle$. It is chosen such that $\langle a|a\rangle$ is always real, $\langle a|b+c\rangle = \langle a|b\rangle + \langle a|c\rangle$, and $\langle \alpha a|\alpha a\rangle = |\alpha|^2\langle a|a\rangle$ for any complex number α. (Note that the vector $|\alpha a\rangle$ is used to designate the vector $\alpha|a\rangle$ and $|b+c\rangle$ designates the vector $|b\rangle + |c\rangle$. The real number $\langle a|a\rangle$ denotes the length squared of the vector $|a\rangle$.)

[6] Shimony (1989).

These vectors in the Hilbert space do not have any direct physical meaning, but they do form an important element in the interpretation of the theory and in the establishment of the relation between the theory and particular realizations of the theory in the real world. The fundamental attributes of the physical world are the linear operators on this Hilbert space. An operator is any function, which takes as input a vector in the Hilbert space, and gives as output a possibly different vector in the same Hilbert space. If we designate an operator by the capital letter A, then we designate the result of the operation by $A(|b\rangle)$ or often by $A|b\rangle$. An operator is linear if the result of the operation on the sum of the two input vectors is the same as the sum of the operations on each of the inputs, i.e., if $A(|b\rangle + |c\rangle) = A(|b\rangle) + A(|c\rangle)$. In the theory there are two types of operators which are of special interest – unitary and Hermitian operators. Unitary operators are operators which do not change the length of a vector. That is, if the length of the vector $U|a\rangle$ is the same as the length of $|a\rangle$ for all vectors $|a\rangle$, then the operator U is unitary. Hermitian operators are those for which their 'expectation values' are real. The expectation value of an operator H in a state $|a\rangle$ is the inner product of the state $|a\rangle$ and the transformed vector $H(|a\rangle)$. In quantum theory all potential physical attributes of the world are modeled by Hermitian operators.

It would be at least conceptually possible that the Hilbert space, and thus the operators that act on the Hilbert space, could change from time to time. One could imagine that certain things which were possible, which were measurable, at least in principle, at one time, did not even exist at some other time, that the set of all physical quantities could be different at one time from the next. However this is not the case in quantum mechanics. The physical variables, the set of operators and the Hilbert space on which those operators act, are assumed to be the same at all times. As with classical physics, the change in the world that we want to describe or explain lies not in a change to the fundamental structures of the world, but in a change to the relation of these structures to each other or in the 'values' that these operators have.

This first role of time might appear trivial. Time plays a role in designating the set of fundamental attributes in terms of which we describe the world. Of course the physicist can decide to focus on some smaller set of fundamental attributes of the world that is of interest at any given time. But these changing simpler structures are not produced by some aspect of time, they are produced by the changing focus of the physicist. The belief is strongly held that at heart there exists some one set of universal operators, some one global Hilbert space, which can be used (barring technical difficulties) to describe everything in the world throughout all time.

New theories may demand new assumptions. The possibility exists that the world could change from time to time in some fundamental way, not just in detail. I will argue below that such genuine novelty may be needed in order to describe the quantum evolution of the universe.

(II) Having defined the mathematical structures used by the theory to describe the world, one must then try to use them to explain the world. In particular, one wishes to use the theory to explain the change we see about us. Since the mathematical structure is constant, the explanation for change must be in terms of changing relations amongst the fundamental entities and changes in the relations between the mathematical structures and the physical world. This explanation uses the equations of motion, equations relating the entities describing the world at one time to those describing the world at another time.

I will work in what is termed the Heisenberg representation in which the identification of the operators on the Hilbert space with some physical attribute changes from time to time. There is an equivalent representation, the Schrödinger representation, in which the Hilbert space vectors transform over time but the identification of certain operators with physical entities remains constant. The two are equivalent in their ultimate predictions, but for various reasons I favor the former. I feel that it makes the distinction between the formal, dynamical aspects of the theory and the interpretative aspects clearer.

Which operator on the Hilbert space corresponds to which physical attribute of the world? This identification can change from time to time. These changes form the essence of the dynamics of the theory, and are expressed in quantum mechanics by a set of equations of motion which take the form of the Heisenberg equations of evolution

$$i\hbar \frac{dA}{dt} = [A, H] \equiv AH - HA.$$

The term A is any dynamical operator representing some aspect of the physical world, H is a special operator in the theory, the Hamiltonian, usually identified with the energy of the system, and $\hbar$ is Planck's constant. It is by means of these dynamical equations of the theory, these changes in identification from time to time of the operators with the physical reality, that one hopes to encode the dual characteristics of the world as envisioned by physicists. That dual character is one of a fundamental identity from one instant of time to the next (at all times one can describe the world by the same set of operators), together with the possibility and reality of change and transmutation, which is so much a part of the world around us.

(III)   Quantum mechanics arose out of, and encodes within its interpretation, a very uncomfortable feature of the world, that the world seems to operate on the basis of insufficient cause. Things just happen, without our being able to ascribe any sufficient cause to explain the details of what occurred. Given two absolutely identical situations (causes), the outcomes (effects) can differ. This feature caused Einstein so much intellectual pain that he rejected the theory, even though he had been a key player in its foundation. It is still one of the most disconcerting aspects of the theory and the world. It seems to call into question the very purpose of physics, and it lies at the heart of the disquiet felt by even some of the best physicists.[7] However, all the evidence indicates that nature operates in accordance with quantum mechanics. One's first reaction would be to say that somehow at heart the universe surely operates with sufficient cause – that when we say that identical situations produce differing results, it is really that the situations were not identical, but that there were overlooked features of the world which caused the differing outcomes. However, the evidence is that this is not how the world operates, that God truly does 'play dice'.[8]

How is this element of insufficient cause encoded in the quantum theory, and how does time, the subject of this volume, enter into this encoding? As mentioned above, each physical attribute of the world is represented by an operator on the Hilbert space. But physical attributes are not seen by us to be operators, rather they are seen to have definite values and definite relations to each other. The position of my car is not some operator which moves abstract vectors around but is some number, say twenty feet in front of my house. How can I relate the operator which represents the position of my car in the theory, with this number which represents the position of my car in the world?

The answer is that Hermitian operators have the property that there are certain vectors (called eigenvectors) in the Hilbert space for which the

---

[7] Bell (1987).
[8] See Bell (1987) and various articles in Greenberger (1986), especially Aspect's.

operation of the operator is simply multiplication by some constant (called the eigenvalue). Quantum mechanics states that the set of numbers, the set of all of the possible eigenvalues of the operator, is also the set of possible values that the physical attribute corresponding to the operator can have. If my car can have any one of the real numbers as a possible value for its position, the operator representing the position of my car must have an eigenvalue corresponding to each of those real numbers. The vectors of the Hilbert space are now used in the following way. If one of the vectors in the Hilbert space represents the actual state of the world, then the theory tells us what the probability is that, given the state of the world, the value of the physical attribute corresponding to that operator takes some given value. That is, the theory does not tell us what the value of the attribute is, it tells us what the probability is that it has some value.

How do those values correspond to our experience of the physical world and our experience with those attributes that the system has? In classical physics any attribute of a physical system is taken to have some unique value at all times. In quantum theory the situation is more difficult. It seems to be impossible to hold to the classical notion of each attribute having a value at all times. However there are certain situations, called measurement situations in which the attribute is taken to have some value, because it has been measured to have a value. Each operator representing a physical value has in general many eigenvalues, and thus the attribute has many potential values. The assumption is that at any given time, if a 'measurement' is made of the attribute, one and only one of these potential values can be realized. The theory does not specify which value will be realized, but rather gives probabilities for the various possibilities. Furthermore, these probabilities are such that the probability of obtaining two (or more) distinct values is zero, and the probability of obtaining any one of the complete set of values is unity. From the definition of probabilities, if a and a' are two separate possible values for an operator A, then the impossibility of having two separate values gives $\text{Prob}(a \text{ and } a') = 0$. This then leads to $\text{Prob}(a \text{ or } a') = \text{Prob}(a) + \text{Prob}(a')$. Furthermore, if we ask for the probability that one of the eigenvalues is realized, we have

$$\text{Prob}(a \text{ or } a' \text{ or } \ldots \text{ or } a'^{\cdots\prime}) = \text{Prob}(a) + \text{Prob}(a') + \ldots + \text{Prob}(a'^{\cdots\prime}) = 1$$

where the set is the whole set of all possible eigenvalues of A.

Again this feature seems obvious, but time enters into this statement in a crucial way. The statement that one and only one value is obtained is true only at some given specific time. It is simply not true without the temporal condition 'at one time'. If we do not specify time, my car can have many different positions (and it did today). It is only at a single time that we can state that the car had one and only one position.

This seemingly trivial fact is encoded into quantum theory at the most fundamental level. The probability of an eigenvalue is given by the square of the dot product between the two vectors, or in symbolic terms, the probability that the operator A has value a is given by the square of the dot product between the eigenvector $|a\rangle$ associated with the eigenvalue a, and the state vector of the system, written $|\psi\rangle$, so that

$$\text{Prob}(a) = |\langle a|\psi\rangle|^2.$$

The statement that only one value can be realized leads to

$$\text{Prob}(a \text{ or } a') = |\langle a|\psi\rangle|^2 + |\langle a'|\psi\rangle|^2$$

and that at least one value must be realized gives

$$\text{Prob}(a \text{ or } a' \text{ or } \ldots \text{ or } a'^{\cdots\prime}) = |\langle a|\psi\rangle| + |\langle a'|\psi\rangle|^2 + \ldots + |\langle a'^{\cdots\prime}|\psi\rangle|^2 = 1.$$

It is because of this physical demand that we can use the eigenvectors of Hermitian operators as the models for physical attributes – the eigenvalues and eigenvectors of Hermitian operators have exactly this required structure.

The third role that time plays in the theory, then, is that, given some physical attribute, that attribute can take one of a whole range of values, but at any single instant in time it can take at most one of those values (the values are statistically independent), and it must take at least one of those values (the values are complete).

As with the first property of time, this property seems almost to be trivial. It is at least very difficult to see how it could be changed without completely altering the structure of quantum mechanics. Furthermore it would seem to reflect a fundamental attribute of the real world. However, as I will argue, quantum gravity may require such a change.

(IV) The fourth role that time plays is in setting the contingencies or conditions for the predictions of the theory. Theories in physics are designed to be broad. They are, especially if they are to be fundamental theories, designed to be applicable in any and all conceivable situations. They are generic and not specific, universal and not particular. But any experiment, any experience of the world is specific and particular. How can the theory be applied to these specific cases?

In classical physics, the answer lies in the 'initial' conditions. Although the theory itself is universal and generic, the specific contingencies of any particular situation can be encoded so as to make the predictions of the theory specific and particular. The theory itself identifies the dynamical variables in terms of which one will describe any situation. The equations of motion specify how the values of these variables at any one time are related to those at any other time. To complete the picture, therefore, one must specify the variables at some one time. Given the values of all of the variables at one time, the values at any other time are completely determined by the theory through the equations of motion. This specification of the values of all of the variables is given the name in classical physics of 'initial conditions'. Although the term 'initial' is used, there is no need that the values be specified at a time earlier than the time of interest, or even that the variables all be specified at one time. For example, instead of specifying the values of the position and momentum of a particle at one time, one can specify the position at two separate times. Both have the effect of completely particularizing the theory. Having specified these initial conditions, the theory provides a complete model for the world in some particular situation and for all times. Any new information gleaned about the system, which takes the form of finding new values for some variables at different times, must either agree in particular with the values predicted by the theory, or the theory must be thrown out as a model for the physical situation. This led to Laplace's famous statement

> An intelligence knowing all the forces acting in nature at a given instant, as well as the momentary positions of all things in the universe, would be able to comprehend in one single formula the motions of the largest bodies as well as the lightest atoms in the world, provided that its intellect were sufficiently powerful to subject all data to analysis; to it nothing would be uncertain, the future as well as the past would be present to its eyes.[9]

This is not true for quantum mechanics, although three hundred years of classical physics still exerts its views on quantum theory, and quantum mechanics is often treated as though it were cast in the same mold.

[9] From the introduction to Laplace (1820) as translated in Nagel (1961), 281–2. Note that the correct initial conditions for most physical theories are not those given by Laplace.

As mentioned, quantum mechanics is a theory of insufficient cause. The complete specification of the theory at one instant of time is not sufficient to completely specify the outcomes of any experiment at all other times. Some things will 'just happen'. As a result the effect of the setting of the conditions on the predictions of the theory are much more subtle, complex, and profound than they are in classical physics.

Let us begin with the simplest text book case. Let us say that at some time t_0, we know[10] that the dynamic variable A has the value a. As I stated, the value a must be one of the eigenvalues of A and has associated a vector in the Hilbert space called the eigenvector, which I will write $|a\rangle$. Since we know the physical variable A to have value a, we need the probability that it has value a to be unity, and the probability that it has value a' different from a to be zero (it can have only one value at a given time). As stated previously, the probability of having a value a' for some vector in the Hilbert space $|\psi\rangle$ is given by the square of the dot product between the eigenvector $|a'\rangle$ and the state vector $|\psi\rangle$. It is one of the fortunate features of Hermitian operators, that $\langle a'|a\rangle = 0$ for eigenvectors associated with different eigenvalues of the same operator. If we choose the state vector to be $|\psi\rangle = |a\rangle$, we will precisely encode the belief that we know the value of variable A to be a. The state vector for the system is thus the way we have of encoding the conditions under which we want the theory to deliver answers to us. In particular we choose the state vector to encode the information that we know that some physical property has some definite value.

Now, for any other physical variable B, at a later time say, the probability that B has value b is given by the square of the dot product $|\langle b|a\rangle|^2$. Note that this does not in general lead to the statement that the system has some value b at that later time, as it would in classical physics. It leads to the statement that there is some probability that it has the value b. However, what value

---

[10] In the interpretation of quantum mechanics the term 'measurement' has two separate meanings. It can refer to the acquisition of knowledge about the physical world, or it can refer to the physical process which results in the acquisition of that knowledge. In the latter case measurements, as a physical process, are part of the physical world and should therefore be analyzable in terms of one's theory of the physical world (quantum mechanics, in this case). I will retain the term 'measurement' for this activity. In the former case, the term plays a more axiomatic, and thus unanalyzable, role in the theory. I will use the term 'knowledge' instead of measurement for this sense in order to avoid confusion between the physical and axiomatic roles that the term 'measurement' plays in quantum theory.

The emphasis that the term 'knowledge' places on the acquisition of information about the world as a human activity is a more accurate portrayal of the *product* of the measurement *process* in quantum theory. If use of the term 'knowledge' bothers the reader, however, it can be replaced almost everywhere by the more traditional term, 'measurement'.

it will actually be found to have if it were measured is unknown. It could have any of the allowed values (i.e., those with non-zero probability). The actual value 'just happens', and the theory can give no further reason as to why that, and not one of the other possible values, is obtained.

This procedure has much of the flavor of classical physics. Knowing that A has value a allows us to assign as an 'initial condition' the 'value' $|a\rangle$ to the wave function of the system. Future predictions now use this 'value' of the state vector. That the 'value' is actually a vector in the Hilbert space rather than the assignment of the value a to some specific variable could be seen as a difference in detail rather than essence. The theory differs from classical physics, but it would seem in this description to fall into the same 'initial condition – equation of motion' framework as classical physics. This is especially true if one works in the Schrödinger, rather than the Heisenberg representation of quantum mechanics where the dynamical evolution is encoded in a time variation of the state vector of the theory rather than in a changing identification of operators with physical attributes. The Schrödinger equation looks like

$$i\hbar \frac{d|\psi\rangle}{dt} = H|\psi\rangle,$$

and the specification of the condition now looks like a specification of the initial conditions for $|\psi\rangle$. This fact together with the three hundred year involvement with classical physics has helped to confuse students into thinking that quantum physics really is little different from classical physics.

The differences between the two are not simply matters of detail, however, but are really a matter of essence. This becomes clear if we now ask a further question. In addition to knowing that the value of A at time t_0 is a, I now also know that at the time t_1 the value of B is b. How do I encode this additional information into the theory? In classical physics you do not. The additional information is either redundant (in that it does not alter the predictions which could have been made using a alone), or it is inconsistent, in which case the theory must be scrapped. (I am assuming that the knowledge of a was complete, in that it specified the value of all dynamic variables in the classical theory. If not, the additional knowledge of b could further refine one's knowledge of the 'initial conditions'. The knowledge of b could not make something that was impossible knowing a alone into something possible with the extra knowledge.)

In quantum mechanics the situation is very different. The knowledge of a predicted only a probability for the various values that B could have. The additional information that B had value b is therefore certainly not redundant – the knowledge of the actual value is obviously a much stronger piece of information than merely a list of probabilities. It is also in general not inconsistent – anything with a non-zero probability could, after all, occur. Since the knowledge of b and a is stronger than the knowledge of a alone, how is this new stronger piece of information incorporated into the theory for the prediction of the values of some other variable C say? The elementary answer is that if B is later than A, then B supersedes A. That is, for all measurements made after that of B, one uses the wave function $|b\rangle$ as $|\psi\rangle$ rather than $|a\rangle$.

The process of replacing the knowledge of A by the knowledge of B is traditionally called the 'collapse of the wave packet'. It has caused much confusion in the literature. In the Schrödinger representation, in which the wave function changes both due to the dynamic evolution and due to such 'collapses', this change in the state due to a change in knowledge has called forth much, in my opinion misplaced, speculation about the dynamics of such a 'collapse'. The collapse is not dynamics. It is the way that new information, new conditions, not contained in old information, is incorporated into a theory of insufficient cause.

This rule, that later knowledge supersedes earlier, has also led to comments that quantum mechanics is, in some sense, inherently time asymmetric. After all, the latter supersedes the former, not the other way around. However, the appearance of time asymmetry is due to the fact that the question being asked is inherently time asymmetric. The latter replaces the former *if* one is asking questions about the system at an even later time. The latter does not supersede the former in other cases. To highlight this point let us ask a different type of question. Given that I know that A had value a at time t_0, and that B had value b at a later time t_1, what are the probabilities that C had value c at an intermediate time t_2 between t_0 and t_1?

As specific examples are often more comprehensible than general arguments, I will present a specific example, but the conclusions drawn will have general applicability. As always in the field of interpretation, one tries to work with as simple a system as possible so as not to obscure the essential point with a forest of technical detail. I will therefore take ubiquitous spin $\frac{1}{2}$ system, with the most trivial of Hamiltonians, namely zero. The dynamics are therefore trivial in that nothing ever changes – the operators at different times are related to each other by being the same at all times. The dynamical equations

of motion are
$$\frac{dA}{dt} = 0.$$

A spin $\frac{1}{2}$ quantum system has the peculiar property that the value of the spin in any direction can only take one of two values, namely $\pm\frac{1}{2}$. We are going to assume that at some time, say one of 9 a.m., 10 a.m., and 11 a.m., we know that the physical system has a value for the x component of the spin of $+\frac{1}{2}$. Similarly at another one of those three times, we know that the system has a value of $+\frac{1}{2}$ for the y component. The question we ask is 'What is the value of the component of the spin along an axis between the x and y axes making an angle of θ to the x axis at the third time?' In classical physics, the answer is simple and straightforward. Because the dynamics is trivial, we then know that the x and y components of the spin are independent, and both had value $+\frac{1}{2}$ at all times. Furthermore since spin is a vector, the midway component, let me call it S_θ, will just be an appropriate sum of the two known vector components, and must have a value of $(\sqrt{2}/2)\cos(\theta - 45°)$. Having specified the values of S_x and S_y the values of all intermediate components are specified uniquely at all times.

In quantum mechanics on the other hand, the situation is more complicated. There are in principle six different answers, depending on the times at which the system was assumed to have had those values in relation to the time about which we are asking the question. Let me write $S_a S_b S_c$ to designate the condition that the spin has value S_a at 9 a.m., S_b at 10 a.m., and S_c at 11 a.m. The various possibilities for the temporal order of the conditions are the six permutations

(a) $S_x S_y S_\theta$ (b) $S_y S_x S_\theta$ (c) $S_x S_\theta S_y$
(d) $S_y S_\theta S_x$ (e) $S_\theta S_x S_y$ (f) $S_\theta S_y S_x$

I will concentrate on the first four of these. Although one can say something also about the last two cases, the potential controversy would hide the point I am trying to make. In the first two cases, the answers quantum mechanically are different, while classically they must be the same. In the first the value of S_y supersedes that of S_x in determining the probabilities for the two possible outcomes of S_θ. In the second the value of S_x supersedes that of S_y. In each case the 9 a.m. condition is irrelevant, because the 10 a.m. condition completely supersedes it. One says that the state of the system is the state determined by the 10 a.m. condition, i.e., it is the $+\frac{1}{2}$ eigenstate of S_y and S_x respectively, and the prediction for the probability that S_θ has value $+\frac{1}{2}$ at 11 a.m. is $\frac{1}{2}(1+\sin(\theta))$ in case (a), and $\frac{1}{2}(1+\cos(\theta))$ in case (b). Note that these are not the same as each other.

The cases (c) and (d) both also have unambiguous answers in quantum mechanics, and both are identical. The probability that S_θ will have value $+\frac{1}{2}$ is

$$P_\theta = \frac{1+\cos(\theta)+\sin(\theta)+\cos(\theta)\sin(\theta)}{2(1+\cos(\theta)\sin(\theta))}.$$

This probability function has at least one peculiar property. We see that the probability of measuring S_θ to have value $+\frac{1}{2}$ is unity (certainty) both when θ has the value zero and when θ has the value 90°. That is, the probability that one will measure $S_x \equiv S_{0°}$ at the intermediate time to have value $+\frac{1}{2}$ is unity and the probability that the measurement of $S_y \equiv S_{90°}$ will have value $+\frac{1}{2}$ is also unity. It is however easy to prove that there exists no state vector whatsoever in the Hilbert space of this spin $+\frac{1}{2}$ particle which could give this result. That is, there exists no $|\psi\rangle$ such that $|\langle+\frac{1}{2},x|\psi\rangle|^2 = |\langle+\frac{1}{2},y|\psi\rangle|^2 = 1$.

We learn from this example that the temporal ordering of the conditions that one places on the question that one asks of the theory are crucial to obtaining answers from the theory. Unlike classical mechanics, the conditions cannot in general be mapped back onto initial conditions. Time, and in particular temporal ordering, is needed in a crucial way not just to set up the dynamical relations, but also to extract sensible predictions from the theory.

Another lesson we can learn from this example is that quantum mechanics does not have an inherent time order to it. If it had, one would have expected the answers to conditions (c) and (d) to differ. After all the temporal order is completely reversed in the latter with respect to the former. However the predictions are identical. The fact that (a) and (c) differ, even though just the temporal order of S_y and S_θ have changed is no surprise since the condition on S_x at 9 a.m. in both cases ensures that these are not the time reverse of each other.

This is of possible relevance to the discussion that Roger Penrose gives in his book, *The Emperor's New Mind* (1989). He argues that quantum mechanics itself implicitly contains a time ordering, that the specification of the conditions leads to a clear and natural time ordering. The relevant section occurs on page 357, where he describes an experiment in which a lamp is placed in front of a half-silvered mirror. One now places a photon detector just in front of the mirror, to detect when the lamp sends a photon toward the mirror, and a detector behind the mirror. One finds that although the probability is only $\frac{1}{2}$ that the second one

1 Time, gravity, and quantum mechanics

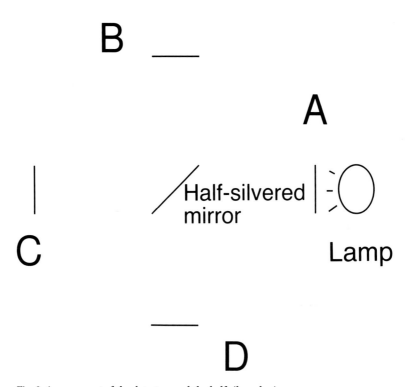

Fig. 6. Arrangement of the detectors and the half-silvered mirror

will have detected a photon when the first one does, the probability is unity that the first one will have when the second one is found to have detected one. He then adduces this as an argument in favor of the position that quantum measurement contains within itself an arrow of time (as well as a time ordering). That is, placing the condition on the photon exiting and asking for the probability that the photon entered is not the same as placing the condition on the photon entering and asking the probability that it exited. Let me rephrase the problem posed by Penrose in a slightly diffcrent way. Consider a half-silvered mirror, with non-destructive photon detectors placed at ports A, B, C, and D (see figure 6). Each detector detects the presence of a photon but lets it pass by unimpeded. Suppose we come across the notebook of an experimentalist in which she has carefully recorded the readings from all the detectors at all the ports. As expected, anytime either A or B gave a reading, so did C or D. Furthermore, A and B never gave readings together, nor did C and D. Beyond these facts however, the readings turn out to be slightly peculiar. The notebook never reported B as giving a reading. It seems that each photon always passed through port A. The results for the C and D ports were however roughly equally split, in that 50% went through C and 50% through D. The experimentalist furthermore stated that all four detectors were working and that all readings in which at least one detector fired were recorded. Do these results imply a time ordering in and of themselves? Does this time ordering imply that the time ordering is an inherent part of quantum mechanics? The naive answer is that the photons must have come in through port A and exited through either of the ports C and D. We know that there is no quantum state which would allow a photon to come in through port C and exit, with unit probability, through port A, while states exist for which the photon could enter port A and exit with 50–50 probability through port C or D. Does this imply a time asymmetry in quantum mechanics as Penrose implies?

The problem is that there is something to be explained in either direction of time. Ports A and B are completely symmetric. Why is there then an asymmetry in the readings at the two ports, i.e., port A is the only one with any detections? The answer that Penrose would give when we interpret A as the input port is that there is a lamp at A and not one at B. The lamp conditions the initial state of the electromagnetic field (i.e., we know that there are photons in the initial state caused by the lamp). However the measurement at A coming later than the conditioning by the lamp supersedes the effect of the lamp, if we are interested in the outcomes at ports C and D. All we need to calculate the measurements at C and D is this knowledge that the photons came through A. If we now regard the system in the time reverse sense, where A is an output port, there is again something to be explained, namely why did all of the photons exit through A? The photons come in at C or D with equal probability (which we would have expected *a priori*). However none exited at B. We cannot explain this from any initial conditions, because the knowledge that the photons came in through C or D will supersede any other initial knowledge. We can however explain it through a final condition, a condition which for some reason or another disallowed any recorded photons at B. Because we are now specifying both initial (photon through either C or D) and final conditions, there will be no wave function which encodes these conditions. However quantum mechanics can still make well-defined predictions even in such an intermediate time case.

Now, one might ask how one could arrange these final conditions in any natural way. After all we all know how to set up the time reversed situation easily – light sources are easily built to send photons into port A rather than port B. The reverse of this experiment with something disallowing photons from exiting from B is not easily arranged. However this takes us into the whole realm of the usual statistical arrow of time. Certain conditions are easy to arrange, and certain conditions are very difficult to arrange in the world we live in. Cups of tea cool but do not heat up when placed in a room-temperature environment. The difficulty in arranging the latter is not generally accepted as a proof that the fundamental laws of physics are time asymmetric. Similarly, the difficulty of arranging an anti-light sink at B does not imply that there is anything fundamentally time asymmetric about quantum mechanics.

The fact that quantum mechanics itself does not pick out a direction in time was recognized over 20 years ago by Aharonov, Bergmann, and Lebowitz.[11] They also suggested the formalism for handling cases in which the conditions do not necessarily precede the times at which the predictions are to be made. That formalism has been independently rediscovered a number of times since by others (including me), which illustrates the lack of impact that this fundamental insight has had at least on the teaching of the subject. It also illustrates the fact that the very different role played by the conditions in quantum mechanics from that in classical mechanics has still not been instilled into the thoughts of most of the practitioners of the subject. In the appendix I have outlined this formalism. For a more detailed exposition I would refer the reader to the literature.

Aharonov and his collaborators have recently been emphasizing the peculiarities of situations in which one sets conditions both before and after the time at which one wishes to discuss the possible outcomes of experiments. Because of the paradoxical nature of some of these results, I will present one of their examples here.[12] This is done to further reinforce the point I am making about the difference in behavior between the conditions in quantum mechanics and the initial conditions in classical mechanics, and in addition illustrates the point that the subject of knowledge and the relation of knowledge to physical measurements can be subtle in quantum mechanics.

The spin of a system in quantum mechanics is a vector operator $\vec{S}$, and represents an internal type of angular momentum for the system. Angular momentum in quantum mechanics has a number of strange properties, one of which is that it cannot take on any arbitrary value. The eigenvalues for the operator corresponding to the total spin, namely $\vec{S} \cdot \vec{S}$, take on only a

[11] Aharonov, Bergmann, & Lebowitz (1964).
[12] See Aharonov, Albert, Casher, & Vaidman (1986).

range of values of the form $s(s+1)$, where s is an integer divided by two. That is, the allowed values are discrete. Furthermore, if the total spin is s (which is the conventional shorthand for saying that the operator $\vec{S} \cdot \vec{S}$ has value $s(s+1)$), then any component of the spin can only have values lying between $-s$ and s and the value must differ from s by a whole number. Consider some system with total spin s. At time t_0 we measure the value of the spin in the x direction (i.e., the x component of the spin), and find it to have value s (i.e., the maximum amount). At time t_1 we measure the spin in the direction lying in the (xz) plane half way between the x and z axes (i.e., we measure the operator $S_{45} = (S_x + S_z)/\sqrt{2}$). This measurement we carry out inexactly, to an accuracy of only of order $\pm\sqrt{s}$. Finally at time t_2 we measure the spin in the z direction, and again find it to be s. What is the probability distribution for the outcomes of the intermediate measurement of S_{45}? The answer turns out to depend on exactly what one means by 'measure inexactly', but there is a perfectly well defined and acceptable meaning in which the answer is that the measurement of S_{45} gives a value of $\sqrt{2}s$, i.e., about 40% larger than the maximum eigenvalue that S_{45} has. Such an answer is obviously silly if at the intermediate time we could imagine the spin system to have some wave function which describes its state, since the mean value of the probability distribution over the eigenvalues must be smaller than the largest possible value. This strange, but true result arises from a conspiracy between the fact that we are asking an intermediate time question (i.e., our conditions are not purely initial or final conditions, but rather are mixed conditions), and the fact that the intermediate measurement is inexact. Note that if the inexactness of the measurement is of order $\sqrt{s}$, then for a sufficiently large s, $0.4s \gg \sqrt{s}$, i.e., the error is much smaller than the deviation from the maximum value.

Let me put a bit more flesh onto the above bones. The initial and final measurements are assumed to be perfect exact measurements. For the intermediate measurement we will institute the requirement that the measurement be inexact by coupling the spin system to another system which will be our measuring apparatus. The measuring apparatus will be assumed to be a free particle of infinite mass and zero potential energy, i.e., I will assume that the free Hamiltonian of the particle is zero. (This is supposed to represent say the dynamics of a massive apparatus pointer.) The coupling will be taken to be such that the interaction between the apparatus and the spin system is instantaneous (i.e., a delta function in time) and is proportional to the

momentum (p) of the particle multiplied by the spin. That is, the full Hamiltonian for the system is
$$H = S_{45}\, p\delta(t - t_1).$$
In the Heisenberg representation, we find that the dynamics of the apparatus is given by
$$p(t) = p(0)$$
$$q(t) = q(0) + S_{45}$$
$$S_{45}(t) = S_{45}(0)$$
$$S_{-45}(t) = \cos(p)S_{-45}(0) + \sin(p)S_y(0).$$
The inaccuracy of the measuring apparatus will be introduced by assuming that it is the value of q which will be used to infer the measured value of S_{45}. That is, we will measure q exactly after the coupling between the particle and the spin has completed and use that value to infer the value of the 45° component. To obtain the value of this component we must subtract the final value of q from the initial value of q since the coupling to the spin causes the value of q to change. To mimic the inaccurate measurement, I will assume that the initial value of q is not exactly known, that the initial state of the apparatus is such that the initial q has a spread in possible values over a range $\Delta q = \sigma \approx \sqrt{s}$. I will take the initial wave function for q to be a Gaussian, centered at 0 with standard deviation of σ.
$$\Psi(q) = \frac{1}{(\pi\sigma)^{1/2}} e^{(q/\sigma)^2}.$$
This measuring apparatus does behave like a proper apparatus should. *If* we assume that the state of the spin system is in fact an eigenstate of S_{45}, the final probability distribution of q values is just a Gaussian, centered around that eigenvalue of S_{45}, with width σ. That is, the best estimate for the value of S_{45} will just be the value measured for q with an uncertainty in the inferred value of $\pm\sigma$. I have carried out the calculation for the situation presented in our problem above, namely that S_x had value s before the measurement and S_z had value s afterwards for a value of $s = 20$, and for various values of σ. These results are presented in figures 7–10. In figure 7, I have taken $\sigma = 0.5$. The intermediate apparatus is sufficiently accurate to distinguish between the various possible values of the spin components (which we recall must be separated by unity from each other). The result is as expected, a series of Gaussian peaks

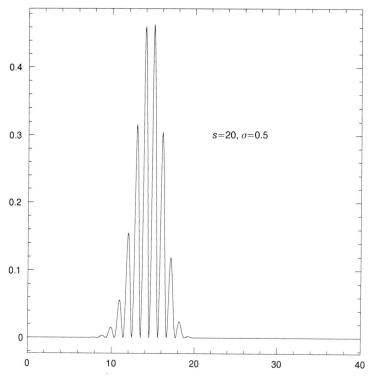

Fig. 7. The probability distribution for the pointer measuring the spin with standard deviation 0.5

centered on the expected possible values for the total spin, with various heights representing the differing probabilities that the spin has those different values.

In figure 8, I have increased σ to 1, and the measurement is not as accurate. There is still a series of peaks, but these are no longer centered on the values we would have expected for the spin, i.e., they are no longer centered on the integers. In figure 9, with $\sigma = 2$ this trend away from the naive expectation has continued. The spacing between the peaks is definitely greater than unity, and the peaks in the probability distribution have begun to extend beyond the maximum possible value for the spin, namely 20. Finally, in figure 10, σ has the value of 5. This is far too coarse to be able to distinguish individual spins, which have a spacing of unity. However we notice that the expected value of q is now about 28, 40% higher than the maximum possible value that the spin is supposed to be able to have. The measurement of q would give an inferred value of the spin higher than it could possibly be. One might at this complain

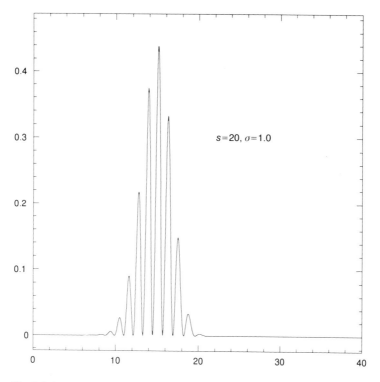

Fig. 8. Pointer measuring the spin with standard deviation 1

that the measuring apparatus is poor, that it does not measure the spin properly. For almost all normal situations it is, however, a good measuring apparatus for the spin. In all normal situations, in which one specified only an initial condition and not a final condition as well, the outcome would have been exactly what one would have expected – namely a sum of Gaussian peaks centered around the allowed values of the spin. It is because one has instituted both initial and final conditions that the measurements have produced the strange result of a value much higher than the maximum allowed value.

This is a wonderful example of the unusual nature of the effect of the imposition of conditions in quantum mechanics. The conditions do not reduce simply to initial conditions. There are no initial conditions whatsoever which could give the results of this *gedanken* experiment. It is precisely because of the insufficient-cause nature of quantum theory that the temporal order of the conditions is crucial. New knowledge changes the results of the theory in a way completely unexpected in classical physics.

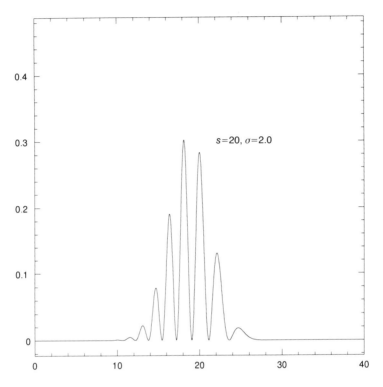

Fig. 9. Pointer measuring the spin with standard deviation 2

3 Quantum gravity

In this section I will argue that attempts to quantize a theory like General Relativity, in which time plays the central role, will potentially meet with problems in all four of the roles that time plays in quantum mechanics. I say potentially, because we do not at present have a quantum theory of gravity. There may be some subtle way in which the difficulties can be avoided which we do not at present recognize. The ultimate theory of quantum gravity may be so different from our present notions that the role of time will not even be an issue. (In the same way that quantum mechanics itself was so different from classical theory that issues which arose in the latter were not even a question in quantum theory. Furthermore, the formalism of quantum mechanics itself suggested interpretations which would never have occurred in classical physics.) Despite the complication of discussing a non-existent theory, it may be worthwhile to look at the difficulties in the hope that a

1 Time, gravity, and quantum mechanics

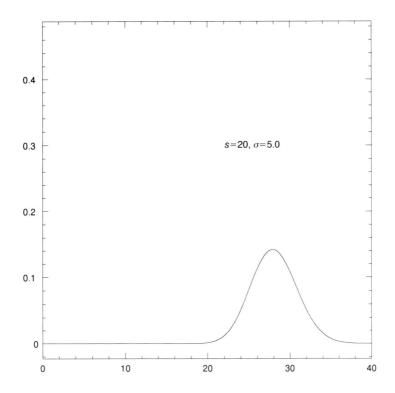

Fig. 10. Pointer measuring the spin with standard deviation 5

clear view of them may suggest a solution to the problems as well. (See also footnotes 13 and 14.)

(I) Mathematical structure: what is the Hilbert space and what are the operators of the theory? And does one expect the Hilbert space and the physical operators to be time independent? I will tackle the second question, and leave the first to paragraph (II). One of the striking predictions of Einstein's theory of gravity was that not only was time intimately involved in gravity but so was space. In particular one of the first sets of solutions to the theory were solutions in which space became dynamic, and the universe as a whole grew in size. These solutions, first discovered by Friedman, and hated at first by Einstein, led to what we now know as the Big Bang theory for the origin of the universe. In the popular imagination, the Big Bang is like an anarchist's

[13] Kuchar (in press).
[14] Unruh & Wald (1989).

bomb, in which at some moment the universe, the cosmic egg, exploded, and what we now see are the fragments of that explosion hurtling away from us. As usual, the popular image misses the most radical part of the theory. It is not that there was some explosion in a preexistent space, but rather that the universe was born very small, and as time went on, the universe created more space for itself to live in. The reason that we see the distant galaxies recede from us is not that they are moving away as the remnants of some inconceivable initial force, but rather that the distance between us and them is increasing due to the creation of new space between us and them at a more or less constant rate. Because of this increase in the amount of space, anything in the universe is continually being diluted. In particular, anything which now has some macroscopic scale, very early on had a much, much smaller scale (by a factor of about 10^{25} on the most naive assumptions, and by a factor of at least 10^{70} if inflation occurred).

Does this also mean that because there was less space early on, the universe also had fewer physical attributes? My suggestion is yes. Modern physics sees the fundamental structures which make up the universe as being fields. Fields, like the electromagnetic field, are entities that have values at each point in space. The number of different physical attributes of the world are thus related to the number of different field values at the different points there could be. At first this would seem to suggest an infinite number, since there are an uncountably infinite number of points in a classical space. However, gravity itself would be expected to put a limit on the number of different values that any field could have. If the field fluctuates at a sufficiently small scale, those fluctuations would have a sufficient energy to collapse into a black hole. That is, we would expect gravity itself to put a limit on the scale of the fluctuations, a limit which very naive estimates would put at scales of the order of the Planck scale – 10^{-33} cm or 10^{-44} sec. One would expect that field variations on scales smaller than these scales simply would not exist. If the above is true (and it is very naive, although it gains a slight amount of support by realizing that the entropy of a black hole is directly related to the number of distinct regions of Planck size that there are on its horizon), then the number of distinct field values might be finite in any finite volume of space, and furthermore, the number would decrease as the universe decreased in size. This would suggest that the critical assumption behind condition (I) in section 2 would be wrong, that one should rather describe the universe with a Hilbert space and a set of operators which could change in time, rather than stay constant. One of the real difficulties is that no-one knows how to implement a theory where the set of operators and the Hilbert space change in time. One would also have to face the question of what the laws would be which would determine such changes. This idea has therefore received almost no study but remains a disquieting possibility.

(II) Dynamics: in the model I presented for the unequable flow of time near the earth, I used the notion that the structure of the spacetime near the earth is essentially constant throughout time. This allowed me to build up the spacetime out of a number of identical pieces. It was also this notion of a time with respect to which nothing changed that allowed me to talk about an unequable flow of time. It was the distance notion of time, the time measured by clocks or other dynamical processes, which 'flowed unequably' with respect to the time, the 'symmetry time', with respect to which the spacetime was the same from time to time. In the generic situation one does not have any symmetry time, any time with respect to which nothing changes. Rather one must introduce an arbitrary notion of time, what is called coordinate time, with respect to which one measures change. Although at first this arbitrariness in the definition of time would seem to make any definite statements impossible, the techniques of differential geometry, which were developed at the turn of the century just before Einstein developed his theory using those techniques, allow one to identify physical features which are independent of the arbitrariness of the choice of this coordinate time. However, this arbitrariness has a consequence for the quantum theory. One of the consequences of the arbitrariness in the choice of coordinate time is that the natural variables in terms of which one would describe the spacetime are themselves partially redundant. There are too many natural variables, and the theory demands that some of them obey an equation known as a 'constraint equation'. Such constraints are relations between the variables at one instant of time rather than relations between the variables at a different instant of time, as in the equations of motion. These constraints are very difficult to implement quantum mechanically. The most natural technique, introduced by Dirac to handle any theory with redundant variables, leads to a restriction of the Hilbert space on which the physical theory is to be defined. This leads as well to a restriction on the operators. Because of the arbitrariness of time in

General Relativity, these constraints on the natural variables lead to a Hilbert space such that the only natural operators defined on this space are the constants of the motion. Only those features of the spacetime geometry which do not change from arbitrary time to arbitrary time seem implementable as operators on the Hilbert space. Thus the dynamical content of the theory seems to be trivial – nothing changes.

How then do we describe or explain the change that we experience in the world around us? This is one of the questions that most bedevils any putative theory of quantum gravity. Although many have thought about the problem, and some have felt that they have solved it to their own satisfaction, there is no generally accepted answer. Moreover I feel that all of the suggested answers (including my own) have severe difficulties. How can change be described in a theory where the only valid physical quantities apparently do not change?

(III) Probabilities: one of the suggested resolutions of the problem mentioned above is to select one of the variables of the unconstrained theory as the time. The physically intuitive reasoning is that time in reality is an unobservable feature of the world anyway. What actually passes for time is the reading on various and sundry pieces of physical apparatus called clocks. If you as a child are late for school, it is not because your arrival at the school is late in relation to any abstract notion of time. It is rather that the reading on the face of your teacher's watch is later than the reading at which school was supposed to start. Note that this approach is in direct contradiction to Newton's approach as stated in the quote at the beginning of this chapter. Time, according to the proponents of this view is exactly the common view, and Newton's non-relationist view is wrong. The key problem with this approach is that it removes the foundation for the third aspect of time in quantum mechanics. At any one time, any variable has one and only one value. It is this which physically justifies the whole Hilbert space structure of quantum mechanics. But any real physical watches are imperfect. It can be proven that any real watch not only has a finite probability to stop, it has a finite probability to run backwards. Now as long as the watch is simply the measure of some outside phenomenon, one could take these probabilities into account. If, however, time is *defined* to be the reading on the face of the clock, the question as to whether or not the clock can stop or run backwards is moot – it cannot by definition.

1 Time, gravity, and quantum mechanics

However the other question now raises itself, namely what basis have we for the assertion that at a time a physical quantity can have one and only one value. For example, at the same readings on a broken watch, a physical quantity can have an arbitrarily large number of readings. There is furthermore no reason why the probabilities should add up to unity if the events are not mutually exclusive. They could add up to far more than unity, and still be in accord with probability theory if they are not independent, mutually exclusive events. What is the mathematical structure which should be used if time no longer plays this third role of defining the sets of mutually exclusive and exhaustive possibilities?

(IV) Conditions: we argued in part (IV) of the last section that the setting of conditions in quantum mechanics was more subtle and rich than in classical theory. In particular, the setting of conditions was not equivalent to the setting of initial conditions. Furthermore, unless the experiments were structured so that the conditions always preceded the questions, the usual use of a state vector to encode the conditions was inapplicable. How does this apply to the attempts to quantize gravity?

One of the approaches is to regard the constraints as a sort of Schrödinger equation, called the Wheeler–DeWitt equation. The solutions to this equation are then supposed to correspond to 'wave functions of the universe'. If we furthermore regard, as in paragraph (III), one of the unconstrained variables as the time, the equation is supposed to represent the probability of measuring various quantities at the 'time' denoted by the value of the variable chosen to play the role of the watch.

In this approach, one of the questions has been: how does one choose which of the possible solutions to the Wheeler–DeWitt equation is supposed to represent the real world? One of Hawking's key contributions to the field was to suggest that there was a natural choice for the 'wave function for the universe'. He suggested that the 'initial value' problem for the wave function could be solved in a natural way. The technique involves a trick. One can rewrite the Wheeler–DeWitt equation in terms of what is known as a path integral in which one integrates the exponential of a function of the geometrical configuration called the action. If instead of using the action in which the time component has the minus sign in the Pythagorian formula, one instead uses the geometries in which the time component is treated in all respects like a space component, one can find a natural solution to the Wheeler–DeWitt equation. This solution is obtained by performing the path integral over all geometries which are completely closed except for a boundary representing the geometry one is interested in. This approach is described in Hawking's best seller, *A Brief History of Time* (1988).

However in light of the discussion I gave in section 2, the problem of setting the conditions in quantum theory is not a problem of initial conditions. In particular in the case in which the definition of time, of before and after, is problematic, one does not expect a wave function to properly encode the setting of conditions. Since the physical answers differ substantially when the time ordering is changed, the lack of such a time ordering in the quantum theory of gravity makes this a critical issue. In what order are the conditions to be set? If a wave function is not the right formalism for describing the theory, what is the correct formalism?

As should be obvious, the conflicting roles of time in gravity and in quantum theory have raised a number of difficult issues. Although the problems seem no nearer resolution than they did forty years ago, it does seem that the effort to understand the problems has given us a much better understanding of the roles played by time in both General Relativity and in quantum theory.

One of the only people to try to come to terms with the above difficulties in the formulation of quantum theory for gravity has been Jim Hartle.[15] In a series of recent papers he has been applying the consistent histories formalism (briefly sketched in the appendix) to the problems of quantum gravity. His attempts are still in an embryonic stage, but the formulation is one in which the usual operators on a Hilbert space approach of conventional quantum mechanics is abandoned. Instead, starting from Feynman's 'sum over histories' approach, he casts quantum gravity directly in terms of histories of observation, histories which need not be tied to particular instants of time or locations in space, incorporates the possibility for the setting of conditions at arbitrary instants and ordering in time, and also allows one to speak of situations in which the possible observables can change in time, as discussed in paragraph (I) above. The exact nature of time, of dynamics and change, has still to be elucidated in my opinion, but it is possible that our present notions of time can arise from the theory as an approximation. But to go into the details of his attempts

[15] Hartle (1991), (1993), (1994).

would make this chapter much longer than it already is, and I will refer the reader to his papers instead.

Appendix

In this appendix, I want to introduce the formalism, developed by Aharonov, Bergmann, and Lebowitz,[11] by Griffiths,[16] by me,[17] and by Gell-Mann and Hartle[18] to describe quantum mechanics under circumstances where the conditions are set at arbitrary times and not solely before the time of interest. This formalism is used by all of the above people in different ways, and Griffiths and Gell-Mann and Hartle have tried to use it to define a new interpretational scheme for quantum mechanics. I will not go into the details of that scheme here but refer the reader to the literature.

To develop the formalism, we must introduce some notation. We will again operate in the Heisenberg representation. For any operator A with a discrete eigenvalue a, we can define a Hermitian projection operator P_a, such that $P_a^2 = P_a$, such that it commutes with A, and such that it obeys $AP_a = aP_a$. To use the language of Gell-Mann and Hartle, we now define a 'history' as a sequence of eigenvalues of various operators at different times, $a_1, b_2, \ldots, z_n$ where a_1 is an eigenvalue of A at time t_1, etc., and $t_1 > t_2 > \ldots > t_n$. They define an operator

$$C_{a_1 b_2 \ldots z_n} = P_{a_1} P_{b_2} \ldots P_{z_n}.$$

This operator represents the successive projection onto the eigenstates of the sequence of operators $A_1, \ldots, Z_n$. Let the 'vector' of eigenvalues

$$\mathbf{v} = (a_1 b_2 \ldots z_n)$$

represent a history, so we can write

$$C_{a_1 b_2 \ldots z_n} = C_\mathbf{v}.$$

Now we define the 'decoherence' function

$$D(\mathbf{v}; \tilde{\mathbf{v}}) = \text{Tr}(C_\mathbf{v} \rho_I C_{\tilde{\mathbf{v}}}^\dagger)$$

where $\tilde{\mathbf{v}}$ is assumed to be another 'history' of possible outcomes of the *same* sequence of operators as in $\mathbf{v}$. The term ρ_I is the initial density matrix, which

[16] Griffiths (1984).
[17] Unruh (1986).
[18] Gell-Mann & Hartle (1991).

I will discuss below and Tr denotes the trace of the operator (the sum of the diagonal elements of the matrix). We will divide the components of **v** into two subsets. One is those values which we know – i.e., those which we wish to use to set the conditions on the questions we ask of the theory, and the other subset is those values which we do not know, i.e., those for which we wish to determine the probability of that particular outcome. Let me designate the first subset ('known') by $K(\mathbf{v})$, and the second by $Q(\mathbf{v})$. Then one finds that

$$\text{Prob}[Q(\mathbf{v})] = \frac{D(\mathbf{v};\mathbf{v})}{\sum_{Q(\tilde{\mathbf{v}})} D(\tilde{\mathbf{v}};\tilde{\mathbf{v}})}$$

where the sum is over *all* possible $\tilde{\mathbf{v}}$ obeying this condition. Notice that we have chosen the same history in both slots of the decoherence function D in order to define the probabilities. It is worth pointing out features of this formula. If all of the conditions imposed occur at a time before any question, then the formula simplifies. That is, if $K(\mathbf{v})$ all occur before $Q(\mathbf{v})$, the above may be written as

$$\text{Prob}[Q(\mathbf{v})] = \text{Tr}[C(Q)C(K)\rho_I C^\dagger(K)C^\dagger(Q)] \Big/ \sum_Q \ldots.$$

We can thus replace the ρ_I by a new $\rho = C(K)\rho_I C^\dagger(K)$. Now ρ_I in the above is supposed to represent the initial 'density matrix' for the system. We note that in reality the initial density matrix represents either some theoretical prejudice about the truly initial state of the system, or represents the cumulative effect of all of those conditions placed on the system at times earlier than the time in question. We also note that we are free either to change the initial density matrix in this way once we set more *a priori* conditions, or to simply include the extra conditions in the formula as parts of the K terms in the history. The first option, changing the density matrix, is the process known as 'reduction of the wave packet' in the conventional approach to the interpretation of quantum mechanics.

It is useful at this point to mention that the decoherence function has been made the center of a new interpretation of quantum mechanics. In my presentation above, the results are all just minor modifications of the standard interpretation of the theory to unusual situations. However, Griffiths and Gell-Mann and Hartle have suggested a more radical use of this decoherence function, called the 'consistent histories approach'. One of the properties of this function is that the probability of outcome of some individual measurement depends on which other measurements are assumed to have been made. That is, the probabilities of the various possible outcomes of say the operator B at time t_2 in the above depends on whether or not Z_n is included in the history, and on precisely which operator (and thus physical quantity) Z_n represents. This just corresponds to the usual quantum condition that the outcome of measurements of say the position of a particle depends on whether or not I earlier measured the momentum of that particle, even if I do not know what the outcome of that earlier measurement was. In classical physics of course, the measurement of a quantity need not in and of itself alter the subsequent behavior of the system. The proposal made by Griffiths and Gell-Mann and Hartle is that ONLY those histories which have the property of decoherence (hence the name 'decoherence function') are histories which one can meaningfully talk about. All other histories are mathematical fictions of the theory. The property of decoherence is that a complete set of histories $\{v\}$ is physically meaningful only if the decoherence function obeys the property

$$C(\mathbf{v}; \tilde{\mathbf{v}}) = 0$$

if $\mathbf{v} \neq \tilde{\mathbf{v}}$. (Griffiths chooses the weaker requirement that only the real part of C need obey this condition.) What this condition means is that for this set of histories one does not need to worry about whether or not some attribute has actually been measured, or included in the history. The probability of the outcomes of one set of measurements is independent of whether or not one includes that other measurement in the history. The system behaves classically in that the measurements do not affect the subsequent behavior of the system *as indicated by the probabilities of the restricted class of measurements included in that set of histories.*

Let us express this condition formally. Define $\mathbf{v}'$ in relation to $\mathbf{v}$ by omitting one of the elements of the history.

$$\mathbf{v}' = (a_1, \ldots, p_h, s_j, \ldots, z_n)$$

where

$$\mathbf{v} = (a_1, \ldots, p_h, r_i, s_j, \ldots, z_n).$$

Assume furthermore that r_i is in $Q(\mathbf{v})$, and that we want the probability of some specific subhistory $Q(\mathbf{v}')$ independent of the value of r_i. Now given

Prob(Q', r_i), the probability of Q' independent of the value of r_i is just

$$\text{Prob}(q') = \sum_{r_i} \text{Prob}(Q', r_i)$$

$$= \sum_{r_i} \frac{\text{Tr}[C(\mathbf{v})\rho_I C^\dagger(\mathbf{v})]}{\sum_Q \text{Tr}[C(\mathbf{v})\rho_I C^\dagger(\mathbf{v})]}$$

$$= \sum_{r_i} \sum (\tilde{r}_i) \frac{\text{Tr}[C(\mathbf{v})\rho_I C^\dagger(\tilde{\mathbf{v}})]}{\sum_Q \text{Tr}[C(\mathbf{v})\rho_I C^\dagger(\mathbf{v})]}$$

$$= \frac{\text{Tr}[C(\mathbf{v}')\rho_I C^\dagger(\tilde{\mathbf{v}}')]}{\sum_{Q'} \text{Tr}[C(\mathbf{v}')\rho_I C^\dagger(\mathbf{v}')]}$$

where $\tilde{\mathbf{v}}$ in the third line is $\mathbf{v}$ with the element r_i replaced by $\tilde{r}_i$. The third line is valid because by assumption $D(\mathbf{v}; \tilde{\mathbf{v}}) = 0$ unless $\tilde{\mathbf{v}} = \mathbf{v}$. (Although the argument is slightly less direct in the Griffiths formulation it is still true there as well.) The consistent histories formulation can therefore be phrased as the statement that only those histories are physically real for which the probabilities are independent of whether or not one of the unknown elements of the history is present or not.

How do the four roles of time (section 2) enter into this formalism? The first two are present in exactly the same way they are in ordinary quantum theory. The third, the exclusivity and completeness, are included by demanding that the sum in $C(\mathbf{v})$ of all of the possible values for one of the elements of $\mathbf{v}$ is the same as the C function excluding that element. That is, we have

$$\sum_{r_i} C(a_1, \ldots, q_h, r_i, s_j, \ldots, z_n) = C(a_1, \ldots, q_h, s_j, \ldots, z_n),$$

and if, say, r_i and s_j are eigenvalues of the same item in the history, at the same time ($t_i = t_j$), then

$$C(a_1, \ldots, q_h, r_i, s_j, \ldots, z_n) = 0$$

unless $r_i = s_j$. That is, at the same moment in time, the same item in the history cannot have two different values.

This decoherence function formalism is clearly designed to implement the fourth condition. Time enters in a crucial way because the answers obtained depend crucially on the time ordering of the projection operators in the definition of C. If we reverse the order, or any pair, the probabilities will not be the same.

It is of interest[19] that one can also generalize the above formula to include the setting of final conditions in a final density matrix. That is, we can generalize the formula by including a ρ_F as a final condition,

$$D' = \frac{\text{Tr}\{\rho_F C(\mathbf{v})\rho_I C^\dagger(\mathbf{v})\}}{\sum_{Q(\mathbf{v})} \text{Tr}[\rho_F C(\mathbf{v})\rho_I C^\dagger(\mathbf{v})]}.$$

Again ρ_F will represent either a prejudice about the probabilities of the final state of the system (e.g., one may want to calculate the probabilities in one's experiment conditional on the laboratory still existing at the end of the experiment) or the accumulated effect of the parts of $K(\mathbf{v})$ which occur after all of the times associated with $Q(\mathbf{v})$.

[19] Gell-Mann & Hartle (1994).

2 Cosmology, time's arrow, and that old double standard

HUW PRICE

A century or so ago, Ludwig Boltzmann and others attempted to explain the temporal asymmetry of the second law of thermodynamics. The hard-won lesson of that endeavour – a lesson still commonly misunderstood – was that the real puzzle of thermodynamics is not why entropy always increases with time, but why it was ever so low in the first place. To the extent that Boltzmann himself appreciated that this was the real issue, the best suggestion he had to offer was that the world as we know it is simply a product of a chance fluctuation into a state of very low entropy. (His statistical treatment of thermodynamics implied that although such states are extremely improbable, they are bound to occur occasionally, if the universe lasts a sufficiently long time.) This is a rather desperate solution to the problem of temporal asymmetry, however,[1] and one of the great achievements of modern cosmology has been to offer us an alternative. It now appears that temporal asymmetry is cosmological in origin, a consequence of the fact that entropy is much lower than its theoretical maximum in the region of the Big Bang – i.e., in what we regard as the *early* stages of the universe.

The task of explaining temporal asymmetry thus becomes the task of explaining this condition of the early universe. In this chapter I want to discuss some philosophical constraints on the search for such an explanation. In particular, I want to show that cosmologists who discuss these issues often make mistakes which are strikingly reminiscent of those which plagued the nineteenth century discussions of the statistical foundations of thermodynamics. The most common mistake is to fail to recognize that certain crucial arguments are blind to temporal direction, so that any conclusion they yield with respect to one temporal direction must apply with equal

[1] Not least of its problems is the fact that it implies that all our historical evidence is almost certainly misleading: for the 'cheapest' or most probable fluctuation compatible with our present experience will always be one which simply creates the world as we find it, rather than having it evolve from an earlier state of even lower entropy.

2 Cosmology, time's arrow, and the double standard

force with respect to the other. Thus writers on thermodynamics often failed to notice that the statistical arguments concerned are inherently insensitive to temporal direction, and hence unable to account for temporal asymmetry. And writers who did notice this mistake commonly fell for another: recognizing the need to justify the double standard – the application of the arguments in question 'towards the future' but not 'towards the past' – they appealed to additional premises, without noticing that in order to do the job, these additions must effectively embody the very temporal asymmetry which was problematic in the first place. To assume the uncorrelated nature of initial particle motions (or incoming 'external influences'), for example, is simply to move the problem from one place to another. (It may *look* less mysterious as a result, but this is no real indication of progress. The fundamental lesson of these endeavours is that much of what needs to be explained about temporal asymmetry is so commonplace as to go almost unnoticed. In this area more than most, folk intuition is a very poor guide to explanatory priority.)

One of the main tasks of this chapter is to show that mistakes of these kinds are widespread in modern cosmology, even in the work of some of the contemporary physicists who have been most concerned with the problem of the cosmological basis of temporal asymmetry – in the course of the chapter we shall encounter illicit applications of a temporal double standard by Paul Davies, Stephen Hawking and Roger Penrose, among others. Interdisciplinary point-scoring is not the primary aim, of course: by drawing attention to these mistakes I hope to clarify the issue as to what would count as an adequate cosmological explanation of temporal asymmetry.

I want to pay particular attention to whether it is possible to explain why entropy is low near the Big Bang without thereby demonstrating that it must be low near a Big Crunch, in the event that the universe recollapses. The suggestion that entropy might be low at both ends of the universe was made by Thomas Gold in the early 1960s.[2] With a few notable exceptions, cosmologists do not appear to have taken Gold's hypothesis very seriously. Most appear to believe that it leads to absurdities or inconsistencies of some kind. However, I want to show that cosmologists interested in time asymmetry continue to fail to appreciate how little scope there is for an explanation of the low entropy Big Bang which does not commit us to the Gold universe. I also want to criticize some of the objections that are

[2] See Gold (1962), for example.

raised to the Gold view, for these too often depend on a temporal double standard. And I want to discuss, briefly and rather speculatively, some issues that arise if we take the view seriously. (Could we observe a time-reversing future, for example?)

Let me begin with a very brief characterization of what it is about the early universe that needs to be explained. There seems to be widespread agreement about this, so that if what I say is not authoritative, neither is it particularly controversial.[3] The question of the origin of the thermodynamic arrow appears to come down to why the early universe had just the right degree of inhomogeneity to allow the formation of galaxies: had it been more homogeneous, galaxies would never have formed; had it been less homogeneous, most of the matter would have quickly ended up in huge black holes. In either case we would not have galaxies as the powerhouses of most of the asymmetric phenomena we are so familiar with. For present purposes the latter issue – that as to why the universe is not less homogeneous – is the more pressing. This is because, as we shall see, there are strong arguments to the effect that if the universe recollapses, the other extremity will be very inhomogeneous. So the homogeneity in the region of the Big Bang would appear to represent a stark temporal asymmetry in the universe as a whole.

The natural mistake

The contemporary cosmological descendant of the problem that Boltzmann was left with – the problem as to why entropy was low in the past – is thus the question: why is the universe so smooth near the Big Bang? Now in effect this question is a call for an explanation of an observed feature of the physical universe, and one common kind of response to a request for an explanation of an observed state of affairs is to try to show that 'things had to be like that', or at least that it is in some sense very probable that they should be like that. In other words, it is to show that the state of affairs in question represents the *natural* way for things to be. Accordingly, we find many examples of cosmologists trying to show that the state of the early universe is not really particularly special. For example, the following remarks are from one of Paul Davies' popular accounts of cosmology and time asymmetry: 'It is clear that a time-asymmetric universe does not demand any very special

---

[3] There is an excellent account of this in Penrose (1989), chapter 7.

2 Cosmology, time's arrow, and the double standard

initial conditions. It seems to imply a creation which is of a very general and random character at the microscopic level. This initial randomness is precisely what one would expect to emerge from a singularity which is completely unpredictable.'[4]

The mistaken nature of this general viewpoint has been ably pointed out by Roger Penrose, however. Penrose asks what proportion of possible states of the early universe – what percentage of the corresponding points in the phase space of the universe – exhibit the degree of smoothness apparent in the actual early universe. He gives a variety of estimates, the most charitable of which (to the view that the early universe is 'natural') allows that as many as 1 in $10^{10^{30}}$ possible early universes are like this![5]

There is another way to counter the suggestion that the smooth early universe is statistically natural, an argument more closely related to our central concerns in this chapter. It is to note that we would not regard a collapse to a smooth *late* universe (just before a Big Crunch) as statistically natural – quite the contrary, as we noted above – but that in the absence of any prior reason for thinking otherwise, this consideration applies just as much to one end of the universe as to the other. In these statistical terms, then, a smooth Big Bang should seem just as unlikely as a smooth Big Crunch.[6]

The fact that this simple argument goes unnoticed reflects the difficulty that we have in avoiding the double standard fallacies in these cases. It deserves more attention, however. Indeed, its importance goes beyond its ability to counter the tendency to regard a smooth early universe as 'natural', and hence not in need of explanation. As we shall see, it also defuses the most influential arguments against Gold models of the universe, in which entropy eventually decreases. It is common to find considerations concerning gravitational collapse offered as decisive objections to the Gold view. However, what the objectors fail to see is that if the argument were decisive in one temporal direction it would also be decisive in the other, in which case its conclusion would conflict with what we know to be the case, namely that entropy decreases towards the Big Bang.

---

[4] Davies (1977), 193–4.
[5] His most recent estimate is 1 in $10^{10^{120}}$; see Penrose (1989), chapter 7.
[6] Penrose himself makes a point of this kind; see Penrose (1989), 339. However, we shall see that Penrose fails to appreciate the contrapositive point, which is that if we take statistical reasoning to be inappropriate towards the past, we should not take it to be appropriate with respect to the future.

In view of the importance of the argument it is worth spelling it out in more detail, and giving it a name. I will call it the *gravitational argument from symmetry* (GAS, for short). The argument has three main steps:

1. We consider the natural condition of a universe at the end of a process of gravitational collapse – in other words, we ask what the universe might be expected to be like in its late stages, when it collapses under its own weight. As noted above, the answer is that it is overwhelmingly likely to be in a very inhomogeneous state – clumpy, rather than smooth.
2. We reflect on the fact that if we view the history of our universe in reverse, what we see is a universe collapsing under its own gravity, accelerating towards a Big Crunch. As argued in step 1, the natural destiny for such a universe is not the smooth one we know our own universe to have.
3. We note that there is no objective sense in which this reverse way of viewing the universe is any less valid than the usual way of viewing it. Nothing in physics tells us that there is a wrong or a right way to choose the orientation of the temporal coordinates. *Nothing in physics tells us that one end of the universe is objectively the start and the other end objectively the finish.* In other words, the perspective adopted at step 2 is just as valid as a basis for determining the natural condition of what we normally call the early universe as the standard perspective is for determining the likely condition of what we normally call the late universe.

The lesson of GAS is that there is much less scope for differentiating the early and late stages of a universe than tends to be assumed. If we want to treat the late stages in terms of a theory of gravitational collapse, we should be prepared to treat the early stages in the same way. Or in other words if we treat the early stages in some other way – in terms of some additional boundary constraint, for example – then we should be prepared to consider the possibility that the late stages may be subject to the same constraint. Failure to appreciate this point has tended to obscure what may justly be called the *basic dilemma* of cosmology and temporal asymmetry. In virtue of the symmetry considerations, it seems that our choices are either to follow Gold, admitting reversal of the thermodynamic arrow in the case of gravitational collapse; or to acknowledge that temporal asymmetry is simply inexplicable – i.e., that the low initial entropy of the universe is not a predictable consequence of our best physical theories of the universe as a whole.

Later on I shall discuss a range of possible responses to this basic dilemma. First of all, however, in order to illustrate how thoroughly contemporary cosmologists have failed to appreciate the dilemma and the symmetry considerations on which it rests, I want to discuss two recent suggestions as to the origins of cosmological time asymmetry.

The appeal to inflation

The first case stems from the inflationary model. The basic idea here is that in its extremely early stages the universe undergoes a period of exponential expansion, driven by a gravitational force which at that time is repulsive rather than attractive. At the end of that period, as gravity becomes attractive, the universe settles into the more sedate expansion of the classical Big Bang.[7] One of the attractions of this model has been thought to be its ability to account for the smoothness of the universe after the Big Bang. However, the argument for this conclusion is essentially a statistical one: the crucial claim is that the repulsive gravity in the inflationary phase will tend to 'iron out' inhomogeneities, leaving a smooth universe at the time of the transition to the classical Big Bang. The argument is presented by Paul Davies, who concludes that 'the Universe ... began in an arbitrary, rather than remarkably specific, state. This is precisely what one would expect if the Universe is to be explained as a spontaneous random quantum fluctuation from nothing.'[8]

This argument graphically illustrates the temporal double standard that commonly applies in discussions of these problems. The point is that as in step 2 of GAS we might equally well argue, viewing the expansion from the Big Bang in reverse, that (what will then appear as) the gravitational *collapse* to the Big Bang must produce *inhomogeneities* at the time of the transition to the inflationary phase (which will now appear as a deflationary phase, of course). *Unless one temporal direction is already privileged, the argument is as good in one direction as the other.* So in the absence of a justification for the double standard – a reason to apply the statistical argument in one direction rather than the other – Davies' argument cannot possibly do the work required of it.

This is close to the point made in reply to Davies by Don Page.[9] Page objects that in arguing statistically with respect to behaviour during the inflationary phase, Davies is in effect assuming the very time asymmetry which needs to

[7] For a general introduction to the inflationary model see Linde (1987).
[8] Davies (1983), 398.
[9] Page (1983).

be explained (i.e., that entropy increases). However, Davies might reply that statistical reasoning is acceptable in the absence of constraining boundary conditions. It seems to me there are two possible replies at this point. One might argue (as Page does) that initial conditions have to be special to give rise to inflation in the first place, and hence that Davies' imagined initial conditions are in fact far from arbitrary. Or more directly one might argue as I have, viz. that if there is no boundary constraint at the time of transition from inflationary phase to classical Big Bang, then we are equally entitled to argue from the other direction, with the conclusion that the universe is inhomogeneous at this stage.

Davies himself argues that 'a recontracting Universe arriving at the big crunch would not undergo "deflation", for this would require an exceedingly improbable conspiracy of quantum coherence to reverse-tunnel through the phase transition. There is thus a distinct and fundamental asymmetry between the beginning and the end of a recontracting Universe.'[10] However, he fails to notice that this is in conflict with the argument he has given us concerning the other end of the universe, a conflict which can only be resolved either (i) by acknowledging that inflation is abnormal (and hence requires explanation, if we are to explain temporal asymmetry); or (ii) by arguing that although the coherence required for deflation looks improbable, it is in fact guaranteed by the reverse of the argument Davies himself uses with respect to the other end of the universe. But to accept (ii) would be to accept that entropy decreases as the universe recollapses, a view that as we shall see, Davies feels can be dismissed on other grounds. As it is, therefore, Davies is vulnerable to the charge that his own admission that collapse does not require deflation automatically entails that expansion does not require inflation. Again, this follows immediately from the realization that there is nothing objective about the temporal orientation. A universe that collapses without deflation just *is* a universe that expands without inflation. It is exactly the same universe, under a different but equally valid description.[11]

[10] Davies (1983), 399.
[11] All the same, it might seem that there is an unresolved puzzle here: as we approach the transition between an inflationary phase and the classical phase from one side, most paths through phase space seem to imply a smooth state at the transition. As we approach it from the other side most paths through phase space appear to imply a very non-smooth state. How can these facts be compatible with one another? I take it that the answer is that the existence of the inflationary phase is in fact a very strong boundary constraint, invalidating the usual statistical reasoning from the 'future' side of the transition.

Hawking and the Big Crunch[12]

Our second example is better known, having been described in Stephen Hawking's best seller, *A Brief History of Time*.[13] It is Hawking's proposal to account for temporal asymmetry in terms of what he calls the *no boundary condition* (NBC) – a proposal concerning the quantum wave function of the universe. To see what is puzzling about Hawking's claim, let us keep in mind the basic dilemma. It seems that provided we avoid double standard fallacies, any argument for the smoothness of the universe will apply at both ends or at neither. So our choices seem to be to accept the globally symmetric Gold universe, or to resign ourselves to the fact that temporal asymmetry is not explicable (without additional assumptions or boundary conditions) by a time-symmetric physics. The dilemma is particularly acute for Hawking, because he has more reason than most to avoid resorting to additional boundary conditions. They conflict with the spirit of his NBC, namely that one restricts possible histories for the universe to those that 'are finite in extent but have no boundaries, edges, or singularities.'[14]

Hawking tells us how initially he thought that this proposal favoured the former horn of the above dilemma: 'I thought at first that the no boundary condition did indeed imply that disorder would decrease in the contracting phase.'[15] He changed his mind, however, in response to objections from two colleagues: 'I realized that I had made a mistake: the no boundary condition implied that disorder would in fact continue to increase during the contraction. The thermodynamic and psychological arrows of time would not reverse when the universe begins to contract or inside black holes.'[16]

This change of mind enables Hawking to avoid the apparent difficulties associated with reversing the thermodynamic arrow of time. What is not clear is how he avoids the alternative difficulties associated with *not* reversing the thermodynamic arrow of time. That is, Hawking does not explain how his proposal can imply that entropy is low near the Big Bang, without equally implying that it is low near the Big Crunch. The problem is to get a temporally asymmetric consequence from a symmetric physical theory. Hawking suggests that he has done it, but does not explain how. Readers are entitled to feel a little dissatisfied. As it stands, Hawking's account reads a bit like a suicide

---

[12] This section expands on some concerns expressed in Price (1989).
[13] Hawking (1988).
[14] Hawking (1988), 148.
[15] Hawking (1988), 150.
[16] Hawking (1988), 150.

verdict on a man who has been stabbed in the back: not an impossible feat, perhaps, but we would like to know how it was done!

It seems to me that there are three possible resolutions of this mystery. The first, obviously, is that Hawking has found a way round the difficulty. The easiest way to get an idea of what he would have to have established is to think of three classes of possible universes: those which are smooth and ordered at both temporal extremities, those which are ordered at one extremity but disordered at the other, and those which are disordered at both extremities. Let us call these three cases *order-order* universes, *order-disorder* universes and *disorder-disorder* universes, respectively. (Keep in mind that in the absence of any objective temporal direction we could just as well call the second class the *disorder-order* case.) If Hawking is right, then he has found a way to exclude disorder-disorder universes, without thereby excluding order-disorder universes. In other words, he has found a way to ensure that there is order at at least one temporal extremity of the universe, without thereby ensuring that there is order at both extremities. Why is this combination the important one? Because if we cannot rule out disorder-disorder universes then we have not explained why our universe is not of that sort; while if we rule out disordered extremities altogether, we are left with the conclusion that Hawking abandoned, namely that order will increase when that universe contracts.

Has Hawking shown that order-disorder universes are overwhelmingly probable? It is important to appreciate that this would not be incompatible with the underlying temporal symmetry of the physical theories concerned. A symmetric physical theory might be such that all or most of its possible realizations were asymmetric. Thus Hawking might have succeeded in showing that the NBC implies that any (or almost any) possible history for the universe is of this globally asymmetric kind. If so, however, then he has not yet explained to his lay readers how he managed it. I shall describe my attempts to find a solution in Hawking's technical papers below. It seems clear, however, that it cannot be shown by reflecting on the consequences of the NBC for the state of one temporal extremity of the universe, considered in isolation. For if that worked for the 'initial' state it would also work for the 'final' state; unless of course the argument had illicitly assumed an objective distinction between initial state and final state, and hence applied some constraint to the former that it did not apply to the latter. What Hawking needs is a more general argument, to the effect that disorder-disorder universes are impossible (or at least overwhelmingly improbable). It needs to be shown that almost all possible universes have at least one ordered temporal extremity – or equivalently, at most one disordered extremity. (As Hawking points out, it will then be quite legitimate to invoke a weak anthropic argument to explain why we regard the ordered extremity thus guaranteed as an *initial* extremity. In virtue of its consequences for temporal asymmetry elsewhere in the universe, conscious observers are bound to regard this state of order as lying in their past.)

The first possibility is then that Hawking has such an argument, but hasn't told us what it is (probably because he does not see why it is so important).[17] As I see it, the other possibilities are that Hawking has made one of two mistakes (neither of them the mistake he claims to have made). Either his NBC does exclude disorder at both temporal extremities of the universe, in which case his mistake was to change his mind about contraction leading to deceasing entropy; or the proposal does not exclude disorder at either temporal extremity of the universe, in which case his mistake is to think that the NBC accounts for the low entropy Big Bang.

I have done my best to examine Hawking's published papers in order to discover which of these three possibilities best fits the case. A helpful recent paper is Hawking's contribution to a meeting on the arrow of time held in Spain in 1991.[18] In this paper Hawking describes the process by which he and various colleagues applied the NBC to the question of temporal asymmetry. He recounts how he and Halliwell 'calculated the spectrum of perturbations predicted by the no boundary condition'.[19] The conclusion was that 'one gets an arrow of time. The universe is nearly homogeneous and isotropic when it is small. But it is more irregular when it is large. In other words, disorder increases, as the universe expands.'[20] I want to note in particular that at this stage Hawking doesn't refer to the stage of the universe in temporal terms – *start* and *finish*, for example – but only in terms of its size. Indeed he correctly points out that the temporal perspective comes from us, and depends in practice on the thermodynamic arrow.

---

[17] This loophole may be smaller than it looks. Hawking's NBC would not provide an interesting explanation of temporal asymmetry if it simply operated like the assumption that all allowable models of the universe display the required asymmetry. This would amount to putting the asymmetry in 'by hand' (as physicists say), to *stipulating* what we wanted to *explain*. If the NBC is to exploit this loophole, in other words, it must *imply* this asymmetry, while being sufficiently removed from it so as not to seem *ad hoc*.

[18] Hawking (1994).
[19] Hawking (1994), 350.
[20] Hawking (1994), 351.

Hawking then tells us how he

> made what I now realize was a great mistake. I thought that the no boundary condition would imply that the perturbations would be small whenever the radius of the universe was small. That is, the perturbations would be small not only in the early stages of the expansion, but also in the late stages of a universe that collapsed again ... This would mean that disorder would increase during the expansion, but decrease again during the contraction.

He goes on to say how he was persuaded that this was a mistake, as a result of objections raised by Page and Laflamme. He came to accept that

> When the radius of the universe is small, there are two kinds of solution. One would be an almost Euclidean complex solution, that started like the north pole of a sphere, and expanded monotonically up to a given radius. This would correspond to the start of the expansion. But the end of the contraction, would correspond to a solution that started in a similar way, but then had a long almost Lorentzian period of expansion, followed by a contraction to the given radius ... This would mean that the perturbations, would be small at one end of time, but could be large and non-linear at the other end. So disorder and irregularity would increase during the expansion, and would continue to increase during the contraction.[21]

Hawking then describes how he, Raymond Laflamme and Glenn Lyons have 'studied how the arrow of time manifests in the various perturbation modes'. He says that there are two relevant kinds of perturbation mode, those that oscillate and those that do not. The former 'will be essentially time symmetric, about the time of maximum expansion. In other words the amplitude of perturbation, will be the same at a given radius during the expansion as at the same radius during the contraction phase.'[22] The latter, by contrast, 'will grow in amplitude in general... They will be small when they come within the horizon during expansion. But they will grow during the expansion, and continue to grow during the contraction. Eventually, they will become non linear. At this stage, the trajectories will spread out over a large region of phase space.'[23] It is the latter perturbation modes which, in virtue of the fact they are so much more common, lead to the conclusion that disorder increases as the universe recontracts.

Let us focus on the last quotation. If it is not an objective matter which end of the universe represents expansion and which contraction, and there is no

---

[21] Hawking (1994), 354.
[22] Hawking (1994), 355.
[23] Hawking (1994), 355.

constraint which operates simply in virtue of the radius of the universe, why should the perturbations ever be small? Why can't they be large at both ends, compatibly with the NBC?

I have been unable to find an answer to this crucial question in Hawking's papers. However, in an important earlier paper by Hawking and Halliwell,[24] the authors consistently talk of showing that the relevant modes *start off* in a particular condition. Let me give you some examples (with my italics throughout). First from the abstract: 'We ... show ... that the inhomogeneous or anisotropic modes *start off* in their ground state.'[25] Next from the introduction:

> We show that the gravitational-wave and density-perturbation modes obey decoupled time-dependent Schrödinger equations with respect to the time parameter of the classical solution. The boundary conditions imply that these modes *start off* in the ground state ...
>
> We use the path-integral expression for the wave function in sec. VII to show that the perturbation wave functions *start out* in their ground states.[26]

How are we to interpret these references to how the universe *starts off*, or *starts out*? Do they embody an assumption that one temporal extremity of the universe is objectively its start? Presumably Hawking and Halliwell would want to deny that they do so, for otherwise they have simply helped themselves to a temporal asymmetry at this crucial stage of the argument. (As I have noted, Hawking is in other places quite clear that our usual tendency to regard one end of the universe as the start is anthropocentric in origin, though related to the thermodynamic arrow – since they depend on the entropy gradient, sentient creatures are bound to regard the low entropy direction as the past.) But without this assumption what is the objective content of Hawking and Halliwell's conclusions? Surely it can only be that the specified results obtain when the universe is small; in which case the argument works either at both ends or at neither.

We have seen that in reconsidering his earlier views on the fate of a collapsing universe, Hawking appears to be moved by what is essentially a statistical consideration: the fact that (as Page convinced him) most possible histories lead to a disordered collapse. However, the lesson of GAS was that in the absence of any prior justification for a temporal double standard, statistical arguments defer to boundary conditions. (The fact that the Big Bang is smooth trumps

[24] Hawking & Halliwell (1985).
[25] Hawking & Halliwell (1985), 1777.
[26] Hawking & Halliwell (1985), 1778.

any general appeal to the clumpiness of the end states of gravitational collapse.) Accordingly, it would apparently have been open to Hawking to argue that Page's statistical considerations were simply overridden by the NBC, treated as a symmetric constraint on the temporal extremities of the universe. Given that he does not argue this way, however, he needs to explain why analogous statistical arguments do not apply towards (what we call) the Big Bang.

Thus it is my impression that Hawking did indeed make a mistake about temporal asymmetry, but not the mistake he thought he made. Either the mistake occurred earlier, if he and Halliwell have unwittingly relied on the assumption that the universe has an objective start; or it occurred when Hawking failed to see that in virtue of the fact that the Hawking–Halliwell argument depends only on *size*, it does yield boundary conditions for both temporal extremities of the universe, and hence excludes the possible histories which Page and Laflamme took to be problematic for his original endorsement of the time-reversing collapse. And if he didn't make either of these mistakes, then at the very least he has failed to see the need to avoid them, for otherwise he would surely have explained how he managed to do so.

The basic dilemma, and some ways to avoid it

The above examples suggest that even some of the most capable of modern cosmologists appear to have difficulty in grasping what I called the basic dilemma of cosmology and time asymmetry. To restate this dilemma, it is that apparently we have to accept either (option 1) that entropy decreases towards all singularities, including future ones; or (option 2) that temporal asymmetry, and particularly the low entropy condition of the Big Bang, is not explicable by a time-symmetric physics.[27]

The writer who has done most to bring to our attention the force of this dilemma is Roger Penrose, who chooses option 2.[28] (We shall look later at his reasons for rejecting option 1.) Accordingly, Penrose suggests that there is an additional asymmetric law of nature that initial singularities obey what amounts to a smoothness constraint, namely that the Weyl curvature of

---

[27] It might be thought that the dilemma only arises if the universe recollapses. It is possible that the universe will not recollapse, as predicted by some inflationary models. But even if the universe as a whole never recollapses to a singularity, it appears that parts of it do, as certain massive objects collapse to black holes. The dilemma arises again with respect to these regions. This point is made for example by Penrose, R. (1979), 597–8.

[28] See Penrose, R. (1979), and particularly Penrose (1989), chapter 7.

spacetime approaches zero in their region. In effect, he is arguing that it is reasonable to believe that such a constraint exists, because otherwise the universe as we find it would be wildly unlikely. Penrose's use of the term *initial singularity* might itself seem to violate our requirement that the terms *initial* and *final* not be regarded as of any objective significance. However in this case the difficulty is clearly superficial: Penrose's claim need only be that it is a physical law that there is one temporal direction in which the Weyl curvature always approaches zero towards singularities. The fact that conscious observers inevitably regard that direction as the past will then follow from the sort of weak anthropic argument we have already mentioned.[29]

Notice however that even an advocate of the time-symmetric option 1 might be convinced by Penrose's argument that the observed condition of the universe can only be accounted for by an independent physical law. In this case the required law would be that the Weyl curvature approaches zero towards *all* singularities. Before we turn to the various reasons that have been offered for rejecting option 1, let me mention some possible strategies towards a third option, which is somehow to evade the dilemma altogether.

The first possibility is the one mentioned above in our discussion of Hawking's proposal. We noted that it is possible that a symmetric theory might have only (or mostly) asymmetric models. This may seem an attractive solution, at least in comparison to the alternatives, but two notes of caution seem in order.

First, as was pointed out earlier,[30] a proposal of this kind needs to distance itself from the accusation that it simply puts in the required asymmetry 'by hand'. It is difficult to lay down precise guidelines for avoiding this mistake – after all, if the required asymmetry is not already implicit in the theoretical premises in some sense, it cannot be derived from them – but presumably the intention should be that asymmetry should flow from principles which are not *manifestly* asymmetric, and which have some independent theoretical justification.

Second, we should not be misled into *expecting* a solution of this kind – predominantly asymmetric models of a symmetric theory – by the sort of statistical reasoning employed in step 1 of GAS. In particular, we should not

---

[29] It seems to me that Penrose himself misses this point, however. For example in discussing the Weyl curvature hypothesis (Penrose (1989), 352–3), he gives us absolutely no indication that he regards the fact that some singularities are *initial* and others *final* as of anything other than objective significance.

[30] See footnote 17.

think that the intuition that the most likely fate for the universe is a clumpy gravitational collapse makes a solution of this kind *prima facie* more plausible than a globally symmetric model of Gold's kind. The point of GAS was that these statistical grounds are temporally symmetric: if they excluded Gold's suggestion then they would also exclude models with a low entropy Big Bang. In effect, the hypothesis that the Big Bang is explicable is the hypothesis that something – perhaps a boundary condition of some kind, perhaps an alternative statistical argument, conducted in terms of possible models of a theory – defeats these statistical considerations in cosmology, with the result that we are left with no reason to *expect* an asymmetric solution in preference to Gold's symmetric proposal. On the contrary, the right way to reason seems to be something like this: the smoothness of the Big Bang shows that statistical arguments based on the character of gravitational collapse are not always reliable – on the contrary, they are unreliable in the one case (out of a possible two!) in which we can actually subject them to an observational test. Having discovered this, should we continue to regard them as reliable in the remaining case (i.e., when oriented towards the Big Crunch)? Obviously not, at least in the absence of any independent reason for applying such a double standard.

It is true that things would be different if we were prepared to allow that the low entropy Big Bang is *not* explicable – that it is just a statistical 'fluke'. In this case we might well argue that we have very good grounds to expect the universe to be 'flukey' only at one end. However, at this point we would have abandoned the strategy of trying to show that almost all possible universes are asymmetric, the goal of which was precisely to *explain* the low entropy Big Bang. Instead we might be pursuing a different strategy altogether. Perhaps the reason that the universe looks so unusual to us is simply that we can only exist in very unusual bits of it. Given that we depend on the entropy gradient, in other words, perhaps this explains why we find ourselves in a region of the universe exhibiting such a gradient.

I am not going to explore this anthropic idea in any detail here. Let me simply mention two large difficulties that it faces. The first is that it depends on there being a genuine multiplicity of actual 'bits' of a much larger universe, of which our bit is simply some small corner. It is no use relying on other merely possible worlds.[31] So the anthropic solution is exceedingly costly in ontological terms. (This would not matter if the cost was one we

---

[31] Unless in David Lewis' sense, so that 'actual' is simply indexical, and denotes no special objective status. See Lewis (1986a).

were committed to bearing anyway, of course, as perhaps in some inflationary pictures, where universes in our sense are merely bubbles in some grand foam of universes.)

The second difficulty is that, as Penrose emphasizes,[32] there may well be much less costly ways to generate a sufficient entropy gradient to support life. Penrose argues that the observed universe is still vastly more unlikely than life requires. However, it is not clear that inflation does not leave a loophole here, too. If the inflationary model could show that a universe of the size of ours is an 'all or nothing' matter, then the anthropic argument would be back on track. The quantum preconditions for inflation might be extremely rare, but this is not important, so long as (a) there is enough time in some background grand universe for them to occur eventually and (b) when (and only when) they do occur a universe of our sort arises, complete with its smooth boundary. Hence it seems to me that this anthropic strategy is still viable – albeit repugnant to well brought up Occamists!

The case against the Gold universe

To choose the first horn of the basic dilemma is to allow that our universe might have low entropy at both ends, or more generally in the region of any singularity. Of course, the issue then arises as to why this should be the case. One option, perhaps in the end the only one, is to accept the low entropy of singularities as an additional law of nature. As we saw, this would be in the spirit of Penrose's proposal, but it would still be a time-symmetric law. True, we might find such a law somewhat *ad hoc*. But this might seem a price worth paying, if the alternative is that we have no explanation for such a striking physical anomaly.[33]

In suggesting the symmetric time-reversing universe in the 1960s, Gold was attracted to the idea that the expansion of the universe might account for the second law of thermodynamics. He saw that this would entail that the law would change direction if the universe recontracts. Thus the universe would enter an age of apparent miracles. Radiation would converge on stars, apples would compose themselves in decompost heaps and leap into trees, and humanoids would arise from their own ashes, grow younger, and become

---

[32] Penrose, R. (1979), 634.
[33] These considerations apply as much to Penrose's asymmetric proposal as to its symmetric variant. Note also that the proposal might seem less *ad hoc* if attractively grounded on formal theoretical considerations of some kind, as might arguably be true of Hawking's NBC.

unborn. These humanoids would not see things this way, of course. Their psychological time sense would also be reversed, so that from their point of view their world would look much as ours does to us.

However, by now it should be obvious that such apparently miraculous behaviour cannot in itself constitute an objection to this symmetric model of the universe, for reasons exactly analogous to those invoked in GAS. Perhaps surprisingly (in view of his tendency to appeal to a double standard elsewhere) this point is well made by Davies. After describing some 'miraculous' behaviour of this kind, Davies continues: 'It is curious that this seems so laughable, because it is simply a description of our present world given in reversed-time language. Its occurrence is *no more remarkable* than what we at present experience – indeed it *is* what we actually experience – the difference in description being purely semantic and not physical.'[34] Davies goes on to point out that the difficulty really lies in managing the transition: 'What *is* remarkable, however, is the fact that our "forward" time world *changes into* [a] backward time world (or vice versa, as the situation is perfectly symmetric).'

What exactly are the problems about this transition? In the informal work from which I have just quoted, Davies suggests that the main problem is that it requires that the universe have very special initial conditions.

> Although the vast majority of microscopic motions in the big bang give rise to purely entropy-increasing worlds, a very, very special set of motions could indeed result in an initial entropy increase, followed by a subsequent decrease. For this to come about the microscopic constituents of the universe would not be started off moving randomly after all, but each little particle, each electromagnetic wave, set off along a carefully chosen path to lead to this very special future evolution ... Such a changeover requires ... an extraordinary degree of cooperation between countless numbers of atoms.[35]

Davies here alludes to his earlier conclusion that 'a time-asymmetric universe does not demand any very special initial conditions. It seems to imply a creation which is of a very general and random character at the microscopic level.'[36] However, we have seen that to maintain this view of the early universe while invoking the usual statistical arguments with respect to the late universe is to operate with a double standard: double standards aside, GAS shows that if a late universe is naturally clumpy, so too is an early universe. In the present

[34] Davies (1977), 196.
[35] Davies (1977), 195–6.
[36] Davies (1977), 193.

context the relevant point is that (as Davies himself notes, in effect[37]) the conventional time-asymmetric view itself requires that the final conditions of the universe be microscopically arranged so that when viewed in the reverse of the ordinary sense, the countless atoms cooperate over billions of years to achieve the remarkable low entropy state of the Big Bang. Again, therefore, a double standard is involved in taking it to be an argument against Gold's view that it requires this cooperation in the initial conditions. As before, the relevant statistical argument is an instrument with two possible uses. We know that it yields the wrong answer in one of these uses, in that it would exclude an early universe of the kind we actually observe. Should we take it to be reliable in its other use, which differs only in temporal orientation from the case in which the argument so glaringly fails? Symmetry and simple caution both suggest that we should not!

A very different sort of objection to the time-reversing model is raised by Penrose in the following passage:

> Let us envisage an astronaut in such a universe who falls into a black hole. For definiteness, suppose that it is a hole of 10^{10} [solar masses] so that our astronaut will have something like a day inside [the event horizon], for most of which time he will encounter no appreciable tidal forces and during which he could conduct experiments in a leisurely way ... Suppose that experiments are performed by the astronaut for a period while he is inside the hole. The behaviour of his apparatus (indeed, of the metabolic processes within his own body) is entirely determined by conditions at the black hole's singularity ... – as, equally, it is entirely determined by the conditions at the big bang. The situation inside the black hole differs in no essential respect from that at the late stages of a recollapsing universe. If one's viewpoint is to link the local direction of time's arrow directly to the expansion of the universe, then one must surely be driven to expect that our astronaut's experiments will behave in an entropy-*decreasing* way (with respect to 'normal' time). Indeed, one should presumably be driven to expect that the astronaut would believe himself to be coming out of the hole rather than falling in (assuming his metabolic processes could operate consistently through such a drastic reversal of the normal progression of entropy).[38]

This objection seems to me to put unreasonable demands on the nature of the temporal reversal in these time-symmetric models. Consider Penrose's astronaut. He is presumably a product of 10^9 years of biological evolution, to say nothing of the 10^{10} years of cosmological evolution which created the

---

[37] Davies (1974), 199.
[38] Penrose, R. (1979), 598–9.

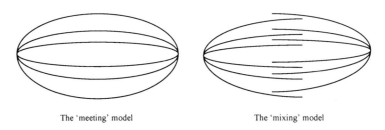

Fig. 1. Two models for a Gold universe

conditions for biology to begin on our planet. So he is the sort of physical structure that could only exist at this kind of distance from a suitable big bang. (What counts as *suitable*? The relevant point is that a condition of low entropy will not be sufficient; at the very least, the bang will have to be massive enough to produce the cosmological structure on which life depends.)

All this means that our astronaut is not going to encounter any time-reversed humanoids inside the black hole, any unevolving life, or even any unforming stars and galaxies. More importantly, it means that he himself has no need of an inverse evolutionary history inside the hole, in addition to the history he already has outside. He need not be a 'natural' product of the hole's singularity. Relative to its reversed time sense, he is simply a miracle. The same goes for his apparatus – in general, for all the 'foreign' structure he imports into the hole.

Notice here that there are two possible models of the connections that might obtain between the products of two low entropy boundary conditions: a 'meeting' model, in which any piece of structure or order is a 'natural' product of both a past singularity and a future singularity; and a 'mixing' model, in which it is normally a product of one or the other but not necessarily both. (See figure 1.) Penrose's argument appears to assume the meeting model. Hawking also seems to assume this model when he suggests[39] that the astronaut entering the event horizon of a black hole would not notice the time reversal because his psychological time sense would reverse. As I say, however, this seems to me to place a quite unnecessary constraint on the time-reversal view. The appropriate guiding principle seems to be that any piece of structure needs to be explained *either* as a product of a past singularity *or* as a product of a future singularity; but that no piece needs both sorts of explanation. The proportions of each kind can be expected to vary from case to case. It may well be that in our region of

[39] Hawking (1985), 2490.

the universe, virtually all the structure results from the Big Bang. This might continue to be the case if in the future we fall into the sort of black hole which does not have sufficient time or mass to produce much structure of its own. In this case the experience might be very much less odd than Penrose's thought experiment would have us believe. The reverse structure produced by the black hole might be insignificant for most of the time we survived within its event horizon.

What if we approach a black hole which is big enough to produce interesting structure – the Big Crunch itself, for example? Will Penrose's argument apply in this case? It seems to me that the case is still far from conclusive, so long as we bear in mind that *our* structure needs no duplicate explanation from the opposite point of view. It is true that in this case we will expect eventually to be affected by the reverse structure we encounter. For example, suppose that we are in a spacecraft that approaches what from the reverse point of view is a normal star. From the reverse point of view we are an object leaving the vicinity of the star. We appear to be heated by radiation from the star, but to be gradually cooling as we move further away from the star, thus receiving less energy from it, and radiating energy into empty space.

What would this course of events look like from our own point of view? Apparently we would begin to heat up as photons 'inexplicably' converged on us from empty space. This inflow of radiation would increase with time. Perhaps even more puzzlingly, however, we would notice that our craft was re-radiating towards one particular direction in space – towards what from our point of view is a giant radiation sink. Whether we could detect this radiation directly is a nice question – which I will discuss below – but we might expect it to be detectable indirectly. For example we might expect that the inside of the wall of the spaceship facing the reverse star would feel cold to the touch, reflecting what in our time sense would be a flow of heat energy towards the star.

These phenomena would certainly be bizarre by our ordinary standards, but it is not clear that their possibility constitutes an objection to the possibility of entropy reversal. After all, within the framework of the Gold entropy-reversing model itself they are not in the least unexpected or inexplicable. To generate a substantial objection to the model, it needs to be shown that it leads to incoherencies of some kind, and not merely to the unexpected. Whether this can be shown seems to be an open question, which I shall take up briefly in the next section.

Penrose himself no longer puts much weight on the astronaut argument.[40] In recent correspondence he says that he now thinks that a much stronger case can be made against the suggestion that entropy decreases towards singularities. He argues that in virtue of its commitment to temporal symmetry this view must either disallow black holes in the future, or allow for a proliferation of white holes in the past. He says that the first of these options 'requires physically unacceptable teleology', while the second would conflict with the observed smoothness of the early universe.[41] However, the objection to the first option is primarily statistical: 'it would have to be a seemingly remarkably improbable set of coincidences that would forbid black holes forming. The hypothesis of black holes being not allowed in the future provides "unreasonable" constraints on what matter is allowed to do in the past.'[42] And this means that Penrose is again invoking the old double standard, in accepting the 'naturalness' argument with respect to the future but not the past. Once again: the lesson of the smooth past seems to be that in that case something overrides the natural behaviour of a gravitational collapse; once this possibility is admitted, however, we have no non-question-begging grounds to exclude (or even to *doubt*!) the hypothesis that the same overriding factor might operate in the future.[43]

Is a reversing future observable? Is it consistent?

Now to return to the issues we deferred above, concerning the observability and hence the coherency of the Gold model. To give the enquiry a slightly less speculative character, let us locate our observers on Earth. Suppose that the

---

[40] Though it appears as recently as Penrose (1989), 334–5.
[41] Penrose (1991).
[42] Penrose (1991).
[43] Also relevant here is the argument of Davies (1974), 96, based on the work of Rees. This too is effectively a statistical argument, for the point is that 'in the normal course of events' radiation will not reconverge on stars. In the normal course of events it would not do so in the reverse direction either, but something seems to override the statistical constraint.

Davies has another radiation-based argument against the Gold universe: 'Any photons that get across the switch-over unabsorbed will find when they encounter matter that the prevailing thermodynamic processes are such as to produce *anti-damping* ... If a light wave were to encounter the surface of a metallic object, it would *multiply* in energy exponentially instead of diminish ... Consistency problems of this sort are bound to arise when oppositely directed regions of the universe are causally coupled together' (Davies (1974), 194). However, Davies apparently no longer regards this as a powerful argument against the time-reversing view, for in Davies & Twamley (1993) he and Jason Twamley canvas other ways in which radiation might behave in a time-reversing cosmos. I turn to the most significant of these arguments in the next section.

actual universe were a Gold universe, and a terrestrial telescope was pointed in the direction of a reverse-galaxy – i.e., a galaxy in what appears to us to be the contracting half of the universe. What effect, if any, would this have on the telescope and its environs?

As in the earlier case, a first step is to consider matters from the reverse point of view. To an astronomer resident in the reverse-galaxy, light emitted from that galaxy appears to be being collected and eventually absorbed by the distant telescope on Earth. This astronomer will therefore expect appropriate effects to take place at the back of our telescope: a black plate there will be heated slightly by the incoming radiation from the astronomer's galaxy, for example. But what does this look like from our point of view? Our temporal sense is the reverse of that of the distant astronomer, so that what she regards as absorption of radiation seems to us to be emission, and vice versa. Apparent directions of heat flow are similarly reversed. Thus as we point our telescope in the distant galaxy's direction, its influence should show up directly in a suddenly increased flow of radiation from the telescope into space, and indirectly as an apparent cooling in the temperature of the black plate at the rear of our telescope. The plate will apparently seem cooler than its surroundings, in virtue of the inflow of heat energy which is to be re-radiated towards the distant galaxy.

As in normal astronomy, the size of these effects will naturally depend on the distance and intensity of the reverse-source. Size aside, however, there seem to be theoretical difficulties in detecting the effects by what might seem the obvious methods. For example, it will be no use placing a photographic plate over the aperture of the telescope, hoping to record the emission of the radiation concerned. If we consider things from the point of view of the distant reverse-astronomer it is clear that the plate will act as a shield, obscuring the telescope from the light from her galaxy. Thus from our point of view the light will be emitted from the back of the photographic plate: the side facing away from the telescope, toward the reverse-galaxy.

The important thing seems to be that the detector effect be one that from the reverse point of view does not depend on any very special initial state – since that would be a special final state from our point of view, which accordingly we could not 'prepare'. Unfogging a photographic plate, for example, is unlikely to work for this reason. We cannot guarantee that the plate finishes in an unfogged condition (in the way that we can normally guarantee that it starts that way). Provided this requirement for a 'non-special' boundary state is satisfied, we appear to be entitled to assume that the detector will behave as

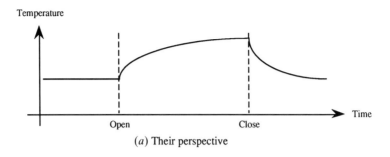

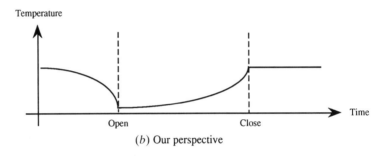

Fig. 2. Two perspectives on a telescope to look into the future

we would expect it to if the boot were on the other foot: in other words, if we were transmitting towards a detector belonging to the inhabitants of the reverse-galaxy.

Let us now look at this behaviour in a little more detail. If we shine a light at an absorbing surface we expect its temperature to increase. If the incoming light intensity is constant the temperature will soon stabilize at a higher level, as the system reaches a new equilibrium. If the light is then turned off the temperature drops exponentially to its previous value. Hence if reverse-astronomers shine a laser beam in the direction of one of our telescopes, at the back of which is an absorbing plate, the temperature change they would expect to take place in the plate is as shown in figure 2(a). When the telescope is opened the temperature of the plate rises due to the effect of the incoming radiation, stabilizing at a new higher value. If the telescope is then shut, so that the plate no longer absorbs radiation, its temperature drops again to the initial value.

Figure 2(*b*) shows what this behaviour looks like from our point of view. As explained above, both the temporal ordering of events and the direction of change of the apparent temperature of the plate relative to its surroundings has to be reversed. One of the striking things about this behaviour is that it appears to involve advanced effects. The temperature falls before we open the telescope, and rises before we close it. This suggests that we might be able to argue that the whole possibility is incoherent, using a version of the bilking argument. Couldn't we adopt the following policy, for example: *open the telescope only if the temperature of the black plate has not just fallen significantly below that of its surroundings.* It might seem that this entirely plausible policy forces the hypothesis to yield contradictory predictions, thus providing a *reductio ad absurdum* of the time-reversing view. However, it is not clear the results are contradictory. Grant for the moment that while this policy is in force it will not happen that the temperature of the plate falls on an occasion on which we might have opened the telescope, but did not actually do so. This leaves the possibility that on all relevant trials the temperature does not fall, and the telescope is opened. Is this inconsistent with the presence of radiation from the reverse-source?

I think not. Bear in mind that the temperature profile depicted in these diagrams relies on statistical reasoning: it is inferred from the measured direction of heat flow, and simply represents the most likely way for the temperature of the absorbing plate to behave. But one of the lessons of our discussion has been that statistics may be overridden by boundary conditions. Here, the temperature is constant before the telescope is opened because our policy has imposed this as a boundary condition. A second boundary condition is provided by the presence of the future reverse-radiation source. Hence the system is statistically constrained in both temporal directions. We should not be surprised that it does not exhibit behaviour predicted under the supposition that in one direction or other, it has its normal degrees of freedom. It is not clear whether this loophole will always be available, though my suspicion is that it will be. If nothing else, quantum indeterminism is likely to imply that it is impossible to sufficiently constrain the two boundary conditions to yield an outright contradiction.

A consistency objection of a different kind to the Gold universe has been raised by Hartle and Gell-Mann.[44] They point out that assuming the present epoch is relatively 'young' compared to the epoch of maximum expansion, it

---

[44] Reported in Davies & Twamley (1993).

Cosmology and time's arrow

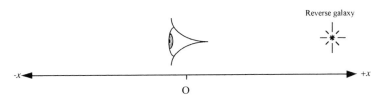

Fig. 3. How not to see the light?

is to be expected that the conversion of matter to radiation in stars will raise the ratio of the energy density of radiation to that of non-relativistic matter considerably by the time of the corresponding epoch in the recollapsing phase. Hartle and Gell-Mann estimate an increase by a factor of the order of 10^3. By symmetry the same should apply in reverse, so that at our epoch there should be much more radiation about than is actually observed. (Indeed, the effect should be accentuated by blue-shift due to the universe's contraction.) As Davies and Twamley put it, 'By symmetry this intense starlight background should also be present at our epoch ... a difficulty reminiscent of Olbers' paradox.'[45]

One thing puzzles me about this argument: if there were such additional radiation in our region, could we detect it? After all, it is neatly arranged to converge on its future sources, not on our eyes or instruments. Thus imagine a reverse-source in direction $+x$ is emitting (in its time sense) towards a distant point in direction $-x$. We stand at the origin, and look towards $-x$. (See figure 3.) Do we see the light which in our time sense is travelling from $-x$ towards $+x$? No, because we are standing in the way! If we are standing at the origin (at the relevant time) then the light emitted from the reverse-galaxy falls on us, and never reaches what we think of as the past sky. When we look towards $-x$, looking for the radiation converging on the reverse-galaxy at $+x$, then the relevant part of the radiation does not come from the sky in the direction $-x$ at all; it comes from the surface at the origin which faces $+x$ – i.e., from the back of our head.

The issue as to whether and to what extent the influence of the low entropy condition of one end of a Gold universe might be expected to be apparent at the other is clearly an important one, if the Gold view is to be taken at all seriously. The main lesson of these brief comments is that the issue is a great deal more complicated than it may seem at first sight. The more general lesson is that because our ordinary (asymmetric) habits of causal and counterfactual reasoning are intimately tied up with the thermodynamic asymmetry, we

[45] Davies & Twamley (1993).

cannot assume that they will be dependable in contexts in which this asymmetry is not universal. To give a crude example, suppose that an event B follows deterministically from an event A. In a Gold universe we may not be able to say that if A had not happened B would not have happened; not because there is some alternative earlier cause waiting in the wings should A fail to materialize (as happens in cases of pre-emption, for example), but simply because B is guaranteed by *later* events.

Indeed, figure 3 illustrates a consequence of this kind. We had a choice as to whether to interpose our head and hence our eye at the point O. Had we not done so, the light emitted (in the reverse time sense) by the reverse-galaxy at $+x$ would have reached $-x$, *in our past*. Our action thus influences the past. Because we interpose ourselves at O, some photons are not emitted from some surface at $-x$, whereas otherwise they would have been. Normally claims to affect the past give rise to causal loops, and hence inconsistencies. However it is not obvious that this need be so here, for reasons similar to those in the telescope case.

These issues clearly require a great deal more thought. Let me therefore conclude this section with two rather tentative claims. First, it has not been shown that the reversing-universe view leads to incoherencies. And second, there seems to be some prospect that the contents of a time-reversing future universe might be presently observable, at least in principle.[46] The methods involved certainly look bizarre by ordinary standards; but in the end this is nothing more than the apparent oddity of perfectly ordinary asymmetries having the reverse of their 'usual' orientation. And the main lesson of this chapter is that unless we have learnt to disregard that sort of oddity, we will make no progress whatsoever with the problem of explaining temporal asymmetry.

Conclusions

What then are the options and prospects for an explanation of the observed cosmological time asymmetry? Let me try to summarize the options. It might

---

[46] The required experiment should not be confused with a well-known test of some cosmological implications of the Wheeler–Feynman absorber theory of radiation. A prediction of that theory was that a transmitter should not radiate in directions in which the future universe is transparent to radiation. Partridge (1973) performed a version of this experiment with negative result. The present discussion does not depend on the absorber theory (which seems to me seriously misconceived; see Price (1991)), and predicts a different result, namely increased radiation in the direction of future reverse-sources.

seem that the most attractive solution would be the possibility mentioned in my discussion of Hawking's NBC, namely a demonstration that although the laws that govern the universe are temporally symmetric, the universes that they permit are mostly asymmetric – mostly such that they possess a single temporal extremity with the ordered characteristics of what we call the Big Bang. But it cannot be over-emphasized that the usual statistical considerations *do not* make this solution intrinsically more likely or more plausible than the Gold time-symmetric cosmology. With double standards disallowed, the statistical arguments concerned are simply incompatible with the hypothesis that the Big Bang itself is explicable as anything more than a statistical fluke. So if anything it is the Gold view which should be regarded as the more plausible option, simply on symmetry grounds – at least in the absence of well thought out consistency objections to time-reversing cosmologies. And failing either of these approaches, the main option seems to be an anthropic account. True, there is also Penrose's view, but here the asymmetry itself is rather unsatisfactory. If we are going to invoke a new physical principle, we might as well have a symmetric one, at least in the absence of good reasons to the contrary.

However, it should be emphasized that none of these alternatives is immediately attractive. As we saw, the anthropic approach involves an enormous ontological cost – it requires that the universe be vastly larger than what we know as the observable universe. (If we take Penrose's calculation as an indication of how unlikely the observed universe actually is, and assume that it is more or less as likely as it can be, consistent with the presence of observers, then we have an estimate of the size of this ontological cost: our universe represents at best 1 part in $10^{10^{30}}$ of the whole thing.) And as for the Gold universe, it is no magic solution to the original problem, even if consistent. We still need to explain why singularities are of low entropy. Unless it can actually be shown that a generic gravitational singularity is of this kind, some additional boundary constraint will again be required. Although in the Gold case this additional constraint need not be time-asymmetric, it may still appear somewhat *ad hoc*. As noted earlier, however, this might seem preferable to having no explanation of what would otherwise seem such a striking physical anomaly. Explanatory power and theoretical elegance are the traditional antidotes to apparent *ad hoccery*. A proposal of sufficient formal merit – perhaps the original symmetric version of Hawking's NBC, for example – might seem a very satisfactory theoretical solution.

What would have to be established to avoid the need for such an additional boundary constraint altogether? To answer this, recall that the heart of GAS was the observation that any gravitating universe still looks like a gravitating universe when its temporal orientation is reversed. It follows that if increasing entropy were the inevitable result of gravitational relaxation, any gravitating universe would look like an entropy-maximizing universe *from both temporal viewpoints*. It was the inconsistency between this conclusion and the observed low entropy Big Bang that showed us that statistical reasoning is untrustworthy in this cosmological context – that undercut our grounds for thinking that gravitational relaxation inevitably or 'normally' produces an increase in entropy.

I want to close by noting that in principle there seem to be two ways to reconcile statistics and low entropy boundary conditions. (I do not suggest that either of these alternatives should necessarily be taken seriously; I am simply interested in sketching the logical structure of the problem.) One way would be to make cosmological entropy, like gravitation, a viewpoint-dependent matter. We have already noted that in whichever temporal orientation one regards the universe, it appears to be gravitating – i.e., subject to the influence of an attractive gravitational force. It is therefore an orientation-dependent matter as to which temporal extremity we take to involve the state of greatest gravitational relaxation (or lowest gravitational potential energy). One way to preserve a strong link between entropy and gravitation would be to suggest that entropy is similarly frame-dependent, and hence that entropy appears to increase monotonically from both temporal viewpoints.

The second way would be to show that gravitational collapse does not naturally lead to a high entropy singularity at all – in other words, to find within one's theory of gravity an argument to the effect that entropy naturally *decreases* in gravitational collapse. In this connection it is interesting to note a recent paper by Sikkima and Israel[47] claiming to show that a low entropy state may indeed be the 'natural' result of gravitational collapse. In one sense this seems to be just the sort of thing that would have to be established, if the boundary conditions are not to require an independent law. However, Sikkima and Israel do not see the argument as supporting the time-symmetric view. For one thing, they say that in a cyclical universe entropy will increase from cycle to cycle. So the old puzzle would re-emerge at this level: how can such an overall asymmetry be derived from symmetric assumptions?

[47] Sikkema & Israel (1991).

These two approaches seek to reconcile statistical arguments in cosmology with the low entropy Big Bang. An alternative is to dispense with the statistical considerations altogether. It might be argued that any such appeal to statistics relies on the assumption that the system in question has the freedom to choose its path from among a range of equally likely futures. Perhaps the root of our dilemma simply lies in this assumption – in the fact that the assumption is itself incompatible with the neutral atemporal perspective we must adopt if we are to explain temporal asymmetry. That is, perhaps we should restore the balance not by seeking a 'natural' endpoint in both directions, but by curing ourselves of our lingering attachment to the idea of natural evolution itself. Perhaps in thinking that low entropy endpoints are anomalous we have been misled by a form of reasoning which is itself grounded in the temporal asymmetry; hence a form of reasoning we should be prepared to discard, in moving to an atemporal viewpoint. This would be a more radical departure from our ordinary ways of thinking than anything contemplated in this chapter. All the same, a possible conclusion seems to be that the project of explaining temporal asymmetry is a hopeless one, unless we are prepared to contemplate a radical departure of some such kind.

Acknowledgements

I am very grateful to Dieter Zeh for patient correspondence on these issues over a number of years, and to Professor Roger Penrose for the correspondence mentioned in footnotes 41 and 42. I would also like to thank Steve Savitt for his comments and encouragement, and several audiences in Vancouver and Sydney for their help in clarifying these ideas.

PART 2 **Quantum theory and time's arrow**

3 Time's arrow and the quantum measurement problem

ANTHONY LEGGETT

More than any other topic in the foundations of physics, the problem of the arrow (or arrows!) of time seems to be characterized by a unique 'slipperiness': it is not only difficult to find answers to the questions once posed, but difficult to find meaningful questions to ask in the first place. Perhaps this is because an asymmetry with respect to the sense of time is built into us at an even deeper level than other 'synthetic a priori' conceptions about the world such as the continued existence of unknown objects. However that may be, in this situation, rather than trying to pose general questions which may turn out in the end to be ill-defined, there may be something to be said for trying to formulate much less speculative, indeed perhaps in retrospect trivial questions to which at least we have a hope of finding a definite answer. This is what I shall do in this essay: to be specific, I shall ask whether a test which is currently being designed concerning the validity of a certain 'common-sense' view of the macroscopic world ('macrorealism') would be affected if we were to relax our common-sense assumptions about the direction of causality in time.

Let me start with a very brief review of established ideas concerning the formal (a)symmetry of the quantum theory with respect to the direction of time. The classic works on this subject are the paper of Aharonov et al.[1] ('ABL') and the book by Belinfante.[2] In these discussions the notion of 'measurement' is taken as primitive and undefined. The argument then goes roughly as follows: starting from either the standard ('textbook') predictive quantum theory in which an initial condition (alone) specifies the state (wave function) of the system, or from a 'retrodictive' version in which the state is specified only by a final condition, we can derive a version of quantum theory which is completely symmetric with respect to the direction of time. In such a version the probability of obtaining a particular value of an observable

[1] Aharonov, Bergmann, & Lebowitz (1964).
[2] Belinfante (1975).

which is measured at some intermediate time t will in general be conditioned by the results of measurements made both before and after time t. Some amusing features of the probabilities obtained in this way have been discussed by Aharonov and Vaidman[3] and verified experimentally by Ritchie et al.[4] Thus, at this stage there is no obvious justification for neglecting what happens after time t, i.e. using the predictive ('textbook') version of the theory. Indeed, in some experiments (for example those reported by Ritchie et al.) one would actually get the wrong answer if one assumed that the results were independent of the result of a 'final' measurement. However, one now introduces the idea of a 'garbling sequence' (or 'coherence-destroying manipulation'), which is a (notional) sequence of 'measurements' such that, if their results are not inspected, it is guaranteed to destroy any conditioning by earlier or later measurements. Thus, if a particular measurement Q is separated from an initial measurement I by a garbling sequence ('pregarbling') then the probabilities of the possible results for Q must be independent of I: while if Q is separated from a final measurement F by a garbling sequence ('postgarbling'), the probabilities for Q must be independent of F. Thus, if a measurement of Q is postgarbled (but not pregarbled) we must use predictive quantum theory (which follows from the symmetric version under these circumstances) while if it is pregarbled (but not postgarbled) we must use the 'retrodictive' version. (If it is both pregarbled and postgarbled, then the probabilities for Q are completely unconditioned, i.e. correspond to no information about the state of the system.)

Why is postgarbling rather than pregarbling the 'natural' state of affairs? At this point we become conscious that the way in which we have framed the argument so far is actually rather unnatural. In real life, experimenters do not normally 'measure' physical quantities and then refuse to inspect the results of the measurement; on the other hand, nature is in effect doing this all the time, since as is well-known the effect of many types of interaction of a quantum system with its environment is to convert the reduced density matrix of the system from that of a pure state to one characterizing a mixture ('decoherence'), which is exactly the result of a 'measurement' whose result is uninspected. Thus, garbling sequences indeed occur naturally and without the intervention of an experimenter. On the other hand, it is a misconception that in interpreting the outcome of a typical real-life experiment we typically

[3] Aharonov & Vaidman (1990).
[4] Ritchie, Story, & Hulet (1991).

assign an initial state (wave function) on the basis of a complete initial measurement; this is in fact very rarely the case. Rather, we typically assign, e.g. to a beam of atoms emerging from an oven, an internal wave function corresponding to the electronic groundstate on the basis of thermodynamic arguments: in effect, we argue that since the atom is in contact with a heat bath whose temperature is very much less than the typical excitation energy of the first electronic excited state, the electrons are overwhelmingly likely to be in the groundstate rather than any excited state.

Why can we not time-reverse this argument? The point is that while the laws of evolution of a complete closed system are presumably invariant under time reversal (I will neglect the complications associated with violation of symmetry under time reversal at the elementary particle level, which are irrelevant in this context), those of a subsystem need not be: in particular, if we consider a process in which a subsystem (e.g. an atom, the 'closed system' being the atom plus radiation field) makes a transition from a set of initial states i to a set of final states j with probability $p(i \to j)$, then we can legitimately make the statement

$$\sum_j p(i \to j) = 1 \tag{1}$$

(since the system must end up in some state!), but we cannot in general make the statement

$$\sum_j p(i \to j) = 1 \quad \text{(wrong)} \tag{1'}$$

and, in fact, in the case considered, if j is the groundstate, we might well have $p(i \to j) \approx 1$ for each i. This argument clearly needs further development (see, for example Watanabe[5]) but in the present context its virtue is that it makes it clear that we can avoid 'pregarbling' by obtaining information in a way which does not require explicit measurement, whereas no similar device can enable us to avoid 'postgarbling'. While this remark does not 'solve' the problem of time asymmetry in quantum mechanics, it shows that it is in some sense only a special case of the more general problem of the so-called thermodynamic arrow of time.

As mentioned above, in discussing the quantum-mechanical arrow of time the references given in footnotes 1 and 2 take the notion of measurement as

---

[5] Watanabe (1955).

primitive and therefore do not have to deal explicitly with the notorious 'quantum measurement paradox'. Nevertheless, it may not be unreasonable to suspect that there may be some deep connection between the two problems. Firstly, it is tempting to regard both the uniqueness of events at the macroscopic level and our every-day sense of the 'direction' of time and causality as pieces of 'synthetic a priori' knowledge about the world. Secondly, and more formally, many alleged 'resolutions' of the quantum measurement paradox rely heavily on the idea of dissipation and thus of irreversibility, which is clearly intimately connected with the thermodynamic arrow. I will explore one very limited aspect of a possible connection.

To motivate the rest of this essay let me review very briefly the familiar argument which leads to the paradox. Consider the 'measurement' of a physical quantity A represented in quantum mechanics by an operator $\hat{A}$ with eigenvalues a_j, and corresponding eigenfunctions ψ_j. An operational definition of 'measurement' is that a 'measuring apparatus' is set up so that it starts in some macroscopic state[6] X_0, and so that if the initial value of A is known with certainty to be a_j, then the final state of the apparatus, as a result of its interaction with the system, is X_j, i.e. schematically,

$$\psi_j X_0 \to \psi_j X_j \qquad (2)$$

where $(X_j, X_k) = \delta_{jk}$, and moreover the different X_j correspond to macroscopically distinct states (otherwise we would not be able to read them off with the naked eye). Note that one (but not more than one) of the X_j may actually be the original state X_0 (provided, of course, that we have independent evidence that the system was there to be measured!). In the case of a two-dimensional Hilbert space of A (only) this leads to the possibility of 'ideal negative result' measurement, a feature we shall discuss more fully and exploit later. Now, as is well-known, the problem is that if the system starts in a *superposition* of the different ψ_j, that is in the state $\sum_j c_j \psi_j$, then the linearity of the laws of quantum mechanics inexorably implies the evolution

$$\left(\sum_j c_j \psi_j\right) \cdot X_0 \equiv \sum_j c_j(\psi_j X_0) \to \sum_j c_j \psi_j X_j, \qquad (3)$$

[6] Purely for notational simplicity, I assume here (unrealistically) that we can describe the apparatus by a pure state. (I also assume that the measurement is 'ideal', i.e. the measured system is left in the state ψ_j as in eqn (2).)

3 Time's arrow and the quantum measurement problem

i.e. the apparatus (or more accurately the system-plus-apparatus complex) ends up in a *superposition of macroscopically distinct states*. Since it is a matter of experience that actual observation, e.g. with the naked eye, reveals a particular macroscopic outcome to have been realized in each case, this poses a *prima facie* problem.

I will not attempt to review here the various alleged 'solutions' of the quantum measurement paradox.[7] Instead I will merely remark (a) that the 'orthodox' solution relies heavily on the idea which is generally known nowadays as 'decoherence', which in turn rests on accepted notions concerning irreversibility and the associated 'direction' of time (thus making contact with the ABL–Belinfante argument reviewed above) and (b) that for reasons given in Leggett[7] and elsewhere, I personally find this 'resolution' fundamentally flawed. Rather, let me go directly to the question of an experimental test of the paradoxical predictions of quantum mechanics versus a 'common-sense' view of the world which I call macrorealism (MR), and define in detail below.

The basic idea of the experiment[8] is to set up a situation where careful application of the standard formalism of quantum mechanics leads to a description of terms of a linear superposition of macroscopically distinct states and then to observe the effects of interference between these two states and compare them with the predictions of MR. What is needed is a single macroscopic system which is effectively restricted to a two-dimensional Hilbert space where the two basis states can be chosen to correspond to *macroscopically distinct* configurations. The system should be characterized by extremely low values of the dissipation (otherwise the usual arguments concerning 'decoherence' apply and rapidly destroy the effects of interference). One wishes, as we shall see, to measure time correlations, and to that end it is necessary to be able, in effect, to switch one's measuring apparatus on and off at will. (This is probably the most difficult aspect of the experiment.)

The most practical realization of the desired set-up is probably obtained by using a single junction SQUID ring.[8,9,10] By biassing this system with a suitable external flux, one can obtain an effective potential energy which as a function of the *total* flux through the ring (or equivalently the current circulating in it) has a symmetric double-well structure; the two wells correspond to states in

---

[7] See e.g. Leggett (1987a).
[8] Leggett & Garg (1985).
[9] Leggett (1987b).
[10] Tesche (1990).

which the current is circulating clockwise (counterclockwise) with a magnitude of the order of 1 µA, so they are reasonably 'macroscopically distinct'. The system is conceptually similar to the NH_3 molecule and we would *prima facie* expect similar 'inversion-resonance' behavior. In preliminary experiments (which have already been carried out on similar rings, see for example, de Bruyn Ouboter and Bol[11]) one verifies that *when actually measured* the current is always found to take one or other of these two distinct values (with small fluctuations around each); this observation forms the basis of our definition of macrorealism for this particular experimental realization, see below. For convenience we label the clockwise (counterclockwise) current as corresponding to a value +1 (−1) of a variable Q. Thus the statement is that whenever Q is measured, we always find either the value +1 or the value −1.

The principle of the experiment is to start the system in a given initial state (e.g. by subjecting the system to a strong bias and turning it off at $t = 0$), and on each run to measure the value of Q at *two and only two* times with the measuring apparatus shut off at all other times. For a reason which will become clear below, these measurements (or at least the first one) should be made by an 'ideal-negative-result' (INR) technique. That is, we first arrange the measuring device so that it interacts physically with the system only when the latter 'has' a value of Q equal to (say) +1, and then throw away all runs in which the device registers; we then repeat the process with the device coupled only to the $Q = -1$ state. The measurement at t_2 may be made in any way we please. In this way we can obtain the complete correlation $\langle Q(t_1)Q(t_2) \rangle$ without ever (*prima facie*) interacting physically with the system at any time before t_2.

Suppose we make four series of runs: in the first we measure the value of Q only at times t_1 and t_2, on the second only at t_2 and t_3, on the third at t_3 and t_4, and finally on the fourth at t_1 and t_4. In this way we determine experimentally the correlations $\langle Q(t_1)Q(t_2) \rangle_{exp}$, $\langle Q(t_2)Q(t_3) \rangle_{exp}$, etc. We now consider the quantity (which may look familiar to those conversant with tests of Bell's inequalities in quantum optics)

$$K_{exp} \equiv \langle Q(t_1)Q(t_2) \rangle_{exp} + \langle Q(t_2)Q(t_3) \rangle_{exp} + \langle Q(t_3)Q(t_4) \rangle_{exp}$$
$$- \langle Q(t_1)Q(t_4) \rangle_{exp} \qquad (4)$$

where $t_{i+1} - t_i = \pi/4\Delta$ with Δ being the frequency of oscillation between the two wells (the 'tunnelling splitting' of the even- and odd-parity combinations).

[11] de Bruyn Ouboter & Bol (1982).

Consider first the predictions of standard quantum mechanics for K_{exp}. If we could consider the system as completely isolated (and moreover could assume the parameters are such that the motion is restricted to the 'ground' Hilbert space spanned by the lowest even- and odd-parity combinations of the single-well states), then everything is in exact correspondence to the 'textbook' example of the NH_3 molecule, and we easily find (this result is actually quite independent of the nature of the measurement process, provided of course that the system is isolated between the measurements)

$$\langle Q(t_i)Q(t_j)\rangle = \cos[\Delta(t_i - t_j)] \tag{5}$$

and hence

$$K_{\text{exp}} = 2\sqrt{2}. \tag{6}$$

For the more realistic case of finite dissipation, we need to do a more involved calculation (see for example, Leggett et al.[12]). As a result we find that for not too strong dissipation we obtain a value of K which while less than that given by (6) is still greater than 2:

$$2\sqrt{2} > K_{\text{exp}} > 2. \tag{7}$$

We shall assume that the experiment can be done in a regime where this prediction is valid, and moreover (for the sake of the present argument) that when it is done the prediction is confirmed. What conclusions can be drawn from this (as yet hypothetical) result?

We will first consider the experiment within the framework of our usual conceptions concerning the arrow of time. We define the class of 'macrorealistic' (MR) theories by the conjunction of three postulates, of which one will be given later:

(P1): a macroscopic system which has available to it two or more macroscopically distinct states always[13] realizes one and only one of these states, *whether or not* it is observed. That is, $Q(t)$ takes one of the values ± 1 for all t.

(P2) (induction): the properties of ensembles depend only on their initial preparation.

The motivation for (P1) is that, as we have seen, whenever the quantity $Q(t)$ is measured, it is found experimentally to have one and only one of the values

---

[12] Leggett, Chakravarty, Dorsey, Fisher, Garg, & Zwerger (1987).
[13] There is a slight complication here connected with the necessity of considering finite 'transit times' between the two states. See Leggett & Garg (1985).

± 1. It is clear that (P1) is insensitive to the 'direction of time' whereas (P2) involves it essentially. Note that (P2) does not exclude the possibility that, if I, Q and F have the same meaning as above, $p(Q|IF) \neq p(Q|I)$ (i.e. the probability of Q given both I and F is not necessarily the same probability of Q given I alone. Such a conditioning on the results of subsequent experiments is perfectly normal in classical physics as well as in 'symmetric' quantum theory!): what it does exclude, however, is for example the possibility that $p(Q(t))$ is sensitive to *whether or not* a measurement of F is made at $t_F > t$ ('causality cannot propagate backwards in time').

We now analyze the experiment from the point of view of MR. Step 1 is that on any given run, whatever measurements are or are not made, the quantity $Q(t)$ exists and takes one of the values ± 1 (this is just (P1)), and therefore by the simple algebra familiar from Bell's theorem we have for any $t_1 \ldots t_4$ the inequality

$$Q(t_1)Q(t_2) + Q(t_2)Q(t_3) + Q(t_3)Q(t_4) - Q(t_1)Q(t_4) \leq 2. \tag{8}$$

Step 2 is that by (P2), provided all expectation values are taken as a *single* ensemble, the corresponding averages (described by the subscript 'ens') satisfy

$$K_{\text{ens}} \equiv \langle Q(t_1)Q(t_2) \rangle_{\text{ens}} + \langle Q(t_2)Q(t_3) \rangle_{\text{ens}} + \langle Q(t_3)Q(t_4) \rangle_{\text{ens}}$$
$$- \langle Q(t_1)Q(t_4) \rangle_{\text{ens}} \leq 2. \tag{9}$$

Now comes the crunch: in order to confront (9) with (7) we need to identify K_{ens} with K_{exp}. But this is in general impossible, since in general the experimental quantities $\langle Q(t_j)Q(t_{j+1}) \rangle$ are *not all measured on the same ensemble*: for example, as regards a measurement conducted at t_4, the ensemble on which Q has been measured at time $t_3 < t_4$ is not in general identical to that on which it has not. To take the argument further at this point we therefore need to supplement (P1) and (P2) by a third postulate.

(P3) ('noninvasive measurability'): it is in principle possible to measure the value of $Q(t)$ on an ensemble without altering the statistical properties of that ensemble as regards subsequent measurements.

The postulate (P3) may be made highly plausible, once given (P1), by considering an INR measurement: if we get a negative result (no change of measuring device states) then we infer that the system was in the state in which it experienced no physical interaction with the device, and hence it seems very unnatural to suppose that its subsequent behavior could have been in any way affected by the carrying out of the measurement. Thus, (P3) is so natural a corollary of (P1) that it is tempting to regard it as implicit in the latter.[14]

Thus, assuming that the first measurement of each pair is indeed of INR type and using (P3), we conclude (step 3 of the argument) that we can indeed identify K_{ens} with K_{exp} and hence any MR theory will make the firm prediction

$$K_{exp} \leq 2 \qquad (10)$$

in contradiction with the result of (7) and, hypothetically, with the experimental data. In other words, given an appropriate parameter regime, an experiment whose results conform to the predictions of quantum mechanics must necessarily exclude the hypothesis of MR as defined by the conjunction of postulates (P1)–(P3).

We now ask: is the implicit use of the 'conventional' arrow of time an essential ingredient in the above conclusion? It seems that in one respect it is, in the sense that it seems impossible, in 'reading off' the results of a measurement from the behavior of the measuring device, to escape the usual conceptions concerning the thermodynamic arrow. However, if we consider the behavior of the isolated system between measurements, it is not obvious that an MR model need respect the same 'arrow of time' as conventional (textbook) quantum mechanics, (or indeed any arrow!). The question is, does this help? i.e. could we escape the refutation of MR by assuming that the conventional assumptions about the arrow of time are violated?

To answer this question, we attempt to generalize the defining postulates of MR so as to avoid any assumption, implicit or otherwise, about the direction of the arrow. Evidently (P1) needs no modification. (P2) however must be replaced by

(P2′): The properties of ensembles may depend on their preparation and/or their 'retroparation' i.e. on the final state to which the system evolves.

Given this modification, step 1 of the above argument is clearly unchanged. Moreover, step 2 remains valid *provided* we define our 'ensembles' by the final as well as initial states (of the system and anything which may have interacted with it). That is, we should make a (further) selection of the experimental runs, such that the final state of the system at some time $t_f > t_4$ is measured on each

[14] Compare with Leggett (1987b).

of the runs and required to be the same for each run kept (all others being discarded).

What, now, of step 3? If we make only the initial measurement of each pair, and not also the final one, by an INR technique, then the final state of the apparatus which has interacted with the system will be different for the different sets of runs (i.e. according to whether Q was measured at time t_2 or time t_3, etc.) and therefore the corresponding ensembles are (or may be) different and step 3 fails. Suppose however we make both the initial *and* the final measurement of each pair by an INR technique, and moreover generalize postulate (P3) to:

(P3'):   It is in principle possible to measure the value of $Q(t)$ on an ensemble without altering the statistical properties of that ensemble as regards either subsequent or previous measurements.

(The justification of (P3') would be similar to that of (P3).) Then it is clear that, even allowing for violations of the conventional arrow of time, we can still identify K_ens with K_exp, and the refutation of MR remains valid.

To summarize this discussion (whose conclusions may well seem in retrospect rather obvious and even trivial!), the postulate which needs to be included in macrorealistic theories in order that their predictions be incompatible with those of quantum mechanics, namely noninvasive measurability, is in some sense 'naturally' time-reversal symmetric (or rather is naturally generalized to be so). As a result, relaxing the conventional assumptions about the 'arrow of time' does *not* mitigate the incompatibility of the predictions of macrorealism with those of quantum mechanics.

It goes without saying that I realize that in this chapter I have barely scratched the surface of the problem. As hinted above, I suspect that there are much deeper connections between these two perennial problems of the foundations of physics, the arrow of time and the quantum measurement paradox. I also suspect that it will be many years, indeed perhaps many generations, before we fully understand what these connections are!

4 Time, decoherence, and 'reversible' measurements

PHILIP STAMP

1 Introduction

In this chapter I attempt to review a few somewhat loosely connected themes, which figure in recent work by physicists on decoherence and irreversibility in quantum mechanics. The looseness is forced upon us by the subject – any proper discussion inevitably drags in questions about thermodynamic reversibility and the various arrows of time, about the foundations of quantum mechanics and quantum measurement, about the nature of quantum coherence, and about the relationship between quantum and classical mechanics. There are inevitably many gaps in our understanding of these questions, both for the physicist and for the philosopher. Some philosophers may feel that the subject is too immature to be interesting to them – that it is not yet ripe for philosophical analysis. Indeed I have recently heard the opinion expressed that quantum mechanics in its entirety was still not amenable to such analysis – that if the physicists could not yet understand it, there was no point in philosophers trying! I hope this feeling is not too widespread amongst philosophers of science – I suspect that it may arise in part because the philosophical community (along with a good part of the scientific community) is unaware of some important advances that have been made in the last decade in understanding the physics of quantum processes at the macroscopic level, as well as the relation between classical and quantum mechanics.

Much of the chapter is therefore unashamedly tutorial in nature; however, given the wide range of topics covered, I imagine that even physicists will find at least some of the material to be novel. What may be of most interest to philosophers is the fact that recent advances in our understanding of quantum mechanics at the macroscopic scale necessitate at least a fresh look at the perplexing philosophical questions surrounding the relationship between the quantum world and our 'common sense' macroscopic world.

I start in section 2 by reviewing the contemporary view of 'time's arrows' held by most physicists (although many might quibble with details, I would be rather surprised if many disagreed with the overall scheme of the arguments). This section was written at the request of the editor of this volume, but it is in fact a useful springboard to the rest of the article, since it sets the phenomenon of decoherence in a very general context. Decoherence is then dealt with – it can only be properly explained in a 'wave-function' view of quantum mechanics, according to which the quantum correlations involved in a coherent wave-function are partially transferred to the 'environment', and cannot be disentangled from the environmental wave-function without great difficulty. The great advances being made in condensed matter physics have nevertheless shown that the disentanglement is by no means impossible, and that the 'barrier' between the quantum and classical worlds can be (and is being) pushed to the macroscopic level with startling speed. In physics these advances are of enormous practical import – they will certainly completely change the lives of citizens of at least the technologically based countries, within the next few decades.

The possibility that the physics and technology of the next century may revolve around quantum coherence might seem reason enough for philosophers to take some interest in these recent developments. But they are also philosophically important – they represent a sharp departure from the usual discussions of quantum mechanics in terms of 'measurements'. The newer picture ignores measurements as much as possible, and treats them simply as a particular kind of irreversible process. No attempt is made to gloss over this break with previous orthodoxy; I rather try to emphasize it, and the advantages of the newer picture over the older 'measurement calculus' (explained in section 4). Of course this does not suddenly solve the well-known 'measurement paradoxes' of quantum theory, although it certainly makes them look rather different (and arguably less serious). Readers will certainly differ as to the importance of this change of perspective – I personally feel that it constitutes the most important advance in the foundations of quantum theory since Bell's inequalities.

In this new light, the 'quantum arrow of time' becomes nothing but the thermodynamic arrow, and the irreversibility involved in 'quantum measurements' becomes no more fundamental, and no less subjective, than in any other thermodynamic process. A very interesting question then arises, because the discretization of energy levels and the 'energy gaps' occurring in some

4 Time, decoherence, and 'reversible' measurements

quantum systems, at low temperatures, permit an extraordinary reduction in the rate at which irreversible processes occur, and a control over them. With each new advance, more and more 'irreversible' processes become reversible in the laboratory, underlining the practical nature of our distinction between reversible and irreversible physics. In classical physics it was (and still is) inconceivable that the division between the two could occur at any but the most microscopic scale. In the quantum mechanics of the 1990s the division is moving very rapidly towards the macroscopic domain – it is already at the micron scale (i.e., visible in optical microscopes).

With a deliberate ambiguity of language it then becomes possible to talk of 'reversible measurements'. This can be done in a reasonably precise way by using a 'quantum computer' to carry out computations which are *at the same time* nothing but measurement operations, in the conventional sense. Quantum computers have not yet been built – indeed their operation depends on the elevation of quantum processes to the macroscopic level, in the way just described. Nevertheless I explore (in section 5) the operation of such a device, in a way which ties together many of the various themes of this chapter. The quantum computer is but one of many speculative ideas for the technology of the next century – needless to say, if it ever gets built, the effect on society will be enormous.

To maintain the flow of the arguments, technical details are kept to a strict minimum, and relegated to two appendices (one on the physics of decoherence, and the other on the quantum computer). Nevertheless these appendices are very important to the understanding of the text, and they are also intended as a bridge to the physics literature, in keeping with the introductory nature of this chapter. If I succeed in bringing philosophical readers at least halfway across the bridge between our two subjects, then the writing of this will have been worthwhile.

2 Time's arrows: the 'orthodox standpoint'

In physics these days one distinguishes (at least) four different arrows of time. The most important is the 'thermodynamic arrow', which refers to the apparently irreversible sense of entropy increase in any non-equilibrium system. For centuries this was the only recognized one. In addition one has the 'radiation arrow' (referring to the existence of only retarded radiation, whether this be, e.g., electromagnetic or gravitational radiation), first seriously discussed at the end of the nineteenth century, and the 'quantum arrow' (describing irreversibility in quantum theory); these two arrows are intimately related to the thermodynamic arrow. Finally there is the 'cosmological arrow', usually discussed in terms of the expansion of the universe.

The debate on these arrows tends to focus on the reasons for their existence, and the relationship between them; this debate is continually stimulated by (and has in return stimulated) new advances in physics. For most practising physicists[1] a number of the questions involved are fairly uncontroversial; the relationship between the thermodynamic and radiation arrows, for example, or the origin of thermodynamic irreversibility, are thought to be understood. Other questions are less clear – for example, the nature of the cosmological arrow, its relation (if any) to the other arrows, and the possible existence of causal loops in general relativity, to name a few. In addition, there is a lot of confusion in most discussions of the 'quantum arrow', even though most physicists probably assume it derives from the thermodynamic arrow.

I will therefore try, in this section, to separate out those arguments that most physicists are reasonably happy with, from those that are controversial. The relatively orthodox arguments involve the connection between the thermodynamic and radiation arrows, and the role of gravity in keeping the universe far from equilibrium, as well as creating and sustaining local regions of high negentropy. The controversies arise (a) near spacetime singularities, where quantum effects must come into play, and (b) when one comes to discuss the quantum arrow, which drags in the subject of quantum measurements and the measurement problem. I will say very little about (a), since its solution presumably requires some theory embracing both quantum mechanics and gravity, which we simply do not have. However I will explain (b) in some detail, particularly as it involves the 'measurement paradox'.

(a) The thermodynamic and radiation arrows, and gravity

As is well known, the time-reversibility of the microscopic Hamiltonian[2] describing any closed system (classical or quantum) implies that its microscopic behaviour must be time-symmetric. To understand the observed

---

[1] I would stress here that 'physicists' are a heterogeneous lot; a condensed matter theorist like me tends to give less weight to, e.g., the speculations coming from quantum gravity than would a cosmologist.

[2] As usual, one ignores here the T-breaking character of the electroweak interaction, assuming it to be irrelevant.

time-asymmetric behaviour requires a chain of arguments that usually begins with the consideration of interacting systems which then separate (so that we now consider open systems). A well-known discussion[3] imagines that in examining a deserted beach, we come across a footprint in the sand. If we consider the beach and its immediate surroundings to be a closed system, then we are faced with something extraordinary – we must accept that the footprint has resulted from a chance thermodynamic fluctuation in an otherwise equilibrium system, and that it must disappear both *before* and *after* we observe it (over a relaxation time characterized by, e.g., the water–beach interactions, sand friction, etc.). The probability of such a low entropy fluctuation is of course quite fantastically small. However it is obvious that the assumption of a closed system is incorrect – the water–beach system interacts with the rest of the universe (even if not now, it has in the past), and apart from the footprint, we see clear signs of this (waves on the water, wind, etc.). Thus we assume automatically that the footprint has been left by a previous visitor; an external system 'from the past' has left its mark.

Why 'the past', and not 'the future'? To understand this, we must now follow the chain of interactions between different sub-systems of the universe, out from the beach and in both directions in time. First, a purely physical point. The footprint is a sign, not only that we are dealing with non-equilibrium systems, but also that the external agent which created the footprint acted as a source of negentropy for the beach. Despite the fact that the *approximately* closed system (beach + water + external agent + surrounding air) has an entropy which can only have increased during the time period involved, nevertheless certain parts of this total system have suffered an entropy *decrease* with increasing t – in particular, the region of the footprint. The source of this negentropy (the external agent) must then clearly have been a system possessing considerable negentropy, and also must have been way out of thermodynamic equilibrium.[4]

But from whence comes this negentropy, and what keeps things out of equilibrium? We must follow the transfer of negentropy and energy back to its source. Although the earth is itself anomalously negentropic compared to its immediate surroundings, it is obviously not the source – as is well known, the source of energy driving almost all processes on earth is the sun, without which it is highly unlikely that life would have evolved, and without which

---

[3] Grünbaum (1973). The example was invented by M. Schlick.
[4] For negentropy, see Brillouin (1962).

the earth would be a much less ordered place.[5] Thus the source of negentropy is the sun – but then what drives the sun? At this point one realizes the inevitable connection between the thermodynamic arrow and the gravitational field, first properly described by Tolman. The sun and stars were formed through gravitational contraction of diffuse matter, and such 'self-organization' necessarily requires the attractive gravitational field. Again, total entropy increases (the gas and dust heat up whilst contracting, and of course once nuclear reactions commence, a massive outpouring of energy and entropy occurs) but it is the gravitational field which produced the highly organized galaxies of stellar systems, and thus acts as a vast source of negentropy – in the same way it keeps the universe very far from equilibrium. It used to be thought that gravity provided an infinite source of local negentropy (as well as global entropy), but now things are not so clear, because of the possibility of black hole formation, and the uncertainties surrounding the behaviour of quantum black holes. In any case gravity alone is responsible for having kept the universe very far from a 'heat death' for a very long time.

Thus we find that the existence of a thermodynamic arrow and of non-equilibrium states apparently derives ultimately from gravity. Moreover, given the coupling of radiation to matter, it becomes clear that the radiation arrow is tied to the thermodynamic arrow – they must have the same sense.[6] And the very long-range (in space and time) nature of radiative coupling means that regions very far apart in spacetime will be coupled together and forced to have the same arrow of time.

At this point it is very tempting to say that the problem has been solved, by saying that the thermodynamic arrow is enforced to be the same throughout spacetime, and that is that. But even if true, this argument is incomplete – there is still nothing to stop us imagining a time t_0 before which entropy *decreases*, and such that the state of the universe at $t = t_0$ is nothing but a massive low entropy fluctuation.

Consideration of a thought-experiment suffices to substantiate this point, and also shows why most physicists pay no attention to it. Imagine the universe at $t = t_0$ as a huge homogeneous mass of particles, throughout space. Now, there is one oft-quoted scenario in which at $t = t_0$, all particles are at rest in

---

[5] It is, however, conceivable that even in the absence of solar energy, primitive life might have evolved on earth, powered by radioactive decay and living deep within the crust.
[6] The Wheeler–Feynman absorber theory (see Wheeler & Feynman (1945) and (1949)) is assumed to be irrelevant here, if for no other reason than that it was rendered obsolete by quantum electrodynamics.

Fig. 1. The typical behaviour in time t of the total entropy $\bar{S}(t)$ of a closed macroscopic system in equilibrium

their co-moving frame – under this circumstance, even in the presence of gravity, entropy increases on both sides of t_0, and the arrow of time reverses at t_0. But consider how ludicrous this scenario is! We have forgotten the photons (and gravitons) which obviously cannot be at rest (presumably at t_0 they have disappeared, having been absorbed by matter!), and the pressure of the gas is zero throughout. This is our massive downward fluctuation of entropy, a footprint on a truly incredible scale! Such a fantastic situation is usually ignored, because of its pathological improbability.

One might try again with a less stringent scenario. We could try supposing, at $t = t_0$, a situation where (particles + radiation) are in *equilibrium*; although their coarse-grained distribution is imagined to be stationary in the co-moving frame, there will now be fluctuations, and the pressure is finite. Then one can apparently argue that for both $t > t_0$ and $t < t_0$, contraction and stellar condensation will occur; entropy will increase, negentropy will be produced, and so on. But this leads to a paradox – we start at $t = t_0$ in equilibrium, where entropy is supposed to be a maximum, and yet it increases in both directions! The resolution of this paradox is that the initial hypothesis of equilibrium at $t = t_0$ is incompatible with the existence of the attractive gravitational field.

Detailed consideration of many such examples of non-equilibrium phenomena, from the sub-nuclear up to the largest cosmological scales, help fill out this remarkable contemporary picture of the universe,[7,8] and the role of gravity in 'enforcing' the thermodynamic arrow therein. However one would like more than examples – we would still like to know whether the thermodynamic arrow really is universal, or whether, e.g., it can reverse itself, in other parts of space (or time).

---

[7] A good text-book on non-equilibrium phenomena is Landau & Lifshitz (1981); this is volume 9 of their 'course' in theoretical physics.

[8] A nice survey of the physics involved in astrophysics and cosmology is given in Zeldovich & Novikov (1971).

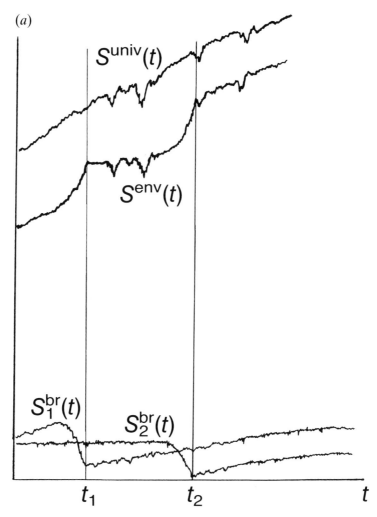

Fig. 2. (a) The observed behaviour of the entropy $S^{\text{univ}}(t)$ of the visible universe, along with the entropy of two 'branch systems' which isolate themselves from the rest of the universe (the 'environment') at times t_1 and t_2.

I think it is probably true that most physicists would still argue along the lines adopted by Boltzmann,[9] Tolman,[10] and Reichenbach,[11] amongst many others, but with one extra ingredient and two main provisos. A sketchy account of this argument goes roughly as follows:

[9] Boltzmann (1964), (1866), (1871), (1872).
[10] Tolman (1934).
[11] Reichenbach (1956).

4 Time, decoherence, and 'reversible' measurements

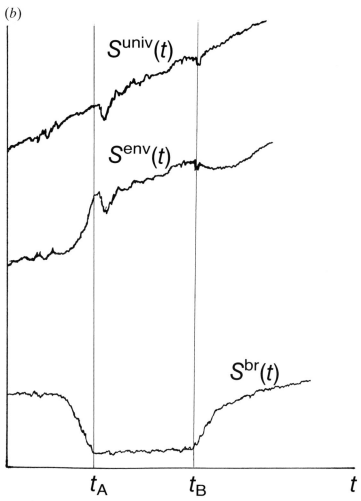

Fig. 2. (b) A branch system splits off from the rest of the universe at time t_A, but rejoins it at time t_B.

(i) We know that for any closed system, the total entropy $\bar{S}$ will behave as in figure 1, with fluctuations $\delta\bar{S}(t)$ much smaller than $\bar{S}$ (for a large system); $\bar{S}$ is the thermodynamic equilibrium entropy. Not only are the fluctuations inversion-symmetric in time (on average), but they are also translation-symmetric (on average) in time; no preferred time exists.

(ii) However one observes that in our present universe, massive disequilibrium exists – not only is the entire universe way out of equilibrium, with rapidly increasing entropy $\bar{S}(t)$ but also it is clear that it has been like this since the big bang, and will continue to be so for a very long time. Moreover, local

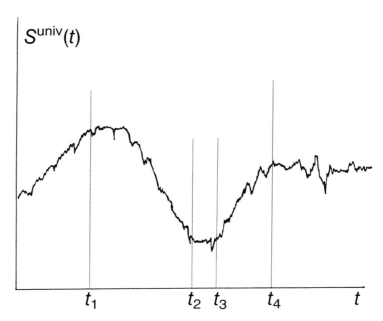

Fig. 3. The behaviour of a closed universe over long time periods (ignoring the effect of gravity). For time $t < t_1$, the arrow of time goes 'forward'; for $t_1 < t < t_2$, it goes backward; for $t_3 < t < t_4$, it goes forward; and for $t_2 < t < t_3$ and $t > t_4$ arrows cannot be meaningfully defined

regions of low entropy are constantly being produced, and 'branching off' from the rest of the universe, at least for some period of time. Such 'branch subsystems'[11] are all around us (in good part because we are constantly creating them ourselves). This situation is illustrated in figure 2. In figure 2(a) we see two branch systems splitting off from the rest of the universe; and to make things clearer, in figure 2(b) the branching is shown for a single system which first separates from, and then is re-absorbed by, the 'environment' (i.e., the rest of the universe).

(iii) Given the symmetry of physical laws under time inversion,[2] there must also be processes corresponding to the time reverse of figure 2, in which the global entropy *decreases* (i.e., $d\bar{S}/dt$ changes sign), over long periods of time. This situation was clearly envisaged by Boltzmann,[9] in the form shown in figure 3. As noted by Reichenbach,[11] to be consistent we must maintain that, when $\bar{S}(t)$ is in a downward stretch, the thermodynamic arrow has *reversed* itself. Thus for a closed universe, it apparently makes no sense to talk about a *universal* thermodynamic arrow of time – one can only talk about such an arrow for those finite intervals of time in which the global entropy changes monotonically.

4 Time, decoherence, and 'reversible' measurements

(iv) However, as we have seen, Boltzmann and Reichenbach missed the crucial role of gravity in all of this. The point is that gravity provides a potentially inexhaustible source of global entropy and local negentropy (at least at first glance), and this makes it possible to imagine the upward rise of $\bar{S}(t)$ continuing forever. Then, given as an 'initial condition' that for example in our present universe things are as they are, one argues $d\bar{S}/dt$ will continue to be positive for all times (positive or negative), and this then *defines* the thermodynamic arrow.

This point was not missed by Tolman[10] and some other early workers in cosmology, and Tolman even gave a nice example of a perpetually oscillating universe (with ever-increasing period) in which $\bar{S}(t)$ increases indefinitely.[12]

Does this mean the problem is solved? Not necessarily; an extra ingredient arises, even in classical general relativity. Since the 1960s it has been understood that both global and local topological properties of the spacetime metric bear directly on the thermodynamic arrow, through their implications for the boundary conditions that we are allowed to choose on the dynamical evolution of the universe.

The global properties[13] enter as follows; it is now known that the universe must possess spacetime singularities, surrounded by an event horizon. These 'black holes' (of which the universe itself may be one) possess a finite entropy S proportional to the area A of their event horizons (in fact $S = \frac{1}{4}A$; a result which comes from adding quantum mechanics on to the background classical black hole metric[14]). The local considerations arise because one observes that for massive stars, galactic nuclei, etc., a black hole is the inevitable result once nuclear and thermal energy have been dissipated.[8] Thus in our present universe, such black holes possess only a future singularity, and a future event horizon, but not one in the past. Of course this would not be true in a universe with a reversed thermodynamic arrow, and so one can imagine 'white holes', with past singularities and event horizons. If one now takes the global point of view that such white holes must be eliminated for all time, one arrives at the 'cosmic censorship' hypothesis,[15] which eliminates such

---

[12] In common with other oscillatory models, Tolman's model has no reversal of the thermodynamic arrow upon contraction. It was suggested in Gold (1962) that the arrow *ought* to reverse with contraction. This suggestion is usually treated sceptically (or even ridiculed – see Zeldovich & Novikov (1971), vol. 2, 670–3), although it has found some support (see Bondi (1962), Ne'eman (1970), and Landsberg (1970)). It obviously requires very special initial or boundary conditions, and one has to explain why, e.g., light from stars which 'leaks between' expansion and contraction cycles does not enforce a uniform arrow.

[13] The standard reference on this subject is Hawking & Ellis (1973).

[14] Hawking (1975, 1976).

[15] Penrose (1969).

naked singularities by fiat – although again, it is by no means clear that this hypothesis is necessary to eliminate white holes (except perhaps for certain examples,[16] which are unstable in the presence of an environment with a well-defined thermodynamic arrow).

A more serious question arises when one considers that the steady accumulation of black holes can limit the rise in total entropy of the universe. A whole range of possible scenarios then arises,[17] and eschatological studies of the end of the universe have become rather popular in recent times, both in the cosmological literature and in science fiction.[18]

Thus we must come to the conclusion that even in classical physics (i.e., including classical general relativity, but excluding quantum mechanics), the problem of the origin of the thermodynamic arrow, and its link with a putative 'cosmic arrow', cannot be regarded as solved, since we simply do not yet know enough about the global properties of our spacetime universe, and possible singularities in it, to decide between the various scenarios that have been envisaged. Lest some theoretical astrophysicists might disagree, I emphasize here that at the present time we do not yet have *convincing* evidence for the existence of even a single black hole in the universe, although there are of course a number of strong candidates.

(b) Conceptual problems: quantum gravity and quantum measurements

These considerations make it clear that at least part of the problem of time's arrows lies in the choice of spacetime boundary conditions. Unfortunately they also bring us to the miasma of current research and the edges of our present understanding. The frontier is clearly signposted by the two provisos mentioned earlier. The first is that the special role played by spacetime singularities inevitably drags in quantum mechanics, and no generally accepted unification or synthesis of quantum theory and general relativity

---

[16] See, e.g., Christodoulou (1984).
[17] I mention here another topic of current interest, viz., the possibility of causal loop solutions both in classical general relativity and in theories of simple quantum fields on a background classical metric. See, e.g., Friedman et al. (1990), Deser, Jakiw, & 't Hooft (1992), Hawking (1992), and 't Hooft (1992). Since these solutions inevitably involve unphysical or improbable boundary conditions and/or wormholes or else do not actually violate causality, it is not clear whether they are relevant to time's arrow. See also Earman's discussion in this volume.
[18] See, e.g., Dyson (1979) and Pohl (1990). These both speculate that the continuing evolution of more intelligent beings could influence the long-term development of the universe.

exists.[19] Thus if we need to understand such 'quantum gravity' problems in order to understand time's arrows, then we must simply twiddle our thumbs for the moment; the reader is referred to the extensive literature,[20] and I shall not further discuss this topic. The second proviso involves ordinary quantum mechanics, and the notorious 'measurement problem'. The idea has been very often put forward that a fundamental irreversibility arises in quantum mechanics (hereafter abbreviated as 'QM') when a measurement takes place, thereby introducing a 'quantum arrow of time'. The measurement problem (or paradox) is thereby dragged into the subject. This problem is usually introduced as follows (following von Neumann[21]). In a measurement one supposes that the system of interest (with initial wave-function ψ_k) interacts with a 'measuring apparatus' (initial wave-function Φ_o) so that the final state of the apparatus is uniquely correlated with the initial state of the system; in the simplest (ideal measurement) case, this implies that $\psi_k \Phi_o \to \psi_k \Phi_k$ during the measurement, with 'macroscopically distinguishable' states Φ_k (otherwise the apparatus would be useless).

Now, however, suppose that the system is initially in the superposition $\Psi = \sum_k c_k \psi_k$, where c_k are constants. We are then forced to conclude that

$$\Psi \Phi_o \to \sum_k c_k \psi_k \Phi_k \qquad (2.1)$$

during the measurement. We are not in the habit of considering superpositions of macroscopically distinguishable states like (2.1) which contradict common sense and classical physics – so we have a paradox.

A number of ways have been proposed to resolve this paradox. Bohr[22] simply asserted the measuring process to be unanalysable, but nevertheless made it the crucial link between the distinct classical and quantum worlds. Von Neumann[21] took this farther, introducing the 'collapse' of the wave-function, according to which

$$\sum_k c_k \psi_k \Phi_o \to \psi_k \Phi_k \qquad \text{(probability } |c_k|^2) \qquad (2.2)$$

must occur before the measurement is finished. This is more neatly expressed in density matrix language, according to which $\rho_{kk'} = c_k c_k^*$ collapses to $|c_k|^2 \delta_{kk'}$

---

[19] There are 'programmes' for unification (such as superstring theory), the practitioners of which are often very confident that they have, or will soon have, the solution. These tend to have a half-life of three to four years and have not yet produced one experimentally verifiable prediction.
[20] See Unruh in this volume and Unruh & Wald (1989).
[21] von Neumann (1935).
[22] Bohr (1936, 1949).

(here $\delta_{kk'}$ is the Kronecker delta-function, and c_k^* is the Hermitian conjugate of c_k). As von Neumann pointed out, the point at which (2.2) occurs is not obvious, and he proposed that it would ultimately take place when a human consciousness (the 'observer') became involved. A similar emphasis on consciousness (whatever this may be!) has been relied on by Wigner[23] and Peierls,[24] in a remarkably anthropocentric view of our most fundamental theory. Quite apart from the feeling that the earth is being placed again at the centre of the universe, these ideas are rather mysterious – one has the impression that the strange core of QM is being replaced by something (consciousness) that is even less well understood – indeed no attempt is even made to define 'consciousness' by advocates of this view. Very few physicists have claimed to understand Bohr's ideas either.

Another early viewpoint, amongst many others,[25] is that the wave-function χ_k embodies only our *knowledge* of the system concerned.[26] This again is remarkably anthropocentric, but it does have the advantage[27] that no physical reality is ascribed to the wave-function (at least when we consider other paradoxes such as the EPR paradox).

All of these early viewpoints are notoriously difficult to understand, and the cures almost seemed worse than the disease – which probably explains why investigations in this area almost ceased for a long time. In the 1950s two other proposed resolutions emerged. One of these, the 'many-worlds interpretation' of Everett,[28] has found some favour amongst cosmologists, but has suffered considerably from some of the wild claims made for it – we shall return briefly to it later. A much more popular resolution is founded on thermodynamic irreversibility, and is thus directly connected to the thermodynamic arrow. This idea (which has a long history[29]) applies the Schrödinger equation throughout, so that (2.2) never occurs, and (2.1) is always valid. However it is then observed that measurements invariably involve some sort

---

[23] Wigner (1963, 1964).
[24] Peierls (1979), pp. 23–34.
[25] A very complete survey of early ideas is given in Jammer (1974).
[26] This view (early espoused by Heisenberg) is usually accompanied by some kind of 'ensemble interpretation', according to which QM and ψ_k are only meaningful with reference to ensembles rather than individual systems.
[27] d'Espagnat (1979).
[28] Everett (1957); reprinted with other articles and commentary in de Witt & Graham (1973).
[29] The history up to 1970 is described in Jammer (1974), pp. 488–96. References to work after this can be found in the material cited in footnote 31. Important early discussions of what is now called 'decoherence' in the measurement process can be found in the works of Ludwig, Daneri et al., Wigner, Simonius, Joos, & Zeh, and Ghirardi et al. These papers are all cited and discussed in Jammer (1974) or the works cited in footnote 31.

of amplification to the macroscopic scale, which involves irreversibility in that it generally couples in many other degrees of freedom to the process in (2.1). Since we cannot follow the behaviour of these other degrees of freedom, the reduced density matrix $\rho_{kk'}$ (obtained by averaging over these other 'environmental' coordinates) is 'for all practical purposes' (FAPP), indistinguishable from $|c_k|^2 \delta_{kk'}$. Another way of saying this is to claim that all 'quantum coherence' or 'quantum interference', embodied in the off-diagonal elements of $\rho_{kk'}$, has been suppressed by the time we reach the macroscopic scale.

Reactions to this argument divide into two camps. Probably a large majority of all physicists like it, and believe it constitutes an 'FAPP solution' to the measurement problem, by showing that QM does not contradict either our experience, or common sense. A smaller but vocal minority, while accepting that there may be no contradiction, FAPP, are not willing to accept such an FAPP solution underpinning our most fundamental (and successful) theory (see, for example, Bell[30]).

In the last 10 years this debate has been turned on its head,[31] by the demonstration that for certain systems, macroscopic interference and coherence can and do exist! This development obviously has very important consequences for the foundations of QM, as well as for our understanding of quantum dissipation and decoherence. At present there are two serious candidates for systems in which macroscopic quantum coherence effects may have been seen – these are superconducting SQUIDs (almost certainly[32]), and certain ferromagnetic systems (where a convincing experimental proof has not yet appeared[33]). However the ideas and techniques involved are now being applied in all areas of physics.[34]

3 Wave-functions and the quantum arrow

In his attacks on 'FAPP' solutions of the quantum measurement problem, Bell sometimes exhorted physicists[34] to ban the term 'measurement' entirely from their lexicon (along with other terms like 'apparatus', observer, etc.). His

---

[30] Bell (1975, 1990).
[31] A simple discussion of the early theory (due to Leggett and collaborators) is in Leggett (1980). A recent discussion of some of the philosophical implications is in Leggett (1987b).
[32] The situation in SQUIDs is described in Leggett (1986, 1988); see also Clarke, Cleland, De Voret, Esteve, & Martinis (1988).
[33] The situation in magnets is described in Stamp, Chudnovsky, & Barbara (1992) and in Stamp (1993).
[34] See Bell (1975, 1982, 1990).

point was very simply that 'measurements' had no place in the foundations of a physical theory, and could only be secondary to it, and refer to what goes on in, e.g., physics laboratories.

Bell's admonition raises a number of philosophical questions about what theories should look like, to which we shall return; but first I attempt, in this section, to follow this line by discussing the quantum arrow from the 'wave-function' perspective just described. Measurements will eventually emerge, but only as a particular kind of messy process, which often tends to occur when physicists are around. Nevertheless a paradox does emerge, as one might expect, although it really now deserves a new name, since it is associated with the existence of macroscopic quantum phenomena *per se*, rather than just measurements. From this new perspective one may take a fresh look at the quantum arrow.

(a) Quantum phases, dissipation, and decoherence

Let us begin with a quick review of the modern view of decoherence and dissipation (leading to irreversibility) in quantum mechanics, which has evolved in the last 10 years, largely under the stimulus of specific but rather generally applicable models, such as the 'Caldeira–Leggett' model.

One starts with some QM object (microscopic or macroscopic), and first isolates a set of degrees of freedom $\{R_j\}$, with $j = 1, \ldots, M$, pertaining to this object, and in which one is particularly interested. Often (but not always) the choice of $\mathbf{R} = (R_1, \ldots, R_M)$ is physically obvious, and governed by symmetries, etc.; this is the art of 'model-building'. We now divide the universe into 'the system' (defined by $\mathbf{R}$) and the 'environment', which comprises not just the rest of our QM object, but the rest of the universe. Everything is treated quantum mechanically, from beginning to end.[35–40]

Now the system and the environment interact, of course, and as we have seen, this effectively destroys phase coherence in the system wave-function; if we start with a *total* (or 'universal') wave-function

$$\Xi_o(\mathbf{R}, \{x_k\}) = \psi_o(\mathbf{R})\Phi_o(\{x_k\}) \tag{3.1}$$

---

[35] Feynman & Vernon (1963).
[36] Caldeira & Leggett (1981, 1983a).
[37] Caldeira & Leggett (1983b).
[38] Leggett (1984).
[39] Caldeira & Leggett (1985).
[40] Leggett (1987b). See also Chakravarty & Leggett (1984).

where the $\{x_k\}$, with $k = 1, \ldots, N$, label the environmental coordinates, then system–environment interactions will cause evolution to

$$\Xi(\mathbf{R}, \{x_k; t\}) = \sum_\alpha \sum_n c_\alpha^{(n)}(t) \psi_\alpha(\mathbf{R}, t) \Phi_n(\{x_k\}, t) \qquad (3.2)$$

after a time t has elapsed (here α labels the system states, n labels the environmental states, and the coefficients $c_\alpha^{(n)}(t)$ vary with time). Since in general we have no access to the environmental state, we 'integrate out' (i.e., average over) the environmental degrees of freedom. Then the density matrix $\rho_{\alpha\beta}(t)$, incorporating our knowledge of the system, is supposed to evolve rapidly to a 'mixed state' in which any interference between states ψ_α and ψ_β is destroyed; this must be the case, it was long felt, if ψ_α and ψ_β were 'macroscopically different'. A simple example of this is provided by the following 'two-state' system. Imagine a system which is allowed to proceed between two points A and B along one or other of two paths (labelled 1 and 2); see figure 4. If we ignore the environment, then after letting the system set off from A, we expect after a while that the state of the system will be

$$\psi(\mathbf{R}, t) = c_1(t)\psi_1(\mathbf{R}, t) + c_2(t)\psi_2(\mathbf{R}, t) \qquad (3.3)$$

with $|c_1|^2 + |c_2|^2 = 1$, and the wave-functions ψ_1 and ψ_2 representing the system on paths 1 and 2 respectively; the corresponding density matrix $\rho_{\alpha\beta}$ takes the form

$$\rho_{\alpha\beta}(t) = \begin{pmatrix} |c_1(t)|^2 & c_1^*(t)c_2(t) \\ c_1(t)c_2^*(t) & |c_2(t)|^2 \end{pmatrix} \qquad (3.4)$$

in which we see the phase coherence embodied in the off-diagonal matrix elements.

However as soon as this two-state system is coupled to its environment, and we let the combined system–environment evolve, to give a total wave-function of the form (3.2), then the off-diagonal terms get 'washed out', once we average over the environmental states. Thus, goes this argument, in the example of (3.3), we ought not to be able to see interference effects between the two paths, if (a) they are widely separated, and (b) the system moving on these paths is 'macroscopic' (by which, roughly speaking, one means 'massive', or at least comprising many particles).

However, as so often seems to be the case in physics, when a detailed investigation of the transition from (3.1) to (3.2) was finally made, the true picture turned out to be rather different (and much more interesting). There

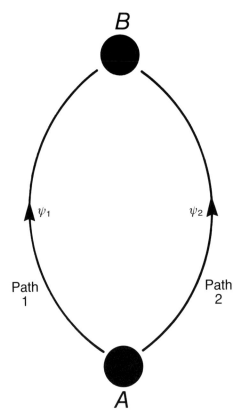

Fig. 4. A 'two-state' system, making transitions from A to B along two different paths (i.e., through two different 'states')

are various ways of looking at the results, and we shall examine several of them.

One useful but rather crude picture is obtained by contrasting two important timescales[41] involved in the system–environment interactions; these are the energy relaxation time τ_E, and the 'decoherence' time τ_ϕ, required for these systems to lose phase coherence. The identification and clarification of the roles of these two times stems from the work of Caldeira and Leggett, who studied simple models in which the initial system state superposed two spatially separated wave-packets.[39,40] After a very short time τ_ϕ, the

---

[41] One may define a number of different characteristic timescales in the theory of transport and non-equilibrium phenomena, referring to the transport, diffusion, and relaxation of various quantities, as in Landau & Lifshitz (1981). Thus we are clearly touching only the surface of a large subject.

interaction with the environment decohered this state, so that no measurements on the system alone could detect any interference between the two initial wave-packets. In density matrix language, the initial system state was $\psi(x) = 2^{-1/2}(\psi_1(x) + \psi_2(x))$, with ψ_1 and ψ_2 normalized, so that the initial density matrix was just $\rho_{\alpha\beta}^{(o)} = \frac{1}{2}\int dx\, \psi_\alpha(x)\psi_\beta^*(x)$; this has off-diagonal interference terms $\rho_{12}^{(o)} = \frac{1}{2}\int dx\, \psi_1(x)\psi_2^*(x)$, in its Hermitian conjugate, as well as diagonal terms, and in fact

$$\rho_{\alpha\beta}^{(o)} = \frac{1}{2}\begin{pmatrix} 1 & e^{i\phi_{12}} \\ e^{-i\phi_{12}} & 1 \end{pmatrix}. \qquad (3.5)$$

However after τ_ϕ, this has evolved, FAPP, to

$$\rho_{\alpha\beta}(t > \tau_\phi) \sim \begin{pmatrix} 1 & 0 \\ 0 & 1 \end{pmatrix}. \qquad (3.6)$$

Now τ_ϕ is very short for most systems – it is roughly the time it takes for one environmental mode to be excited by the system. It is then obvious that τ_E will be much longer; in order for the system energy to relax, many environmental modes will need to be excited, as the system 'cascades down' to a state (actually a mixture of states) in equilibrium with the environmental 'bath'. Notice also that τ_ϕ has a rough analogy with the collision time in a classical system (between, say, a Brownian particle and a background gas of particles).

Now the situation envisaged above is of course very general; as Feynman emphasized, the sort of 'two-slit interference' going on in (3.5) is at the very heart of the mystery of QM. Of course this interference has not been destroyed after τ_ϕ has elapsed – it has simply been 'destroyed FAPP', since we would now have to look both at the system *and* the environment together, in order to see interference again.

The heuristic picture given above may be elaborated (in the form of a time evolution equation[37] for $\rho_{\alpha\beta}(t,T)$, at some bath temperature). However it does miss some very important physics, and this is often not understood in the literature.

The first crucial point is that a discussion in terms of characteristic timescales automatically implies a monotonic decay of off-diagonal matrix elements, and indeed usually an *exponential* decay – I suspect that most readers, when first reading equation (3.4), simply assumed that the off-diagonal matrix elements looked something like $\rho_{12}(t) \sim \frac{1}{2}\exp(-t/\tau_\phi)\exp(i\phi_{12})$. Yet again, when one looks at the details, one finds that this assumption (which is of course a crucial assumption for those many physicists who 'resolve FAPP' the quantum measurement problem) can be completely wrong!

The problem is that whereas the relaxation of quantities like energy, current, etc., is constrained by local conservation laws, decoherence refers to phase randomization, and there are usually no such conservation laws to help with this. Thus the decoherence process is often not an exponential relaxation of off-diagonal matrix elements – in fact these may relax quite differently, or even grow or oscillate, before they are finally suppressed. This should make it quite clear that even though system–environment interactions are responsible both for decoherence and thermodynamic irreversibility, we should not equate these two – quantities such as the system entropy do not generally exhibit any of the oscillatory behaviour of the off-diagonal matrix elements (except in odd cases like the spin echo[42]).

The second important point has already been mentioned – that despite the usual assumption that by the time any quantum process has reached the macroscopic scale it must have 'decohered' and become quasiclassical, nevertheless the existence of macroscopic quantum processes has now been demonstrated. Of course the systems for which this is discussed (superconductors and magnets) show all the usual thermodynamic irreversibility. The key point is that there exist certain 'macroscopic' degrees of freedom (the phase across a Josephson junction in a SQUID ring, or the position of a magnetic domain wall) which couple extremely weakly to their environment.

Again, the details of this are a little complicated (appendix A gives an introduction to these), but the consequences may be discussed fairly simply. First note that by 'macroscopic' we really mean rather big – for example the number of spins (or Cooper pairs) taking part in the macroscopic quantum tunnelling of a magnetic domain wall (or superconducting SQUID) is expected to be as high as $\sim 10^{10}$, at low temperatures.[33] The 'effective mass' of such objects is comparable with small bacteria (and the domain walls are just visible through ordinary optical microscopes). Coherence on such a scale is quite inconceivable for a classical system; what allows it here is (a) the existence of an *energy gap* in the environmental spectrum (a uniquely quantum phenomenon), and (b) effective interactions between the spins (Cooper pairs) which force them to act in concert.

We may summarize the picture as follows. In QM one has irreversibility and dissipation in much the same way as in classical mechanics (indeed, QM is full

[42] Hahn (1950).

of various kinds of Boltzmann equations[7]). However the destruction or decoherence of phase information is governed by different processes and different timescales, even though both dissipation and decoherence arise from 'system–environment' interactions. Moreover decoherence is a much more subtle process than dissipation, and can behave quite differently. It can also be dramatically reduced under certain circumstances (as in superconductors and magnets). The interested reader is urged to read appendix A, to fill out this picture.

(b) The transition to the macroscopic world

Let us now stop to consider some more general implications of all of this. First, for the 'quantum arrow'. It will be manifestly clear that, within the 'wave-function' description just given, the quantum arrow is no different from the thermodynamic arrow (although the details are now quite different from any classical theory). This will be true no matter which position one adopts towards the deeper questions of thermodynamic irreversibility, at least until one is driven eventually to look for a unification of quantum theory and gravity, in the way described in section 2.

Second, what implications are there for measurements, apparati, and observers? As promised, we did not *need* these concepts in the discussion above; and one may now try to argue (as indeed many have[29]) that measurements are simply one particular kind of (apparently) irreversible physical phenomenon, still described in terms of a wave-function as in eqns. (3.5) and (3.6), in which FAPP transitions like (3.3) → (3.4) take place. It is then argued that there is no longer any measurement problem or paradox, and that the classical world of Bohr is *defined* as the world of phenomena in which decoherence has already taken place. As already mentioned, this argument is now at least 40 years old, although it seems to get reinvented every five years or so.[43] Moreover, it is probable that most physicists with an opinion on the subject find this view, if not congenial, at least better than any alternative.

However there is strong opposition to this view as well. As noted by Bell[34] and Leggett,[31] the quantum interference is still *there*, embodied in system–environment correlations; and who is to say that present-day 'FAPP limitations' on our ability to recover this interference, will not be later overcome? The work on macroscopic quantum phenomena[31–33] really rubs this point in,

---

[43] The latest reinvention is embodied in the 'decoherence functional' of Gell-Mann & Hartle (1991).

as do the rapid advances currently taking place in 'mesoscopic physics' (i.e., the physics of systems between the macroscopic and microscopic scales, in which the QM energy level density is high, but fluctuations in it still have important physical consequences).

At this point the argument starts to become as much philosophical as physical. As far as I can see, there is no obvious *contradiction* involved in the orthodox position just described (i.e., that the classical world is classical because of decoherence). It must surely make proponents of this view uncomfortable to know that the 'FAPP barrier' between the quantum and classical worlds is shifting inexorably towards the classical macroscopic realm with every advance in condensed matter physics! Nevertheless, provided one is prepared to follow this barrier wherever it goes, I can see no way of faulting the logic of the argument. In fact I would maintain that this is the case even if we let the FAPP barrier invade our minds. For what evidence can we possibly have against it, given that at present we have *no scientific evidence whatsoever* that we ourselves are not described by the QM laws of physics, including our mental processes? Arguments against the application of QM to the mind usually start from the idea that it is 'obvious' (by, say, introspection) that our minds, or thoughts, or personal state, are *never* in a superposition of states.[44]

I myself cannot see why this is obvious at all! In fact, all known sentient beings operate under conditions of such high temperature and irreversibility that it is difficult to imagine any collective coordinate describing 'thought processes' being sufficiently decoupled from its environment to show macroscopic phase coherence – indeed it is *obvious* that minds and mental processes are very strongly coupled to their environments (otherwise they would be useless!). This point (which should really be elementary in the light of the description of decoherence just given) seems never to have been made, despite the well-known fact that the operation of all biological and genetic processes (as opposed to their atomic components) can be described *classically*. From this point of view, we see that minds (as we know them), even if they are described by QM, are still going to operate, FAPP, according to classical physics – all phase coherence will be irretrievably (FAPP) entangled with the environment. One might imagine some superbeing (or ourselves in a few centuries) being able to push the FAPP dividing line (between the quantum and classical

---

[44] One can also argue, as Bohr did, that any attempt (even by a superbeing) to probe quantum interference in the mind would prevent normal functioning – a kind of 'decoherence of the mind' which would transform the FAPP barrier into a permanent barrier of principle.

worlds) so far into human minds that it might be possible to observe, say, quantum interference in mental processes. However we are obviously very far away from this now, and so it is difficult to see what evidence introspection can presently provide about the applicability of QM to the human mind, or to 'consciousness' (whatever this is taken to be), or to known sentient beings.[45]

Another argument often levelled against the 'decoherence → classical' explanation of the macroscopic world, as applied to sentient beings, is that it ignores free will. This argument is of course not specific to QM; it applies to any physical theory having some deterministic macroscopic limit. Some proponents of free will are even prepared to argue that QM must be somehow involved in mental acts, since QM is 'not deterministic'.

This argument seems to me to be even flimsier than the last. A contradiction between free will and determinism can only occur, for some sentient being, if that being has *complete knowledge* of its own state (and the states of other entities or objects acting on 'his/her' state). Otherwise there is no problem – our being can be under the illusion that his/her acts are not predetermined, provided he/she is not aware of those elements which are determining these acts. Now, since Gödel we are well aware of the contradictions accompanying completeness in formal systems – and complete self-knowledge on the part of a physical system seems equally contradictory. And again, it is certainly very far removed from anything humans presently have!

Having said all of this, it is none the less important to remember why we have a 'measurement paradox' in the first place. It arises because many physicists are so sure that the macroscopic world of common sense must be described classically, or at any rate in terms of definite states, that they are prepared to go to almost any lengths to preserve the 'reality' of this world of definite states. Hence, e.g., the wave-function collapse, or Bohr's mysticism. In this light, one simply cannot ignore the existence of macroscopic quantum tunnelling and interference phenomena[32,33] – they constitute a direct assault on this 'common sense' view.

One final issue (which I will not discuss here at any length) is whether, if one adopts a 'wave-function' point of view, one should also assume some ontological status for the wave-function, i.e., whether it then implies that the wave-function is 'real' in some sense. My personal feeling is there is no such necessary

---

[45] A typical example is Albert (1992), who cannot understand (pp. 99–100) why so many physicists cannot see this introspective argument. Other examples (somewhat less dogmatic) include Wigner (1963, 1964); and the evidence from introspection is widely accepted.

implication, and that in fact any such assumption is fraught with difficulties. Interestingly, Aharonov et al.[46] have recently tried to claim that under certain circumstances the wave-function can be 'measured' unambiguously, and that consequently it should be regarded as real. This claim has been criticized by Unruh,[47] to which I can only add the further criticism that Aharonov et al. have not shown how superluminal propagation of changes in the wave-function can be made compatible with special relativity, if the wave-function is physically real.

In the same context one should mention the Everett 'many-worlds' interpretation of QM, which is held by some to resolve the foundational problems in QM. Again, it holds the wave-function to be real, but this time different elements of a superposition are held to exist in different 'universes'. The great objection (to my mind a quite fatal one) to all of this is that the real structure of the wave-function evolution in time is not the 'branching' one described by proponents of this view, but rather one of branching and 'recombination' of paths, as embodied in the path integral. If, on the other hand, the branching is supposed to describe irreversible separation of different elements of a superposition, after a measurement, then the many-worlds interpretation avoids none of the problems of conventional QM, since no means of saying when such a measurement takes place is given to us by this interpretation either.

Let me now summarize this section by saying that the wave-function point of view, *if consistently maintained*, leads to no obvious contradictions, and merely subsumes measurements under the general heading of (apparently) irreversible processes. However it does have the worrying feature (essentially the measurement problem in another form) that phase correlations are *always there*, and can in principle (and increasingly in fact) be recovered, even on the macroscopic scale. To put this another way, one can begin to think about '*reversible measurements*' in this framework – and I return to this theme in section 5, together with its implications for the thermodynamic arrow.

4 The 'measurement calculus' and the quantum arrow

In the previous section the occurrence of decoherence and irreversibility was described without any reference to measurements in QM, except as a

---

[46] Aharonov, Anandan, & Vaidman (1993); Aharonov & Vaidman (1993).
[47] See Unruh (1994) and his chapter in this volume.

4 Time, decoherence, and 'reversible' measurements

special kind of irreversible process. In this section I briefly deal with a quite contrary view of QM, in which measurements occupy a central role, so central that sometimes papers which take this view omit all mention of the wave-function!

A representative example is provided by the well-known discussion[48] of Aharonov, Bergmann, and Lebowitz of time-symmetric measurements (more recently developed into a 'two-vector' formulation of QM by Aharonov et al.[49]). In these papers (and similar subsequent papers[50,51]) one never deals with the wave-function at all, but instead with measurements and with the time evolution of operator equations (in particular, with the time evolution of the density matrix of the system of interest). This methodological dichotomy goes all the way back to the beginnings of QM; the wave-function point of view originated with Schrödinger, and the operator/measurement point of view with Heisenberg.

The first point that needs to be made about the time-symmetric formulation is that, in so far as it can be straightforwardly applied to real physical situations (and this is actually very rarely – see below) it demonstrates that there is no particular connection whatsoever between the thermodynamic arrow and quantum measurements. Thus consider a sequence of measurements of operators of the form

$$\hat{I}\hat{B}_1, \hat{B}_2, \ldots, \hat{B}_i, \ldots, \hat{N}, \ldots, \hat{A}_1, \ldots, \hat{A}_j, \ldots, \hat{A}_{m-1}, \hat{A}_m \hat{F} \equiv \hat{I}\{\hat{M}_k\}\hat{F} \quad (4.1)$$

in which an operator $\hat{O}_k$ in eqn (4.1) is associated with a projection operator $\hat{P}_k$ (in state-vector language, $\hat{P}_k = |k\rangle\langle k|$, where $|k\rangle$ is one of the set $\{|k\rangle\}$ of eigenvectors of $\hat{O}_k$). In (4.1), $\hat{I}$ is an initial measurement whose value we know, $\hat{F}$ is a final measurement whose value we also know, whereas the results of the $\{\hat{B}_i\}$, before $\hat{N}$, and the $\{\hat{A}_j\}$, after $\hat{N}$, are unknown to us, as is the result of $\hat{N}$ (made 'now'). It then follows that the conditional probability $p(\{A_j\}, N, \{B_j\}/F, I)$ of getting the values $\{A_j\}, N, \{B_i\}$ for the intervening measurements, given initial and final values I and F for the operators $\hat{I}$ and $\hat{F}$, is

$$p(\{M_k\}/F, I) = \frac{\text{Tr}(\hat{I}\hat{B}_1\hat{B}_2 \ldots \hat{N} \ldots \hat{A}_{m-1}\hat{A}_m\hat{F}\hat{A}_m\hat{A}_{m-1} \ldots \hat{N} \ldots \hat{B}_2\hat{B}_1)}{p(F/I)} \quad (4.2)$$

---
[48] Aharonov, Bergmann, & Lebowitz (1964).
[49] See Aharonov & Vaidman (1990, 1993) and also Aharonov, Albert, & Vaidman (1988).
[50] Griffiths (1984) and Gell-Mann & Hartle (1991).
[51] Unruh (1986) and his chapter in this volume.

where Tr represents the trace of the variables. Now (4.2) is clearly time-symmetric; moreover, as emphasized by Unruh, the information in $p(\{M_k\}/F, I)$ is not *directly* embodied in either a wave-function or a density matrix description of the system at any given time (e.g., 'now'), as can be seen by playing around with (4.2) (see also Unruh's chapter in this volume).

Thus (4.2) shows there is no necessary connection with any arrow of time in this 'measurement calculus' formulation of QM. Moreover, it is possible to do experiments which measure conditional probabilities of this type; they are not merely theoretical abstractions.[52]

However what is not always emphasized is that it is nevertheless *extremely difficult* to measure such conditional probabilities, because in the overwhelming majority of situations, decoherence enters into the sequence $\{\hat{M}_k\}$ (in the form of interactions between the system of interest and its environment). The net result is to fuzz out the coherence existing between successive measurements in the sequence $\{\hat{M}_k\}$ (in this picture, the environment itself is repeatedly making measurements, often of unknown operators, and certainly with unknown results). Then the traditional time-asymmetric picture of QM emerges if we assume this fuzzing or 'garbling' occurs after $\hat{N}$ (if we are really interested in $\hat{N}$) rather than before. But this assumption is of course nothing but the introduction of the thermodynamic arrow of time, *by hand*, into the quantum formalism. From this point it should be clear that the wave-function discussion of decoherence and the thermodynamic arrow is preferable to this discussion in the measurement calculus – we did not have to put anything in by hand *because* we did not have to rely on physically ill-defined (or undefined) concepts such as measurements. It is also amusing to notice how artificial the discussions of environmental decoherence appear when they are framed in terms of 'measurements' being made by the environment on the system of interest. The simple fact is that any attempt to go beyond vague generalities in discussing this decoherence, to discuss it quantitatively, requires a reversion to calculations of matrix elements, etc., starting from wave-functions (or related functions). This underlines the way in which the adoption of a measurement calculus approach has blocked progress in understanding the physics underlying measurements and decoherence.

This brings us to the second point I would like to make, and it again concerns the measurement paradox, this time in a rather simpler form. As we have previously noted, one difficulty with the discussion of QM exclusively in the

[52] Ritchie, Story, & Hulet (1991).

wave-function form is that it is not at all clear what is the ontological status of the wave-function itself; it is hard to maintain that it is 'real' in the same sense that we talk about the reality of tables and chairs. This problem is not supposed to arise in the measurement calculus, since measurements and their results are supposed to be objectively real. However in QM, as is well known, we cannot easily maintain this view either, since no prescription is given in a given physical situation for deciding when or even if a measurement has yet taken place. We have the same problem as before – if we can recover phase correlations that were thought to be lost, then we can 'disentangle' the measurement, or 'reverse' it. The measurement calculus, which treats measurements as primal and unanalysable, is of course incapable of dealing with this – and again we see that if we wish to make progress, as physicists or philosophers, we *must* pursue our analysis of measurements in terms of something more fundamental, i.e., the wave-function.

5 The quantum computer – reversible measurements?

In section 3, I explained how, within a wave-function formulation of quantum mechanics, modern physics can be viewed as steadily pushing back the frontier between the quantum and classical domains. This merely underlines the fact that irreversibility, at least on a fairly local scale, is *subjective*; if one is capable of penetrating behind the coarse-graining that is involved in most laboratory experiments (as well as our everyday experience of the world), then the reversibility of the underlying dynamics becomes visible.

This then raises the important questions of whether it makes sense to talk of *reversing* physical processes (at least locally). If these physical processes involve what are traditionally thought to be *measurements*, then one can speak of 'reversible measurements'.

I immediately emphasize that for those physicists who *assume* that measurements must be irreversible, particularly in QM, the idea of reversible measurements is contradictory. However, as we have seen, all attempts to investigate what measurements *actually are* in QM (or any other theory for that matter), merely end up with some kind of FAPP definition. For most physicists (most particularly those who actually carry out measurements), a real measurement in QM involves the interaction of a system with some 'apparatus', in which information about the system is acquired, via the interaction, by the apparatus, and encoded in its wave-function. That the apparatus/system complex appears afterwards to be in some 'definite state' is then usually attributed to decoherence, irreversibility, etc. Now there is no question that experimentalists do not consider the measurement to have taken place *until* decoherence has occurred – all our experience, as well as 400 years of 'scientific method', encourages us to believe that the results of measurements must be objective and immutable, for all to see.

However the obvious problem with this kind of definition is that it is manifestly FAPP; and as we have seen, the last 10 years have made it clear how rapidly FAPP definitions of measurement are beginning to evolve. I would now like to bring all of the various aspects of irreversibility, measurements, and decoherence into sharp focus by looking at a rather interesting model, viz., the 'quantum computer'. What makes this model example particularly useful from our present perspective is that *computations* are very similar to measurements, and some writers have even identified the two. One can, roughly speaking, think of a computation as a 'transformation of information'; information in some form (the 'input') is transformed into an 'output' by the computer. The analogue in QM is with the transformation of the QM wave-function, by some operator, into another form (perhaps another wave-function; certainly another wave-function if we assume the wave-function formulation of section 3). Thus if that special class of transformations in QM, which constitutes a measurement, is considered to be the analogue of the computation operation, we see the potential value of deploying some of the methodology of mathematical logic and computer science to look at QM; and this idea has indeed been pursued by many writers.

Here I shall not insist too much on this analogy, but rather assume that the object that is carrying out the measurements *is a computer*, whose essential operation is quantum mechanical. The idea of the quantum computer was first formulated by Benioff,[53] and somewhat later by Feynman,[54] in a similar way; Feynman's article is extremely readable, and also contains a lot of examples of ways in which such a machine might operate.[55] Since this work a number of other writers have also looked at quantum computers in various ways,[56,57] most particularly from the viewpoint of computer science.[58]

[53] Benioff (1980, 1982, 1986).
[54] Feynman (1986).
[55] Feynman has also explored the more general question of whether one might analyse the universe itself as a computer. See Feynman (1982).
[56] Margolus (1986) or Zurek (1984).
[57] Deutsch (1985, 1989) and Deutsch & Jozsa (1992).
[58] Berthiaume & Brassard (1993).

4 Time, decoherence, and 'reversible' measurements

Here I shall explain enough of the basic idea to give some foundation for the discussion that follows. Let us imagine the construction of a 'quantum Turing Machine' by means of a set of finite-state QM systems, coupled together so that information may be transferred between them. For definiteness one can think of a set of QM spins, arranged in one or more 'spin chains'; the similarity to the 'Coleman–Hepp' model of QM measurements[30] is very revealing, and it is clear that with a computer built in this way, one can model a very large variety of measuring apparatus (whether *all* kinds of measuring apparatus can be so described – at least to an arbitrarily good approximation – is a more difficult question, which I ignore here). As sketched in appendix B, and described in great detail by Feynman, it is sufficient to use two-state or 'two-level' systems for each element; we are already familiar with the physics of their operation from section 3 (and appendix A).

We therefore imagine information encoded in the form of a multi-spin input wave-function $\psi_i(\{\sigma_j\})$, with $j = 1, 2, \ldots, M_s$ labelling the spins; the computer will have the initial state $\Phi_o(\{s_\mu\})$, with $\mu = 1, 2, \ldots, M_a$ labelling its spins; and their interaction is going to lead to a computation, by which we mean that the final state is going to look like

$$\Xi_f(\{\sigma_j, s_\mu\}) = \psi_f(\{\sigma_j\})\Phi_o(\{s_\mu\}) \tag{5.1}$$

in which not only has the computer decoupled completely from the output wave-function ψ_f, leaving no trace of the computation in any kind of 'entanglement' between the computer wave-function and the output wave-function, but also it has returned to its original state! This is the 'reversible quantum computer', and comparison with the general solution (3.2) to the equations of motion shows how improbable is a solution like (5.1), even before we consider decoherence and dissipation.

Nevertheless ways of implementing (5.1) were found by both Benioff and Feynman which use perfectly ordinary spin Hamiltonians; provided one works in an ideal world without dissipation, imperfections, etc., the properties of reversibility and 'disentanglement' are satisfied. This is nicely shown by Feynman's model Hamiltonian, which reads, for the *computer*:

$$\hat{\mathcal{H}} = t\left[\sum_{j=1}^{m}(\beta_{j+1}^+ \beta_j A_j + H.c.) + \sum_{j=m+1}^{n}(\beta_{j+1}^+ \beta_j + H.c.)\right] \tag{5.2}$$

in which the 'program sequence', $\{\hat{M}\} = \hat{A}_m \hat{A}_{m-1} \ldots \hat{A}_1$, of operators acts on the input wave function; the computer is driven by the 'control loop'

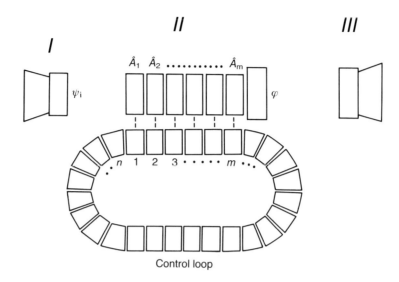

Fig. 5. A schematic diagram of a quantum computer. The input wave-function ψ_i is presented to the program (system II); the output wave-function φ, at the output head of II, is read by system III

consisting of the closed sequence $\beta_n \beta_{n-1} \ldots \beta_m \ldots \beta_1$, connected with periodic boundary conditions (i.e., $n + 1 \equiv 1$). In (5.2), β_j^+ is the Hermitian conjugate of the control loop operator; and '$H.c.$' signifies Hermitian conjugate. The idea is that any transition in the control loop (from, e.g., the jth to the $(j + 1)$th site), causes the computer program to make a similar transition; the coupling between the control loop and the computer program, in the first part of (5.2), ensures this. One can think of (5.2) intuitively as follows – we imagine a 'particle' (actually this particle will be a quantized spin wave, or 'magnon', in the case where our system is built out of spins), which shunts around the control loop; this loop is joined for part of its course to the program sequence, in such a way that its motion drives transitions in the program sequence. The operation is shown schematically in figure 5, and described in more detail in appendix B.

Now it is important to note that this device can operate forward or in reverse (i.e., the 'particle' can move in either direction around the loop); the Hamiltonian is Hermitian. Thus the device is *reversible*, in principle (notice we have not yet started to worry about the environment). We can now begin to see how a computation, strongly resembling a 'measurement', may be carried out with this device. A typical sequence of operations, making up the computation, will proceed as follows:

(i) We imagine a spin-system I, with wave-function ψ_i, which encodes the input information, interacting with the computer (spin-system II); the computer has a spin wave-function Φ_o. It will be useful to separate the computer wave-function into two parts, one describing the pre-existing (program + control) part, which we call χ^α (α labelling the particular program, described by the program sequence $\hat{M}$), and an 'output head', described by the wave-function φ; where the final result of the computation will appear. Then the initial state of the computer will be $\Phi_o \equiv \Phi_o^\alpha = \chi^\alpha \varphi_o$.

We then let systems I and II interact, and the following transition takes place:

$$\psi_i \Phi_o^\alpha \rightarrow \psi_i' \Phi_i^{\alpha'} \equiv \psi_i'(\chi^{\alpha'} \varphi_k). \tag{5.3}$$

The notation in (5.3) is as follows: the final state $\Phi_i^{\alpha'}$ of II is uniquely correlated with the initial state ψ_i of I, in that it depends only on ψ_i and on which program (labelled by α) we happen to use (thus suppose ψ_i described the number '6', and program M operates on it; then the 'answer', 36, is uniquely determined by ψ_i and M). The label α' tells us that the computing part of II has changed its state (from χ^α to $\chi^{\alpha'}$); it is then convenient to write the output head wave-function not as φ_i, but as φ_k, since it is determined not just by ψ_i but also by χ^α. The spin wave-function φ_k encodes the 'answer', i.e., the output of the computation. Notice also that the wave-function ψ_i' of the input system I has also changed. The resemblance of the transition (5.3) to a measurement (made by II and I) will be obvious, and we return to this below; for now we continue with the computation.

(ii) Let a third spin system III interact with the output head, reading its answer; its initial wave-function will be Ω_o, and we have

$$\varphi_k \Omega_o \rightarrow \varphi_k \Omega_k \tag{5.4}$$

we then separate III from II; we have now safely stored the answer.

(iii) We now run the *combined* system I and II backwards, simply reversing (5.3). Thus we may characterize the entire computing operation by the transition

$$\psi_i \Phi_o^\alpha \Omega_o \rightarrow \psi_i \Phi_o^\alpha \Omega_i^\alpha \equiv \psi_i \Phi_o^\alpha \Omega_k \tag{5.5}$$

in which the 'answer' Ω_k is uniquely determined by i and α; we can just write Ω_k as Ω_i^α (the single index k is useful simply to remind us that, in (5.4) and (5.3), Ω_k and φ_k are spin wave-functions just like ψ_i).

It should be noted at this point that while the above sequence of operations is easy to describe, it is not so easy to start with a realistic Hamiltonian for a set of spins, to get it to look something like (5.2), and then to run it in the way discussed. Detailed simulations[59] show that internal reflections of the wavefunctions in the presence of disorder or randomness in the couplings have a very serious effect on its operation, and the transition (5.4) is not so easy to organize. However these problems do not stop the Hamiltonian being Hermitian, and thus reversible – they simply slow down the computation, and make it necessary to 'filter' the output.[59]

A much more serious problem, which has apparently been totally ignored in the quantum computing literature, is of course that of decoherence and dissipation caused by interaction with the environment; and this brings us back to the main theme of this chapter. We have, in (5.3) and (5.4), carried out operations which look just like QM measurements, and then reversed all but one of the sequences of transitions involved in these measurements (the non-reversed sequence being (5.4)).

However the big difference between a typical QM measurement, and the quantum computing operation, is that whereas dissipation, decoherence, and irreversibility are usually desirable features of the measurement (in so far as they establish its result once and for all), they are clearly anathema to the quantum computer. This is all the more obvious if we try to run 'parallel quantum computations', in the manner suggested by Deutsch[57] (i.e., replace the input wave-function ψ_i in (5.3) by a superposition $\sum_k c_k \psi_i^{(k)}$ of input wave-functions). It is quite clear that a very small amount of decoherence would be quite fatal to a parallel quantum computation – detailed simulations[59] show that we are still far from building any such device, with present technology.

Nevertheless the advances made in recent years certainly encourage one in the belief that a quantum computer will be built some time in the next century. If we take this possibility seriously, it should then be clear that we will have to rethink equally seriously our ideas about the nature of the wave-function formulation of QM.

There is no question that the problems involved are partly philosophical. To see this, consider that point of view which considers a sufficiently complex computer as being indistinguishable from a sentient, conscious being, capable of thinking, and if necessary reproduction. This point of view is almost guaranteed to raise philosophical hackles, but I would maintain that the advances in

[59] Unpublished work of Angles d'Auriac (1992).

modern microbiology have shifted the onus of the argument very much against those who maintain that 'life' or 'consciousness' is somehow beyond any purely physical description. But now notice that the possibility of a quantum computer introduces a quite new ingredient into this debate, which could hardly be better designed to exacerbate it. This is because, unlike all presently known examples of living sentient beings, a quantum computer will operate at low temperatures and in the absence of dissipation and decoherence. Conventional computers and (as noted in section 3) all known examples of sentient beings (including humans) operate *classically*, and this arguably protects them from all the philosophical problems surrounding QM. No such protection exists for quantum computers.

Thus the mere idea of an operating quantum computer (let alone its practical implementation!) raises all sorts of fascinating questions for the future. Amongst these one may imagine a 'self-conscious' computer, carrying out 'self-referential measurements', in which, for example, one part of the computer observes another part, and then conveys information about its observations to that other part, as well as obeying instructions from it.[60,61] How closely could such a machine monitor its own state (within the limits imposed by the uncertainty principle)? Is there some version of Gödel's theorems, applicable to such a device? There is almost no limit to the speculations that can arise here (for example, if we take seriously Feynman's idea that the universe is a computer, then to what extent can it be 'self-conscious'?). Previously, such speculation has surfaced only in the wilder reaches of cosmology (or the science fiction literature); but it may be only a matter of a few decades before we shall be forced to deal with some of these questions in the corridors of physics laboratories – and perhaps of philosophy departments.

Appendix A: Decoherence and the emergence of classical properties

In this appendix I fill out some of the qualitative discussion of the main text, in a way which gives the interested reader a window into the extensive literature, without being too technical. I start by looking at some two-state systems, which can be understood in a simple intuitive way – moreover, as made

---

[60] Compare, in this context, Albert (1983, 1986).
[61] The physics of self-referential measurements becomes particularly interesting when described concretely in terms of a quantum computer and will be dealt with in a separate publication.

beautifully clear by Feynman, most of the essential features of QM can be seen most easily in the two-state system.[62] I go on to describe some features which do not appear in the two-state case, and round up with some notes on the relation between quantum and classical mechanics for both closed and 'open' systems.

A.1 Two-state systems: In a remarkably large number of physically interesting cases, it is possible to truncate the Hilbert space of the system to two states only. This is obviously partly because physicists tend to concentrate on systems that are easy to understand, but an even more important reason is the 'natural truncation' that occurs when the influence of all other states, on the two states in question, becomes either adiabatic or just weak. The question that interests us here is the destruction of the coherence between the two states, caused by their interaction with the environment. In what follows I will describe the results for two important examples, and then summarize some of the general points that emerge.

(i) *The 'spin-boson' problem*: Any two-state system can be formally represented in terms of spin-$\frac{1}{2}$ Pauli matrices. As pointed out by Leggett et al.,[36,40] the low-energy properties of the environment acting on such a system can very often be represented in terms of a set of bosons, and this leads to a 'spin-boson' Hamiltonian whose simplest form is[40]

$$\hat{\mathcal{H}} = \frac{\Delta}{2}\hat{\sigma}_x + \frac{1}{2}\sum_{k=1}^{N} m_k(\dot{x}_k^2 + \omega_k^2 x_k^2) + \sum_k c_k x_k \hat{\sigma}_z \qquad (A.1)$$

in which the first term represents the two-state system, the second term the 'environmental bath' of bosonic oscillators (with coordinates x_k; the index k labels the N different oscillators, having frequencies ω_k), and the last term describes the coupling between the system and the environment. Without the environment, we have a 'free' Hamiltonian matrix

$$\hat{\mathcal{H}}_0 = \frac{1}{2}\begin{pmatrix} 0 & \Delta \\ -\Delta & 0 \end{pmatrix} \qquad (A.2)$$

with eigenstates $|\pm\rangle = (1/\sqrt{2})(|\uparrow\rangle \pm |\downarrow\rangle)$, where $|\uparrow\rangle$ and $|\downarrow\rangle$ are the eigenstates of $\hat{\sigma}_z$; the eigenstates of (A.2) are split by an energy Δ. We now notice that the coupling term, proportional to $\hat{\sigma}_z$, causes transitions between the eigenstates of

---

[62] See Feynman, Leighton, & Sands (1965), particularly chapters 6–11.

(A.2). Contrast this with the effect of an environmental coupling $\sim \sum_k c_k x_k \hat{\sigma}_x$, proportional to $\hat{\sigma}_x$; this would simply cause the splitting Δ to change (and to fluctuate), but cause no transitions between $|+\rangle$ and $|-\rangle$. Thus we see that the coupling term in (A.1) has a *dissipative* effect on the system. To see the decoherence clearly, we imagine starting the system off in the state $|\uparrow\rangle$ at time $t = 0$; this is a superposition of the two eigenstates (we have $|\uparrow\rangle = (1/\sqrt{2})(|+\rangle + |-\rangle)$, from above), and so the system state will oscillate in time at a frequency $\Delta/\hbar$ (the difference between the energies of the two eigenstates – this is nothing but a 'beat' phenomenon). In the absence of the bath, this means that at some time $t > 0$, the *probability* $P_o(t)$ that the system will be found in the state $|\uparrow\rangle$ is $P_o(t) = \cos^2(\Delta t)$. However as soon as we add the environmental coupling, these oscillations become *damped* (and for strong enough coupling, they can cease altogether – the environment 'localizes' the spin system).[40] The existence of the oscillations is a direct consequence of the coherence between the states $|\uparrow\rangle$ and $|\downarrow\rangle$ (the eigenstates $|+\rangle$ and $|-\rangle$ are just coherent superpositions of these two states), and their suppression is an example of decoherence. As mentioned in the text, the destruction of interference between $|\uparrow\rangle$ and $|\downarrow\rangle$ is most clearly displayed in the density matrix $\rho_{\alpha\beta}(t)$ (where in this case the indices α and β take the values $\uparrow$ or $\downarrow$). In the absence of the bath, $\rho_{\alpha\beta}(t)$ has the form

$$\rho_{\alpha\beta}(t) = \begin{pmatrix} \cos^2 \Delta t & -i \cos \Delta t \sin \Delta t \\ i \cos \Delta t \sin \Delta t & \sin^2 \Delta t \end{pmatrix}. \quad (A.3)$$

Now the detailed treatment of the decoherence and damping effect of the bath on $\rho_{\alpha\beta}(t)$ is complex; it depends on the temperature of the bath, as well as the details of the couplings c_k and the oscillator frequencies ω_k. However the general effect is shown in figure 6, both for the diagonal elements $\rho_{\uparrow\uparrow}(t)$ and $\rho_{\downarrow\downarrow}(t)$, and also the off-diagonal elements. In these graphs we notice not only the damping of the oscillation, but also the reduction of the oscillation frequency, to a 'renormalized' value $\hat{\Delta}/\hbar$; unless the coupling to the environment is rather small, one has $\tilde{\Delta} \ll \Delta$, if indeed the system still oscillates at all. Inevitably one expects that if the two states involved refer to some 'macroscopic' coordinate, then the coupling to the environment will be stronger than if the system is microscopic – compare the case of a macroscopic superconducting SQUID ring (with two different flux states) with the microscopic examples of a nuclear spin or an ammonia molecule. However this is not inevitable – for example, the damping of coherent motion of a macroscopic

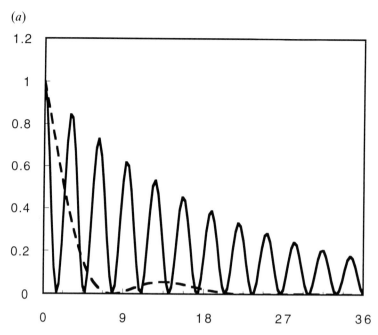

Fig. 6. The density matrix for a dissipative two-state system. In figure 6(a) is shown the difference ($\rho_{\uparrow\uparrow} - \rho_{\downarrow\downarrow}$) between the two diagonal matrix elements. This is done for two different systems; the solid line is for a system with very weak damping, and the dashed line for stronger (but still sub-critical) damping; notice how the oscillation frequency decreases as one raises the damping

domain wall in a magnet can be extremely small,[63] whilst the effect of a 'Fermi liquid' environment (such as a metal) can be so strong as to completely destroy coherence, even for atomic-sized defects or impurity spins.[64] As noted in the main text, the principal reason that some macroscopic systems are weakly coupled to their environments is the existence of an 'energy gap' in the spectrum of the environmental oscillators – all oscillators have frequencies greater than some 'gap energy' E_0, and this removes the low-frequency oscillators that are most effective in destroying coherence. It also helps if the macroscopic object involved is a soliton, since solitons do not couple linearly to their environment.[33] Thus we arrive at the possibility of 'macroscopic quantum coherence'.

[63] A recent summary of the situation in magnets is given in Stamp (1993).
[64] Recent reviews of quantum diffusion are given in Kagan (1992) and Kagan & Prokof'ev (1992).

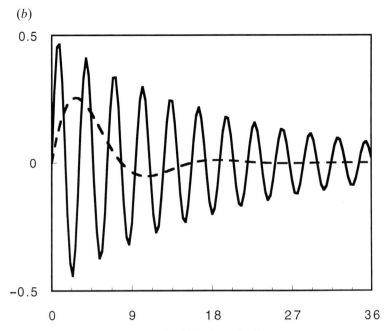

Fig. 6. (b) The off-diagonal quantity $|\rho_{\downarrow\uparrow}(t)|$ is shown for the same two systems

Another useful point illustrated by (A.1) is the way in which decoherence between $|\uparrow\rangle$ and $|\downarrow\rangle$ is caused by an environment which can 'tell the difference' between the two states (via the operator $\hat{\sigma}_z$ in the coupling). In this way the decoherence can be said to be caused by 'measurements' of the system by the environment. All this really means, of course, is that the operator $\hat{\sigma}_z$ in the coupling allows correlation or 'entanglement' to be set up between the states $|\uparrow\rangle$ and $|\downarrow\rangle$, and their environment states, in the way described at length in the text. Needless to say there are many physical examples of 'effective Hamiltonians' like (A.1) or simple generalizations of it.[40,64]

(ii) *'Two-slit' systems*: Another kind of two-state system involves interference between two different 'paths' for the dynamic evolution of the system. A famous example, discussed from the very beginning of QM, is the two-slit experiment, in which, e.g., photons or electrons are forced to pass from a source A to a target B through one or other of two slits; as is well known, the resulting quantum interference is quite inexplicable using the concepts of

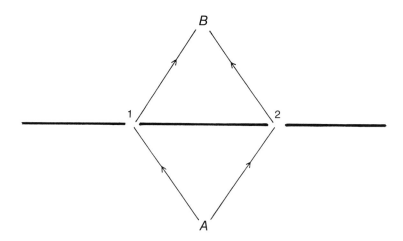

Fig. 7. The two-slit system. Light (or a stream particle) passes through slits 1 or 2 going from A to B (compare with figure 4)

classical physics. The general situation is illustrated in figure 7; again, the variety of physical systems described by this model is enormous, and familiar to any student of physics, ranging from the cosmological to the sub-nuclear scale. Apart from the two-slit example, the reader may find it helpful to think of the example of a ring geometry (e.g., a metal or a superconducting ring) around which particles (e.g., electrons) may propagate; one is interested in the interference between clockwise and anti-clockwise paths, or between left and right paths.

Again, the influence of the environment on the dynamics of such systems has been extensively studied. The most popular way of doing this is in a 'Caldeira–Leggett' kind of model, in which the environment is again modelled as a set of oscillators. For example, in the case of a ring geometry, when the angular coordinate is labelled θ (with a corresponding conjugate momentum p_θ), the following kind of Hamiltonian is often appropriate, for a particle of mass M and charge e:

$$\hat{\mathscr{H}} = \frac{1}{2M}(p_\theta - e\mathbf{A}(\theta))^2 + \sum_{k=1}^{N} c_k x_k\, p_\theta + \sum_{k=1}^{N} \frac{1}{2} m_k(\dot{x}_k^2 + \omega_k^2 x_k^2) \qquad (A.4)$$

in which we again have N oscillators, now linearly coupled to the particle momentum. The effective Hamiltonian (A.4) also allows for the effects of enclosed magnetic flux in the ring (the Aharonov–Bohm effect,[62,65] for example), through the vector potential $\mathbf{A}(\theta)$.

As mentioned in the text, models such as (A.4) are useful in showing (amongst many other things) that the rate at which coherence or interference between the two paths is lost (or alternatively, between clockwise and anti-clockwise motion) is much faster than the rate at which the system relaxes (the relaxation here might be relaxation of the energy of the particle, or of its momentum, these being two different kinds of relaxation). The reader is referred to some of the interesting studies[66] of models such as (A.4).

One very important feature of 'two-slit' or ring geometries is that they involve closed paths, and thereby allow some of the important topological and non-local aspects of QM to appear (in this case, the 'winding number' of the QM phase around the path). In recent years some rather remarkable topological features of QM and quantum field theory have been uncovered – in particular the Berry phase and its generalizations, and the phases involved in the QM treatment of angular momentum and spin, encapsulated in the 'Wess–Zumino' topological term which was used to provide a path integral for spin[67] in the early 1980s. These ideas have consequences for the topic of this chapter as well, as we can best see from another example, shown in figure 8. We imagine a magnetic grain[33] containing some 10^8 individual paramagnetic spins; the strength of the exchange interaction between these spins is strong enough that they are all forced to line up together, and we have what is sometimes called a 'giant spin' $\mathbf{S}$, with quantum number $S \sim 10^8$. Now the direction $\hat{\mathbf{S}}$ in which the spin can lie is not arbitrary, since there exist residual magnetic anisotropy forces acting on the giant spin, which try to orient it. A typical example of an 'effective Hamiltonian' acting on $\mathbf{S}$ is

$$\mathcal{H}_{\text{eff}} = -K_{\parallel} S_z^2 + K_{\perp} S_y^2 \qquad (A.5)$$

where $K_{\parallel}$ and $K_{\perp}$ are positive, and $K_{\parallel} > K_{\perp}$. Thus, as shown in the figure, $\mathbf{S}$ likes to lie 'up' or 'down' along the z-axis. Now the crucial point is that $\mathbf{S}$ can *tunnel* between these two states (which we will call $|\uparrow\rangle$ and $|\downarrow\rangle$). During the course of this tunnelling motion, we can imagine $\mathbf{S}$ moving either clockwise or anti-clockwise between the two degenerable minimum energy states $|\uparrow\rangle$ and $|\downarrow\rangle$, and so again we have a two-slit system in operation.

[65] Aharonov & Bohm (1959).
[66] Applications in different areas of physics are described in Shapere & Wilczek (1989), Aronov & Sharvin (1987), and Altshuler & Lee (1988).
[67] See e.g. Haldane (1983).

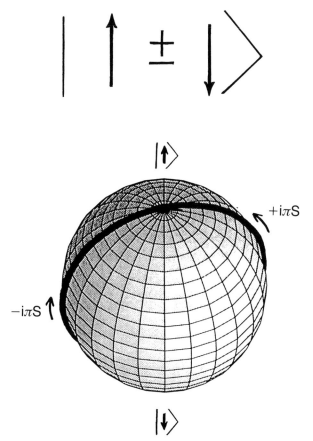

Fig. 8. A 'giant spin' must choose between two paths (clockwise or anti-clockwise) in making a transition between $|\uparrow\rangle$ and $|\downarrow\rangle$. The two paths carry opposite phases. The two energy eigenstates are still given by the sum or difference of $|\uparrow\rangle$ and $|\downarrow\rangle$

The interference between these two paths is important; they carry[68] opposite phases $\pm \pi S$, and so summing the two amplitudes in the usual way, we get a total tunnelling amplitude from, say, $|\uparrow\rangle$ to $|\downarrow\rangle$ of

$$\psi \sim \Delta_0(e^{i\pi S} + e^{-i\pi S}) = 2\Delta_0 \cos \pi S \qquad (A.6)$$

so that the final energy splitting between the state $|+\rangle = (1/\sqrt{2})(|\uparrow\rangle + |\downarrow\rangle)$ and $|-\rangle = (1/\sqrt{2})(|\uparrow\rangle - |\downarrow\rangle)$ is $\tilde{\Delta}_0 = 2\Delta_0|\cos \pi S|$. That this splitting is *zero* for S a half-integer (i.e., $S = n + \frac{1}{2}$), but finite for integer S (i.e., $S = n$) is an example of Kramer's theorem.[69]

[68] See Haldane (1983) or Affleck (1989) and the recent application to magnetic grains by Loss, di Vincenzo, & Grinstein (1992) and von Delft & Henley (1992).
[69] Kramer's theorem is described in Landau & Lifshitz (1965), section 60.

4 Time, decoherence, and 'reversible' measurements

But now this immediately leads us to ask – what if our giant spin couples to (microscopic) spins in the environment? If these are thereby caused to flip, or otherwise change their state, then they will add their own phase contribution to the transition amplitude involved in (A.6). Now since there is no reason why, on different transitions of the giant spin, these environmental spins should undergo the same changes of states,[70] the general effect[71] of these environmental spins is to add a 'random phase' to (A.6), thereby destroying the coherence between $|\uparrow\rangle$ and $|\downarrow\rangle$. This kind of decoherence is called 'topological decoherence', and it can be very destructive. What is interesting here is not only the illustration that the phase of the *environmental* wave-function can enter the picture, but also the way in which a microscopic spin (usually a *nuclear* spin, it turns out) can completely control the quantum dynamics of an object far larger than itself, such as a macroscopic ferromagnetic grain. In fact one finds that this occurs if the interaction energy between the giant spin **S** and microscopic environmental spin σ (call this energy ω) satisfies $\Omega_0 > \omega > \tilde{\Delta}_{\text{eff}}/\hbar$, where Ω_0 is the (very high) small oscillation frequency of the grain, and $\tilde{\Delta}_{\text{eff}}$ is related to (but less than) $\tilde{\Delta}_0$. In this 'intermediate coupling' regime, one environmental spin will convert the coherent oscillations between $|\uparrow\rangle$ and $|\downarrow\rangle$ (at frequency $\tilde{\Delta}_0$) into a more complicated motion; and several more spins will make the giant spin move in an apparently quite incoherent manner. This is reminiscent of the measuring process (in fact, it is reminiscent of the Stern–Gerlach measurement of spin quantization), but where measuring apparatus and system have swapped roles[72] – a kind of 'reverse Stern–Gerlach' measuring operation!

A.2 Quantum to classical mechanics: The preceding discussion in terms of simple examples makes one wonder how much is known in general about the mechanisms of decoherence, and its role in the 'emergence' of classical mechanics (CM) from QM. It should be said immediately that this is a subject in its infancy, with only limited results so far – but it is clearly of the utmost relevance to the theme of this chapter, and a promising area of future research activity. As before, the following is meant mainly to bridge the gap between the main text of this chapter, and the physics literature.

[70] For example, on one transition of the giant spin three environmental spins might change their state (e.g., flip), but on the next transition four (in general different) spins might change their state.
[71] Prokof'ev & Stamp (1993).
[72] Similar 'reverse Stern–Gerlach' set-ups are described in Stamp (1988) and Sidles (1992).

Before discussing any environmental effects, it is important to understand the relation between QM and CM in their absence. This is not a simple topic, and important advances have been made only recently. Traditionally one takes the classical limit of CM by letting $\hbar \to 0$. Thus, e.g., in the Feynman path integral formulation of QM, the amplitude to go from a state $\psi_A(x_A, t_A)$ to state $\psi_B(x_B, t_B)$ is given by

$$\psi_B(x_B, t_B) = \psi_A(x_A, t_A) \int_{x(t_A)=x_A}^{x(t_B)=x_B} Dx(t) \exp\left[\frac{i}{\hbar} S(x, \dot{x}; t)\right]$$

$$\equiv K(x_B, t_B; x_A, t_A)\psi_A(x_1, t_1) \qquad (A.7)$$

where we sum over *all possible* paths $\{x(t)\}$ leading from the initial to the final state – as $\hbar \to 0$, only one or a few classical paths is supposed to emerge from the mess of interfering paths in (A.7), the paths of 'extremal action'. However the limit $\hbar \to 0$ is clearly subtle (since the phase $S/\hbar$ clearly diverges for any path whatsoever unless $S = 0$).

Amusingly, recent progress in understanding the relation between CM and QM, and the classical limit, has come by studying how QM emerges from CM, i.e., going in the opposite direction! One of the more remarkable features of this 'semiclassical' approach[73] is the crucial role played by *closed* classical orbits in understanding the QM behaviour, even for systems displaying classical chaos. This means that intuitive ideas developed in the course of studying the two-slit problem (with its single closed orbit) are still very relevant for any closed quantum system possessing a large (in general infinite) number of closed orbits.

For example, suppose we are interested in the density of energy levels, as a function of energy E, for some closed quantum system with Hamiltonian $\hat{\mathcal{H}}$. It is convenient to study this with a 'resolution' ϵ, i.e., counting the number of levels in a window of width ϵ, centred about E. One then finds[74] that the density of states $N_\epsilon(E)$ is given by

$$N_\epsilon(E) = \bar{N}(E) + \sum_{r=1}^{\infty} e^{-\epsilon t_r(E)/\hbar} A_r(E) \cos\left(\frac{1}{\hbar} S_r(E) + \gamma_r(E)\right) \qquad (A.8)$$

in which $\bar{N}(E)$ represents a 'smoothed' (ergodic) density of states (having nothing to do with, e.g., the shape of the system), and the summation in the

---

[73] Recent results of semiclassical theory are described in Gutzwiller (1990) and Heller & Tomsovic (1993).
[74] See Callan & Freed (1992) and Chen & Stamp (1992).

second term is over all closed orbits of the equivalent *classical* system, labelled by the index r. We see that each orbit adds 'fine structure' to the density of states, but their contributions are damped by the energy resolution factor $\exp(-\epsilon t_r(E)/\hbar)$, in which $t_r(E)$ is the time required for the classical system to accomplish the closed orbit r, given it has energy E. Finally, $S_r(E)$ is the action incurred in making this closed orbit, and $\gamma_r(E)$ is a phase correction coming in when one passes focal points or reflections in the classical paths. The appearance of the cosine factor will be intuitively obvious from our look at two-state systems (compare with eqn (A.6)) as arising from interference between forward and backward paths – the great difficulty with the practical implementation of (A.8) is the computation of the amplitude factor $A_r(E)$. However great advances have been made in recent years in the calculation of semiclassical formulae for many important QM quantities (including the Feynman propagator K in (A.7)).

With all this in mind, we turn now to environmental decoherence. The important point to be made here is that whereas decoherence also causes a suppression of interference between different paths or amplitudes, one should not jump to the conclusion that there is a straightforward 'quantum–classical' transition, caused by decoherence, which is directly analogous to the passage between QM and CM just discussed. In fact decoherence really works very differently, and moreover it does not really lead to a transition to CM, but rather to a diffusive dynamics, often called 'quantum diffusion'. Since this point is often not appreciated (it is not really clear in studies of two-state systems, and still less in studies of the dissipative harmonic oscillator), it is perhaps worth sketching a few details.

Consider some object of mass M moving in one spatial dimension, in some potential $V(R)$, and coupled to an environment. This might be a simple Caldeira–Leggett environment of oscillators, with a coupling of the form $\sum_k F_k(R) x_k$, linear in the oscillator coordinates x_k (but with some arbitrary $F_k(R)$); it might be some higher non-linear coupling, such as that important in quantum soliton motion at low temperatures (of the form $\sum_{k_1} \sum_{k_2} \sum_{k_3} F_{k_1 k_2 k_3}(\dot{R}) x_{k_1} x_{k_2} \dot{x}_{k_3}$, in fact); or it might be something really awful, such as a coupling to an environment of spins. Nevertheless studies of models of this kind do indicate that the general effect of the environment is as described in section 3; interference between different paths or states, like those just described, is suppressed. However the result is *not* to 'strip the quantum flesh from the classical bones', leaving the classical paths of eqn (A.8). Instead one ends up looking at a density matrix $\rho(x, x'; t; T)$, now a function of continuous coordinates x and x' (as well as environmental temperature T) which has the diffusive form

$$\rho(x, x'; T) \sim (\pi\sigma^2(t))^{1/2} \exp[-(x-x')^2/\sigma(t)] \qquad (A.9)$$

in which the variance $\sigma(t, T)$ increases both with time (as t), and temperature; in fact $\sigma(t, T) \sim 4D(T)t$, where $D(T)$ is the diffusion coefficient.

One might think that it must be possible to relate this behaviour to some *damped* classical system at finite T. Unfortunately this does not work either, because quantum diffusion and classical diffusion behave quite differently with temperature. Classical diffusion is driven by environmental thermal fluctuations, and therefore is suppressed by *lowering* T, whereas quantum diffusion rates are suppressed by *raising* T, since this increases the decoherence rate. This striking difference between classical and quantum diffusion is seen in many experiments. For more details on the theory, readers can consult, e.g., the references given in footnote 64. It should also be noted that the long-time stochastic motion of quantum systems undergoing decoherence is not even necessarily diffusive, particularly at zero temperature – good examples are provided by topological decoherence, and by the $\ln t$ spreading of the density matrix interacting with a bosonic environment at $T = 0$.

We thus see that the sense in which one can say that classical properties 'emerge' from a suppression of interference between states of paths is a rather restricted one. Although this point is well understood by at least a part of the condensed matter physics community (e.g., by those working on quantum diffusion!), it is apparently not understood by those working on the foundations of QM – and it constitutes yet another example of why it is so important to look at the *details* behind the sweeping assertions that have characterized much of QM measurement theory and the foundations of QM for so long.

I finally note that this field is likely to hold future surprises in store for us – both semiclassical theory and the fields of 'quantum dissipation' and decoherence are still furrowed by open questions. One example is the recent work on a 'quantum fractal system', the so-called 'WAH' or 'Hofstadter' problem, which has a nested 'Chinese box' spectrum, in which energy structure level is repeated to ever smaller energy scales (the corresponding classical orbits being self-similar to ever larger spatial scales); the Hamiltonian can be written as

$$\hat{\mathcal{H}}_{\text{WAH}} = -t \sum_{\langle ij \rangle} (c_i^+ c_j\, e^{iA^0_{ij}} + H.c.) \qquad (A.10)$$

in which c_j^+ creates a fermion on a two-dimensional lattice site, so that $c_i^+ c_j$ describes hopping between sites, which are restricted in the sum (by the symbol $\langle \ldots \rangle$) to be nearest neighbour; the phase A^0_{ij} accumulated in this hopping corresponds to a constant applied magnetic field perpendicular to the lattice plane. Now one might assume, on the basis of everything said above, that the effect of coupling an environment to the fermions in (A.10) would be to destroy the subtle long-range phase coherence involved in producing the beautiful fractal spectrum of (A.10). Remarkably, recent theory[74] indicates exactly the contrary – further structure is produced, and one finds an infinite fractal sequence of phase transitions as the coupling to the environment is switched on. We are far from understanding all the details, or even the reasons for this, but it does certainly show that the last word has yet to be said on the subject!

Appendix B

Here I give a few more details on the Feynman–Benioff quantum computer; as before, the reader is urged to go to the original literature to supplement the following.

The important point made by Benioff[53] was that it was perfectly possible to build a 'quantum Turing Machine' out of a set of spin systems, which we can choose (with no loss of generality) to be two-state spin-$\frac{1}{2}$ systems. How the computer operates will then depend on how the spins are arranged to interact with each other. As discussed in the text, quantum computers behave very similarly to measuring devices in QM, and so one should note that a lot can be learnt about the measurement operation by modelling the measuring systems by a set of interacting spin systems as well; one nice example is the 'Coleman–Hepp' model,[30] for which the Hamiltonian is

$$\hat{\mathcal{H}} = -i\frac{\partial}{\partial x} + \frac{1}{2}\sum_{n=1}^{\infty} V(x - x_n)\hat{\sigma}_x(n)(1 - \hat{s}_z) \qquad (B.1)$$

where $V(x)$ is short-ranged, and $\int dx V(x) = \pi/2$. If the spin-$\frac{1}{2}$ s is sent down the infinite chain of spins $\sigma(n)$, they end up being aligned with s (or in a superposition, if s is not quantized along $\hat{z}$).

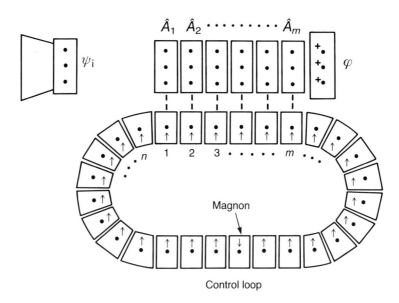

Fig. 9. A close-up of the program part of the quantum computer. We imagine that the computer is built from spin-$\frac{1}{2}$ 'two-state' units, which are in states $|\uparrow\rangle$ or $|\downarrow\rangle$; each spin is represented by a dot. A 'magnon' (i.e., a reversed spin) travels around the control loop; in this way it can drive the program

As noted by Feynman,[54] one simple way to build up a quantum computer is just to build various logic gates from spin-$\frac{1}{2}$ systems; their operation will be represented by a set of Pauli matrices, operating on one or more spins.

Thus a 'NOT' gate can be represented as the operator $(\hat{\sigma}_+ + \hat{\sigma}_-)$, operating on an incoming spin-$\frac{1}{2}$ wave-function (so if $|\psi_{\text{in}}\rangle = |\uparrow\rangle$ then $|\psi_{\text{out}}\rangle = |\downarrow\rangle$, and vice versa). Feynman found that all logical operations required for a computer could be accomplished by three logic gates; a NOT gate (operating on a single spin-$\frac{1}{2}$ wave-function); a 'NOT SWITCH' gate, acting on a *pair* of incoming spin-$\frac{1}{2}$ wave-functions; and a 'DOUBLE NOT SWITCH' gate, acting on a triplet of incoming spin-$\frac{1}{2}$ wave-functions. We may now imagine a computation proceeding according to figure 9, in which three parallel lines of spin-$\frac{1}{2}$ systems carry the information in the computation; as one passes down the line, the three-spin wave-function is operated on by successive logic gates. One may also add extra parallel lines of spin, to carry extra information. This is a spin version of Turing's 'ticker tape' program; here, the program is represented by an operator $\hat{\mathbf{M}}$, acting on the three-spin wave-function, which is just the product

$$\hat{M} = \hat{A}_m \hat{A}_{m-1} \ldots \hat{A}_1 \qquad (B.2)$$

of successive logic operations (the $\hat{A}_j$), acting on the input spin wave-function. This is similar to the Coleman–Hepp model and one should also note that there is nothing to stop us from doing a 'superposition of computations', by presenting the computer with a superposition of input wave-functions.

One must also drive the computer along; in Feynman's model this can be accomplished by adding the 'control loop' shown (also made up of spin-$\frac{1}{2}$ systems); one can imagine a 'magnon' (i.e., a reversed spin in an otherwise parallel array of spins in the loop) circulating around the loop. As the magnon arrives at sites in the loop which are connected to the program line, the program is set in motion – this was accomplished by the Hamiltonian

$$\mathcal{H} = t\left[\sum_{j=1}^{m}(\beta_{j+1}^{+}\beta_j A_j + H.c.) + \sum_{j=m+1}^{n}(\beta_{j+1}^{+}\beta_j + H.c.)\right] \qquad (B.3)$$

also described in section 5.

Now there is no obvious reason why such a machine could not be built. However problems arise when we consider that a more realistic Hamiltonian would be

$$\hat{\mathcal{H}} = \left[\sum_{j=1}^{m} t_j(\beta_{j+1}^{+}\beta_j A_j + H.c.) + \sum_{j=m+1}^{n} t_j(\beta_{j+1}^{+}\beta_j + H.c.)\right.$$
$$\left. + \sum_{j=1}^{n}\sum_{k=1}^{N} \mathcal{H}_{\text{int}}(\beta_j, \sigma_k)\right] \qquad (B.4)$$

in which we allow some randomness in the couplings t_j between neighbouring sites (or triplets of sites) in the lines of spins, as well as having interactions with a set of N 'stray' or 'environmental' spins. The effect of disorder in the set $\{t_j\}$ of couplings will slow down the operation of the computer, but the effect of the environmental spins is much more serious – it will cause 'topological decoherence', of a kind similar to that discussed for macroscopic quantum phenomena in magnets,[71] and only a few environmental spins could quite rapidly decohere the computer wave-function completely. A proper treatment would also have to include the effect of external uncontrolled magnetic fields on the operation of the computer. Preliminary simulations[59] of some of these effects show them to be extremely deleterious, so we are clearly some distance yet from building the first operating quantum computer.

Acknowledgements

Parts of this chapter have been influenced by discussions with A. J. Leggett, S. Savitt, and W. G. Unruh, particularly in section 3. I also thank Y. Aharonov and N. V. Prokof'ev for discussions on the 'reality of the wave-function', and the reader is urged to read the recent articles of Aharonov and Unruh on this subject. Part of this chapter was written during a stay at Princeton, where a number of helpful remarks were made by B. Doucot and P. W. Anderson; and I found the critical remarks of the staff at NEC in Princeton to be useful. Last but certainly not least, J. C. Angles d'Auriac has considerably sharpened my ideas about the quantum computer. The work was supported in part by NSERC of Canada.

5 Time flow, non-locality, and measurement in quantum mechanics

STORRS McCALL

Does time flow? It will be shown in this chapter that *if* the spacetime structure of the world has a certain branched dynamic form, then time flows. In addition to the flow and direction of time, two issues in quantum mechanics, those of non-locality and the definition of 'measurement', are shown to be illuminated by the hypothesis that the world has the spatio-temporal form described. I call the form the *branched model*, and the interpretation of quantum mechanics to which it gives rise I call the *branched interpretation*.

1 Objective time flow

The branched model is a four-dimensional spacetime model in the shape of a tree, each branch of which is a complete Minkowski manifold in which are located objects and events.[1] The trunk represents the past, the first branch point is the present, and the branches constitute the set of all physically possible futures. The scheme is shown in figure 1.

Of the many possible futures which split off at the first branch point, one and only one is selected to become part of the past. The unselected branches vanish, so that the first branch point moves up the tree in a stochastic manner and the tree 'grows' by losing branches. This progressive branch attrition is what in the model constitutes the flow of time.

Suppose for example that 1000 lottery tickets have been sold to 1000 different purchasers. Then at the time of the draw, assuming the procedure is completely fair, there will be at least 1000 different kinds of branch at the first branch point: branches on which A wins, branches on which B wins, etc. Branch attrition ensures a unique winner (unlike the many-worlds interpretation of quantum mechanics, in which if the drawing involved a measurement-like interaction then all branches would be equally 'actual' and everyone would be a winner).

---

[1] A detailed description of the model may be found in McCall (1994).

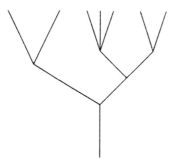

Fig. 1. Schematic form of the spacetime model

The branched model is doubly indeterministic. Firstly, the fact that a given set of initial conditions at a branch point can give rise to two or more physically possible future outcomes contradicts determinism, which permits only one such outcome. Secondly, indeterminism occurs again in the random or chance selection of the branch which becomes actual. It is this dynamic aspect of the model, the fact that at no two moments is its shape the same, all later trees being sub-trees of earlier trees, that enables it to represent in a perfectly objective and realistic way the passage of time. The branched model is a model of objective time flow.

It is true that in the opinion of many philosophers and physicists time flow is not a characteristic of the physical world, but is instead a purely subjective phenomenon. In what would probably be the standard view of scientists today, time's arrow takes flight not in the cosmos, but in the consciousness of intelligent beings. In the words of Paul Davies:

> Relativity theory has shifted the moving present out from the superstructure of the universe, into the minds of human beings, where it belongs. Present day physics makes no provision whatever for a flowing time, or for a moving present moment. Those who might wish to retain these concepts are obliged to propose that the mind itself participates in a novel way in some form of physical activity that is not manifest in the laboratory, a suggestion that meets with a great deal of reserve among the scientific community. Eddington has written that the acquisition of information about time occurs at two levels: through our sense organs in a fashion consistent with laboratory physics, and in addition through the 'back door' of our own minds. It is from the latter source that we derive the customary notion that time 'moves'.[2]

[2] Davies (1974), 3.

5 Time flow and measurement in quantum mechanics

However, despite the current popularity of this view, time flow need have nothing to do with minds and consciousness. The branched model shows this, since the process of branch attrition was a property of the branching structure billions of years before the appearance of living beings, and will presumably continue to characterize it long after their departure.

One objection to the idea of the branched model's being a dynamic entity – the idea of its changing by losing branches – is that it conceals at best a confusion, and at worst a contradiction. If a three-dimensional object like a poker changes, it does so in four dimensions. Time is the dimension of change. But the branched model is already four-dimensional, so how can it change? To do so would appear to require a second time dimension, which would measure the rate of branch attrition. But if the idea of a second time dimension seems absurd, as it does, then the branched model cannot really change after all, and its supposed dynamic quality is illusory. To this objection it may be replied that although the branched model changes in the sense that it suffers progressive branch loss, so that later universe-trees are sub-trees of earlier ones, it does not change *in* time. Branch loss is instead what *constitutes* the flow of time. In the branched model there is only one time dimension, and this dimension runs through every branch and measures the temporal distance of point-events from some assumed origin. No second time dimension is needed to measure the rate of branch attrition. When a branch drops off, it does so at the time (measured on the original time dimension) that it joins the trunk at the first branch point.

A more serious objection is this. The special theory of relativity tells us that there is no unique world-wide class of all the instants simultaneous with 'here–now'. Instead there are many such classes, one for each coordinate frame. Orthogonal to the time axis of each such frame there is a spacelike hyperplane which intersects here–now, and which forms the division between unbranched past and branched future. Therefore there is not just one branched structure relative to 'now', but many. We should speak not of a single branched *model*, but of *models*.

All this is true. But when it comes to modelling temporal passage, and to explaining the distant correlations of outcomes in the EPR experiment (discussed in section 3), the multiplicity of different hyperplane-dependent models is exactly what is needed. To see this in the case of passage, suppose we were to try to represent the 'happening' or 'becoming' of events in a single (unbranched) Minkowski manifold by imagining a tiny light which flashed on as each event 'happened'. Could this device represent time flow? Plainly no haphazard or random illumination would do: the march of time could be represented only by a regimented pattern of lights which swept up the manifold. But to coordinate the tiny lights requires that they be placed in simultaneity classes, and this in turn requires a frame of reference. No frame-independent or hyperplane-independent pattern of illumination could possibly represent temporal becoming. For this reason it is appropriate that its representation in the branched model by branch attrition should require a multiplicity of frame-dependent models, in each of which the branching is along three-dimensional spacelike hyperplanes. Temporal becoming is frame-dependent.

Granted that all branched models are frame-dependent or hyperplane-dependent, they can still be understood as representations of the same underlying spatio-temporal reality. If we take the set W of all future point-events which were possible at the beginning of the world and call this set *the universe*,[3] then the triple $[W, 0, \leqslant]$ is a branched topological space, the structure of which is determined by formal conditions imposed on the so-called 'causal' relation $\leqslant$ of special relativity.[4] The passage of time is marked by progressive point-loss in W.

Given any frame of reference f, and an equal-time hyperplane S in f passing through here–now, consider the subset W_S of W restricted to members of W which are related by $\leqslant$ to some member of S. For each frame f, W_S constitutes one of the corresponding frame-dependent branched models referred to two paragraphs earlier. W_S branches along equal-time hyperplanes in f, and branch-loss in W_S represents time flow in the appropriate frame of reference. But underlying all such frame-dependent models is the dynamic set W which constitutes the frame-independent 'matter' of the four-dimensional physical world.

To return to time flow, this process is objectively represented by branch attrition in frame-dependent models. These models also provide objective correlates of the notions of past, present, and future (the trunk, the first branch point, and the branches), and furnish a clear picture of temporal anisotropy or temporal asymmetry. In the branched model, temporal asymmetry does not depend on knowledge, or causation, or the second law of thermodynamics, or the failure of the principle of the excluded middle to apply to statements about the future, but is a built-in structural feature of spacetime. The model can be

---

[3] Belnap (1992) refers to the set as 'Our World'. The present discussion owes much to him.
[4] See appendix 2 of McCall (1994).

used to provide a consistency proof for the use of tenses and the categories of 'past', 'present', and 'future'. Such notions do not, despite McTaggart's claims about the A-series, engender a contradiction, and a formal proof of this can be given using the branched model as a semantic model structure or 'frame'. In a world which had the structure of the branched model, and which underwent branch attrition, the concepts of temporal becoming and temporal asymmetry, and the categories of past, present, and future, would possess as exact a scientific meaning as the relations of earlier and later.

2 Quantum probabilities

The feature which represents time flow in the branched model, i.e. branch attrition along spacelike hyperplanes, also furnishes an explanation of the much-discussed statistical correlations observed in the EPR experiment. In both the EPR and GHZ experiments, discussed below, a measurement performed at one point on a system of two or more particles in an entangled state can seemingly compel the outcomes of other, simultaneous measurements on distant particles to assume certain values.[5-9] As will be shown in section 3, these instantaneous non-local effects may be explained by the model. We begin by examining how future branching gives an objective physical meaning and a precise value to quantum probabilities.

When a vertically polarized proton enters a two-channel polarization analyzer inclined at an angle ϕ to the vertical, it has a probability $\cos^2 \phi$ of emerging in the $\phi+$ channel, and a probability $1 - \cos^2 \phi$ $(= \sin^2 \phi)$ of emerging in the $\phi-$ channel. These probabilities are represented by the relative proportions of sets of branches in which the photon exits in the $\phi+$ or $\phi-$ channels, and the proportions in question take fixed, precise values for each experimental situation. If $\phi = 30°$, then $\cos^2 \phi = 3/4$. The fact that the photon has a 75% probability of being measured $\phi+$ is represented in the model by the fact that 3/4 of the future branches above the hyperplane in which the photon enters the analyzer are $\phi+$ branches.

There is a potential difficulty lurking here, as was pointed out to the author in 1987 by Bas van Fraassen. If $\cos^2 \phi$ is an irrational number, as it may well be,

---

[5] Clauser & Shimony (1978).
[6] d'Espagnat (1979).
[7] Greenberger, Horne, Shimony, & Zeilinger (1990).
[8] Mermin (1985) (1990).
[9] Clifton, Redhead, & Butterfield (1991).

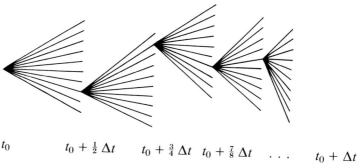

Fig. 2. A decenary tree that branches within a time interval Δt

then no finite set of branches, no matter how large, can represent its value. This would indicate that infinite sets of branches are required. But Cantor showed long ago, or is believed to have shown, that there exist no fixed proportionalities among infinite sets. (Think of the proportion of even numbers in the set of all natural numbers, which varies according to how the numbers are ordered. If we count 1, 2, 3, 4, ... the proportion appears to be 1/2, but it changes to 1/3 if we count 1, 2, 3, 5, 4, 7, 9, 6, 11, 13, ...) These considerations constitute a dilemma for anyone who wishes to represent probability by relative proportions of sets.

Fortunately, in the case of sets of branches on trees, the dilemma can be resolved. Consider what will be called a *decenary tree*, which branches in ten, and each branch of which branches in ten, at each of a denumerably infinite number of levels. Suppose that the vertically polarized photon enters the polarization analyzer at $t = t_0$, and that by $t_0 + \Delta t$, where Δt may be as small as we please, the photon must exit in the $\phi+$ channel on exactly $\cos^2 \phi$ of the branches. This will be so if at t_0, at the base node of the model, there is located a decenary tree, which branches first at t_0, then at $t_0 + 1/2\Delta t$, then at $t_0 + 3/4\Delta t, \ldots$, as shown in figure 2.

By $t_0 + \Delta t$, the branched model will contain $10^{\aleph_0}$ branches, and in this set there exist subsets of any desired proportionality. For example, a decenary tree in which exactly 3/4 of its $10^{\aleph_0}$ branches are $\phi+$ branches (i.e. branches on which the photon exits in the $\phi+$ channel) has the following form. Of the first ten branches, seven are $\phi+$ and two are $\phi-$. The remaining branch is neither $\phi+$ nor $\phi-$, meaning that on it the photon enters neither channel. At the next level however this branch splits in ten, and of these five are $\phi+$ branches and five are $\phi-$. At this level 75 of the 100 branches are $\phi+$ and

25 are $\phi-$, and since all branches are now labeled this 75% to 25% proportionality holds for the entire set of $10^{\aleph_0}$ branches.

Suppose that an irrational proportionality is desired, say $\pi/10 = 0.314159\ldots$ Then at the first level three branches are $\phi+$ and six $\phi-$, with one unlabeled, at the second level one is $\phi+$ and eight are $\phi-$, etc. Given any decimal between 0 and 1 there exists a decenary tree corresponding to it, and conversely given a decenary tree with some branches having property P and some not, the exact proportion of those with P can be read off it. There is of course no reason to think that nature favors decenary trees over binary trees or ternary trees. Any proportionality-preserving tree, each branch of which divides into the same finite number of offspring branches at each level, would serve the purpose of giving the exit probability of the photon as it enters the analyzer. A tree of this kind, of temporal height Δt, I shall call a prism.

The passage of the photon through the analyzer is not the only indeterministic quantum event taking place at time t_0. On the same spacelike hyperplane there will be a radioactive atom on Venus which may or may not decay, a hydrogen atom on Jupiter whose electron may move to a lower orbit and emit a photon, etc. All these events have, according to quantum theory, a certain probability of occurring or not occurring, and in the branched interpretation these probability values will be built into the same prism with base node at t_0. One prism does for all.

It may be asked, and indeed *has* been asked on several occasions by colleagues, whether the branched model provides an explanation of quantum probabilities in any genuine sense, or whether this explanation is not an instance of *obscurum per obscurius*. A related question is whether future branches, being pure *potentia* or even worse *phantasmata*, are too insubstantial to play any explanatory role. A third question is where the probabilities 'come from' – are they not taken over directly from the mathematics of quantum theory? These questions reflect a skeptical attitude towards the ontological status of future branches. Are they real? Or are they instead something imagined, or something with a purely abstract or mathematical existence, for example maximal consistent sets of propositions?

On this point it must be said that in the branched interpretation future branches are as real, concrete, and solidly four-dimensional as the present and the past. All that the branched model adds to the ontology of special relativity is a multiplicity of four-dimensional manifolds above any such 'present'. Given this branching ontological structure, quantum probabilities are not merely redescribed in an esoteric way, they are *explained* in the sense that a physical correlate is provided for them. In quantum mechanics, probability values have only a theoretical or mathematical existence. In the branched interpretation, these values correspond to something real and physical, namely branch proportionality. This physical characteristic explains quantum probabilities in the sense that it gives the term 'quantum probability' an objective, physical, spacetime reference.

Returning to the notion of a 'prism', the branched model is in effect an enormous stack of prisms. At the tip of every branch of every prism there sits the base node of another prism. From this stack, the probability value of any future event, not only quantum events, may be extracted. For example, what is the probability of snow falling next March 16th at 4 p.m.? To determine this, we must find the proportion of branches of the base prism which have at least one offspring branch on which snow falls at 4 p.m. on March 16th, then we repeat this process for each prism at each subsequent level until March 16th is reached. The final result is obtained by multiplying the individual prism probabilities up each snow-containing branch and then summing to get the proportionality of the disjunctive set of such branches at the March 16th level. This last total to be sure may involve summing an uncountable infinity of elements, which rules out the operation as a practical possibility. But whether the operation can be performed or not is irrelevant, and the existence of the prism stack guarantees that, computable or not, the value of the uncountable sum will lie between zero and one. In the branched model, every future event has an exact probability.

3 Non-locality

It is a fundamental principle of relativity theory that no event E_1 can influence or causally affect another event E_2 unless E_1 is located within E_2's backward light cone. If E_1 and E_2 are spacelike separated, meaning that neither lies within the light cone of the other and hence that no light signal can connect them, then they cannot be causally related in any ordinary way. And yet the results of the EPR and GHZ experiments, as well as the non-local character of other quantum phenomena which troubled Einstein,[10] indicate

---

[10] Howard (1990) argues that one of the principal reasons which led Einstein to repudiate quantum mechanics was its denial of the statistical independence of spatially separated systems.

5 Time flow and measurement in quantum mechanics

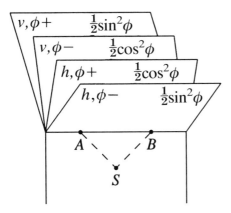

Fig. 3. The relative proportions of the four sets of branches that occur when the two photons enter their respective devices at the same time

that somehow it is possible for reciprocal influences between spacelike separated events to exist. In the branched model, owing to the fact that the branches split along spacelike hyperplanes, statistical and other quasi-causal connections can exist among spacelike separated events, and a natural explanation of non-locality is forthcoming which does not contradict relativity theory.

Consider a version of the EPR experiment in which two photons in the singlet state are emitted from a source S and pass through polarization analyzers some distance apart. Let us suppose that the photons are created by positronium decay and have anti-correlated polarizations, so that if one is observed to be vertically polarized, the other will be horizontally polarized. For dramatic effect, imagine that one analyzer is close to the photons' source on earth and that the other is on the moon. Let the terrestrial device be a two-channel horizontal–vertical plane polarization analyzer, and let the lunar one be similar but slanted at an angle ϕ to the vertical. For each pair of photons, there will therefore be four possible joint measurement outcomes, namely $(v, \phi+)$, $(v, \phi-)$, $(h, \phi+)$, $(h, \phi-)$. On the branched model, these four possible outcomes are represented by four different kinds of branch in a prism. The base node of the prism is the hyperplane on which the entry of the two photons into their respective devices occurs at the same time, and the relative proportions of these four sets of branches, which in the model constitute the physical correlate of the joint probabilities $p(v, \phi+)$, $p(v, \phi-)$, $p(h, \phi+)$, and $p(h, \phi-)$, are represented schematically as on figure 3. (Since the photons have orthogonal plane polarizations, if $\phi = 0°$ then $p(v, \phi+) = 1/2 \sin^2 \phi = 0$ and there will be no $(v, \phi+)$ or $(h, \phi-)$ branches.)

On this diagram, A and B are the two measurement events, which in the coordinate frame chosen are simultaneous. From the proportions of the different branch sets, it is plain that the outcomes of the two measurement events cannot be independent. For the outcome on earth to be independent of the outcome $\phi+$ on the moon, the joint probability $p(v, \phi+)$ must be factorizable into the product of $p(v)$ and $p(\phi+)$ taken separately:

$$p(v, \phi+) = p(v) \times p(\phi+).$$

But in the case where $\phi = 30°$, $1/2 \sin^2 \phi = 1/8$ and $1/2 \cos^2 \phi = 3/8$. Then $p(v) = p(v, \phi+) + p(v, \phi-) = 1/2$, $p(\phi+) = p(v, \phi+) + p(h, \phi+) = 1/2$, and $p(v) \times p(\phi+) = 1/4$, whereas $p(v, \phi+) = 1/8$. Therefore the two outcomes v on earth and $\phi+$ on the moon are not independent, despite the fact that the measurement events lie outside each other's light cones.[11] How can two events be statistically correlated without being causally related?

They could be, of course, if there existed a local hidden variable or common cause explanation of the two events; for example, if each photon were accompanied by a set of instructions telling it how to react to measuring devices oriented at different angles. But Bell's theorem rules out any explanation of this sort. As things stand at present, we seem to have no choice but to accept the correlations revealed by the EPR experiment as brute fact, lacking any further explanation.[12]

The branched model *does* however provide an explanation. The future branches are spacetime manifolds arranged in sets of fixed and definite proportionality, and correlations of the EPR variety are explicable by the physical structure of the model. Since branch selection is random, the probability of obtaining the joint result $(v, \phi+)$ will be exactly $1/2 \sin^2 \phi$, no matter how far apart the two observations are made. And except where $\phi = 45°$, $p(v, \phi+) \neq p(v) \times p(\phi+)$, so that future branch proportionality explains the fact that the two measurement results v and $\phi+$ are not independent. To

---

[11] The results violate what Jarrett (1984) calls 'completeness', that is to say *outcome independence*, while at the same time satisfying the requirement that each measurement should be independent of the *state* or *orientation* at the opposite measuring device, which Jarrett calls 'locality'. Violation of the latter condition would permit signalling at speeds faster than light (Ballentine & Jarrett (1987)).

[12] Cf. Fine (1989).

underline this point, consider a frame of reference in which the earth measurement takes place before the moon measurement.

Imagine the two photons being emitted, and shortly afterwards the earth photon being measured as v vertically polarized. Once this result is obtained, because the photons have anti-correlated polarizations, it is certain that the moon photon will behave exactly like a horizontally polarized photon. If $\phi = 0°$, it will never pass $\phi+$; if $\phi = 90°$, it will always pass $\phi+$; if $0 < \phi < 90°$ its probability of passing $\phi+$ will be that of a horizontally polarized photon. Three questions concerning this state of affairs arise.[13]

(i) Was the moon photon horizontally polarized *prior* to the earth measurement? The answer seems to be no. If the earth photon had been measured h instead of v, the moon photon would not have been horizontally polarized. If at the last moment the earth analyzer is removed, so that no measurement at all is made, then the moon photon exhibits no consistent angle of polarization, passing + or − through analyzers oriented at any angle with equal frequency. All the evidence indicates that before the earth measurement, the moon photon was not horizontally polarized.

(ii) Was the moon photon horizontally polarized *after* the earth measurement? Yes. After the earth measurement all tests confirm that the moon photon behaves in every respect like a horizontally polarized photon.

(iii) If the moon photon was not polarized horizontally *before* the earth measurement, but was *afterwards*, how did it acquire the property of horizontal polarization, given that it was then travelling to the moon at the speed of light? Alternatively, how was the information concerning the result of the earth measurement conveyed to the moon in time to affect the subsequent measurement? On the face of it, there would seem to be no way of answering this question consistently with relativity theory.

Consider however the branched picture of the EPR experiment in a frame of reference in which the earth measurement A precedes the moon measurement B, shown schematically in figure 4. At the moment of the earth measurement, the progressive vanishing of all branches but one forces a choice between a v-branch, in which the earth photon is vertically polarized, and an h-branch. But on every v-branch, the moon photon has probability $\sin^2 \phi$ of being measured $\phi+$, and probability $\cos^2 \phi$ of being measured $\phi-$. On every one of those branches it behaves exactly like a horizontally polarized photon. One could say that on those branches it *is*

[13] Cf. Mermin (1985), 46–7.

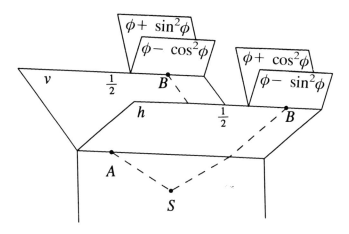

Fig. 4. The branched model picture of the EPR experiment when the earth measurement A precedes the moon measurement B

a horizontally polarized photon.[14] Since the 'moment of decision' for the moon photon in this model is the earth measurement, many miles away, the vanishing of all but a single v-branch as a result of the earth measurement instantaneously affects the moon photon. On the branch selected by the earth measurement, the moon photon has the property of being horizontally polarized. In this way branch attrition along spacelike hyperplanes can produce instantaneous non-local effects, a kind of effortless action-at-a-distance.

It is characteristic of the kind of non-local influence permitted by the branched model that it is reciprocal: X affects Y and Y also affects X. Suppose that a coordinate frame is chosen in which the moon measurement precedes the earth measurement. (Since A and B are spacelike separated, this will always be possible.) In this case the model takes the shape shown in

[14] It is true that a more sophisticated experimental situation might be imagined, in which the earth photon after passing through its analyzer was made to interfere with itself in a reversed apparatus and emerge once more in a superposition of v and h states (cf. Wigner (1963)). In these experimental conditions the moon photon would presumably not be horizontally polarized, and this would be reflected in the branched model by the fact that the proportion of branches on which it was measured $\phi+$ would be 1/2, not $\sin^2 \phi$. There is in the branched model what John Earman has called 'pre-established harmony' between initial or experimental conditions on a hyperplane and the proportions of sets of branches in the prism stack above that hyperplane, which in turn entails that branching above the more sophisticated 'reversed measurement' apparatus differs from that given earlier in the diagrams of the text. McCall (1994), 138 and 216–17 proposes a large-scale hypothesis which is highly speculative, but which accounts for this pre-established harmony between initial conditions and branching.

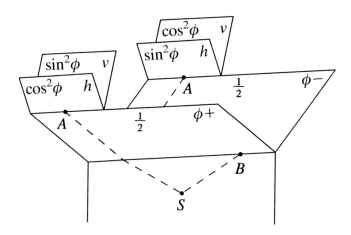

Fig. 5. The branched picture of the EPR experiment when the moon measurement B precedes the earth measurement A

figure 5, and the 'moment of decision' for the earth photon is the moon measurement.

Plainly no ordinary causal relationship can exist between A and B if the question whether A influences B or conversely depends upon the choice of reference frame, and for this reason it is better to describe non-local influences as 'quasi-causal' rather than causal. Nevertheless, whether causal or quasi-causal the influences in question are real enough, once the branched interpretation is accepted. At the same time they are entirely consistent with relativity theory. Each complete path through the tree (trunk plus a single branch) is a Minkowski spacetime, with the geometry characteristic of special relativity, and the principle that nothing travels faster than light is preserved by the fact that though the outcome of one measurement in the EPR experiment can instantaneously affect the other, nothing at all travels between the two events. Hence relativity is not violated. Attrition in the branched model is inherently non-local, and provides a relativistic account of EPR action-at-a-distance.

In the case of the GHZ experiment, the connection between distant measurements and the way in which an observed result at point X can instantaneously influence a result at point Y is equally striking. Suppose that a system of three spin-1/2 particles is emitted from a central source in an entangled state consisting of the following superposition:

$$\Psi = \frac{1}{\sqrt{2}}(|1,1,1\rangle - |-1,-1,-1\rangle).$$

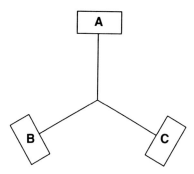

Fig. 6. GHZ experiment: devices to measure the spin orientations of three particles moving in different directions in a plane

Here the vector $|1,1,1\rangle$ represents the state in which each of the three particles is spin-up in its direction of propagation (along its z-axis), and the vector $|-1,-1,-1\rangle$ denotes spin-down for the three particles along their z-axis.[15] The three particles move off on a plane along different directions and pass through Stern–Gerlach devices which can be set to measure their spin orientation perpendicular to their motion in either the x or the y directions, as shown in figure 6. If all three instruments are set to measure spin in the x direction then quantum mechanics predicts that the product of the three spin measurements will be -1. Consequently, if the measurement at apparatus **A** is 1 (spin-up) then the outcomes at **B** and **C** must be different, whereas if **A** yields the result -1 (spin-down) then the outcomes at **B** and **C** must be the same (i.e. both spin-up or both spin-down). Since the experiment can be arranged so that all three measurement events lie outside each other's light cones, the question arises how the information about the outcome at **A** can be communicated to the other devices in time to bring about the desired correlation of their results. As in the case of the EPR experiment, relativity theory permits no answer to this question since such information transfer would require faster-than-light signaling.[16]

The branched model provides the following interpretation of what is going on in the GHZ experiment.[17] If the three instruments are all set to measure spin in the x direction, then the only permitted joint results for the three

---

[15] Mermin (1990).
[16] Stapp (1993).
[17] The experiment is still only a thought-experiment, but Greenberger, Horne, Shimony, & Zeilinger (1990) offer suggestions about ways in which it could one day be performed.

particles are $(1, 1, -1), (1, -1, 1), (-1, 1, 1)$, and $(-1, -1, -1)$. In the model in which all three measurement events are simultaneous there are only four different kinds of branch above the relevant hyperplane. That is, there are no branches containing the joint outcomes $(1, 1, 1), (1, -1, -1), (-1, 1, -1)$, or $(-1, -1, 1)$. Therefore if the outcome at **A** is 1, the outcomes at **B** and **C** must differ. Again if the outcome at **A** had been -1 and the outcome at **B** had been 1, the outcome at **C** would have had to have been 1. And so forth. The truth-maker for all these conditionals and counterfactuals is the *absence* of certain kinds of branch in the branched model. As in the case of the EPR experiment, one can also consider frames of reference in which the **A** measurement occurs before the **B** and **C** measurements, or in which **A** and **B** are simultaneous followed by **C**, etc., and again the phenomenon of a measurement outcome apparently bestowing a definite spin orientation upon a particle a long distance away will be observed. As before, the explanation is branch attrition along hyperplanes.

4 The definition of measurement

In his last published paper, 'Against "measurement"',[18] the late John Bell lists some words which in his opinion have no place in any exact formulation of quantum mechanics. Among them he mentions *system, apparatus, microscopic, macroscopic, observable,* and *measurement,* of which the last is the most notorious and raises the most problems. Bell cites Landau and Lifshitz as expressing the traditionally accepted quantum mechanical view of measurement:

> the 'classical object' is usually called *apparatus* and its interaction with the electron is spoken of as *measurement*. However, it must be emphasized that we are here not discussing a process ... in which the physicist-observer takes part. By *measurement*, in quantum mechanics, we understand any process of interaction between classical and quantum objects, occurring apart from and independently of any observer. The importance of the concept of measurement in quantum mechanics was elucidated by N. Bohr.

In this passage we encounter what Bell describes as the 'shifty split' between system and apparatus, between what is measured and what does the measuring, between the quantum and the classical domains. Underlying the approach is the assumption, as Bell puts it, that apparatus may be 'separated off from the

[18] Bell (1990).

rest of the world into black boxes, as if it were not made of atoms and not ruled by quantum mechanics' (p. 33). The picture is that of a world in which quanta lead their own lives, moving freely and described by wavefunctions which expand according to the time-dependent Schrödinger equation, until from time to time they encounter a classical object and undergo 'measurement'. But the so-called measurement process then becomes entirely mysterious, without any agreement as to whether and under what conditions it involves a 'collapse' or narrowing of the wavefunction, or whether on the contrary Schrödinger evolution spreads from the microscopic to the macroscopic domain, involving the measuring apparatus itself in a superposition of macroscopic states.

To this debate the branched interpretation can make a contribution, namely a definition of measurement which is precise and unambiguous, and which in no way relies on the concept of apparatus, or on the microscopic/macroscopic distinction, or on the presence or absence of conscious observers. The definition preserves a clear dividing line between measurement and Schrödinger evolution, and supports von Neumann's view (now somewhat unfashionable) that both processes are seen in the physical world. The distinction between the two is stated in terms of characteristics of branching.

In the branched interpretation, measurement is a process which a *system S* undergoes with respect to a *property P*. It is true that Bell recommends not using the word 'system', but some word is needed to denote the object or collection of objects which undergoes measurement. No mention will be made of the system/apparatus dichotomy, and the system S may be microscopic, macroscopic, or whatever. Since S ranges in size from a single photon to an elephant or larger, the property P may denote anything from a quantum spin state to 'weighing ten tons'. In the quantum domain, P represents a state, not an observable or an operator. It will be clearer in what follows if P is understood to be a *dated* property, i.e. *P-at-t.*

In addition to *system* and *property*, the third concept needed for the definition of measurement is that of an **R**-*type prism stack*. To recall, a prism stack of temporal height t is a branched structure with a single prism at its base, at the tip of each branch of which there sits another prism, and so on up to t. If a system S exists at the base node and throughout a prism stack T, and if S has property P on some of the branches of T and lacks P on others, then we shall say that T is an **R**-*type prism stack* for S with respect to P. If on the other hand S has P on all the branches of T, or fails to have P on any

5 Time flow and measurement in quantum mechanics

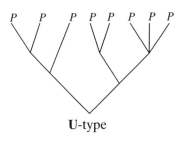

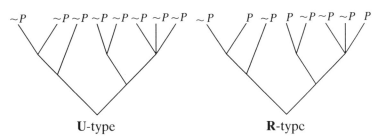

Fig. 7. **R**- and **U**-type prism stacks

branch, then T will be a **U**-*type prism stack* for S with respect to P,[19] as shown schematically in figure 7.

Defined in this way, it is plain that for any S, any P, and any prism stack T which contains S, T is either a **U**-type or an **R**-type stack for S with respect to P. We may now proceed to define measurement. A system S exists in the present, or 'now', if a hyperplane containing it constitutes the first branch point of the branched model.[20] If, with respect to some property P, the prism stack immediately above 'now' is an **R**-type prism stack for S, then when branch attrition takes place S undergoes *measurement* with respect to P. But if on the other hand the stack is a **U**-type stack, then S undergoes *Schrödinger evolution* with respect to P. According to this definition every presently existing physical system, with respect to any property, undergoes either Schrödinger evolution or measurement with respect to that property. Which of these two processes it undergoes is an objective, precisely defined fact about the branched model.

In the definitions of Schrödinger evolution and measurement, the sharp differences between the two processes are preserved. Schrödinger evolution is

---

[19] The letters **U** and **R** come from Penrose (1989) where they stand for 'unitary evolution' and 'reduction'.

[20] More accurately, if the hyperplane consists of limit points of the model's trunk. This covers both the 'lower cut' and 'upper cut' branching described by Belnap. See McCall (1990), 233.

smooth, continuous, and deterministic. Measurement is abrupt, discontinuous, and indeterministic. Branch attrition within **U**-type as compared to **R**-type prism stacks bears out these differences. Since a system at the base node of a **U**-type stack has property P on all or none of the branches, branch attrition produces no surprises. The system evolves towards its eventual state in a deterministic way, like Newton's apple falling to the ground. But if the system finds itself at the base node of an **R**-type stack, then at some point within that stack it will suddenly either 'jump' into some state P, or into not-P, in a completely indeterministic manner. Whether a system undergoes measurement and 'jumps' into some state P, or whether it evolves smoothly and predictably into some other state Q, depends upon the branched structure of the universe. It is characteristic of the branched interpretation that something very small, such as the transition of an electron from one energy state to another in an atom, is accounted for by something very large, namely the overall structure of spacetime.

This concludes, in outline, the 'branched' analysis of measurement. Many questions remain open, notably the difficult one of determining in exactly what circumstances Schrödinger evolution takes place, and in what circumstances measurement or collapse takes place. In terms of the branched interpretation, this is the problem of which type of prism stack, **U**-type or **R**-type, is associated with which type of initial conditions on a hyperplane. Conceivably, it might turn out that **R**-type prism stacks are associated with *no* permissible initial conditions, so that measurement never takes place and all is Schrödinger evolution. But this seems unlikely. In any case, whatever the answer may be to whether there are in fact quantum jumps, the branched interpretation answers Bell's criticisms in providing an exact definition of measurement.

6 Stochastically branching spacetime topology

ROY DOUGLAS

1 Introduction

In this chapter the modelling of spacetime is discussed, with the aim of maintaining a clear distinction between the local properties of a model and those attributes which are global. This separation of attributes (into local and global type) logically motivates the construction of the non-Hausdorff branched model for spacetime. Of course, much of the physical motivation for this construction is derived from the stochastic nature of quantum mechanics. The *Many-World Interpretation* is seen to be (at least, topologically) a consistent and complete interpretation of quantum mechanics.

In any serious inquiry into the nature of *spacetime* (on either the quantum or cosmological level), mathematical models will be constructed which are in substantial agreement with the collective empiricism of experimental physics. If such a mathematical model is to be more than just a prescription for prediction, then it is necessary to give careful consideration to the appropriate level of mathematical generality of the model. Common sense, as well as the history of science, appear to indicate the need to maintain as general (i.e., unrestricted) a model as possible, constrained only by empirical data on one side, and the limits of our mathematical sophistication and imagination on the other.

In the spirit of generalization, I will construct a model of the *time parameter*, which extends the usual concept of a *time-line*, and can be used to model spacetime locally, with the usual product structure of Minkowski spacetime. Our imagination will be constrained only by the mathematical discipline of topology and a clear view of all underlying assumptions. As such, it seems to be a worthwhile exercise, no matter what one may (or may not) believe to be true of our universe.

In this chapter I present two kinds of argument for a Many-World Interpretation of quantum mechanics. The first type of argument is partly mathematical (section 2 and the appendix). In this approach, certain standard results in topology are assembled, which I hope may convince the reader that (a) imposing the (global) structure of a *manifold* on a model of spacetime is unreasonable, while (b) it is reasonable to assume that spacetime models are *locally Euclidean* (but *not* necessarily *Hausdorff*). This viewpoint seems to require that the *branching-time*, locally Euclidean model for spacetime be seriously considered, for strictly abstract reasons, such as maximal generality.

The second type of argument (section 3) is more physically motivated, and relies on the necessity to explain the probabilistic nature of quantum mechanics. The point of view taken here is not the only reasonable way to regard the concept of *probability*; however, it is based on the following abstract principle: probability can only be meaningful, if there exists a *population* which has a *distribution* over some measurable parameter space. In the branching-time model, the population consists of a specific *set of time-lines* embedded in a larger time parameter space (see definition 16). In this context, *experiments* should be viewed as sampling procedures.

If our models are to reflect both the subtlety and locality of our physical experience, rather than our prejudices (such as the unfortunate, ubiquitous bias toward mathematical simplicity for its own sake), then we must examine the foundations of our modelling technique. I wish to argue that any model of spacetime (or more generally, any kind of associated phase space) will have at least the structure of a *topological space*, if only to retain the usual, useful concept of *continuity*.[1]

What can we expect to know about this topological space? There are quite accurate and subtle theories which predict our physical experience (experimental data). However, our physical experience is entirely *local*, with the possible exception of the cosmic microwave background radiation. Thus, it seems reasonable to *want* to specify *only local properties*, when building a topological model of spacetime.

Of course, it is not possible to construct effective models of spacetime without any global (i.e., non-local) restrictions. In fact, we require two such constraints, where the first is the assumption that spacetime can be modelled as a topological space.

The second global assumption, which might be viewed as a *Topological Cosmological Principle*, asserts that local conditions are *topologically equivalent*

---

[1] More general settings and ideas that might relate to *fluctuating topologies* have been considered by C. J. Isham (at least in private conversation with the author).

everywhere in the non-singular part of spacetime. For example, in this chapter, we will assume that spacetime is *locally Euclidean*, since this is a *local property*, which is useful for computation, and corresponds well to our intuitive sense of our neighborhood in spacetime.

In this chapter, no mention is made of a differentiable structure, curvature, or any other structure which is obviously related to the mass–energy density. Thus, even in a 'lumpy' universe (where the geometry is determined by this density), we can assume the Topological Cosmological Principle, since *local density fluctuations* do not seem to alter the *local topology*, at least until one approaches the event horizon of a black hole.

As will be discussed in later sections, the various *separation axioms* for topological spaces behave quite differently when imposed locally, as opposed to globally. Moreover, when models are locally Euclidean, it is important to understand the separation axioms in this context.

In section 2 we will briefly review the topological separation axioms (or properties) denoted T_0, T_1, T_2, T_3, and T_4, in order of increasing constraint. The main observation is that while locally T_1 is the same as T_1 (globally), none the less locally T_2 is very much weaker than the global constraint, T_2.

Consistency with our earlier discussion demands that we consider appropriate spacetime models which are locally Euclidean, but are not globally Hausdorff. Such a model is explicitly constructed in section 3, in order to represent the stochastic branching implicit in quantum mechanics. Branching-time models seem to have been first considered by H. Everett;[2] they are also discussed at length in B. S. de Witt and N. Graham,[3] and in J. A. Wheeler and W. H. Zurek,[4] as well as other sources. These and several other interpretations of quantum theory are presented and discussed by Bell.[5] Sections 2 and 3 of this chapter consist almost entirely of mathematics and are therefore unlikely to be controversial. The author's opinions are discussed in section 4.

2 Topological separation axioms: locally and globally

The mathematics in this and the next section is presented in a way that is nearly self-contained. Every technical term is defined simply and precisely (with one minor exception). All deductions (or proofs) in this discussion

---
[2] Everett (1957).
[3] de Witt & Graham (1973).
[4] Wheeler & Zurek (1983).
[5] Bell (1987).

remain at the level of only two or three logical steps (with the exception of one argument, which is a few steps longer, and is explained fully in the appendix). I make an apology to topologists, for there is nothing here that should surprise you. If novelty there be, it resides in the strict application of these ideas to the study of time.

For completeness, this section will begin with a very brief review of the concepts from *general (point-set) topology* which are required for this discussion. The general definition of a *topological space* is given, from which the entire subject of *topology* follows logically. Of course, this short section is no substitute for a proper study of topology; the curious reader can find a clear and comprehensive treatment of this subject in various texts.[6]

Definition 1 *Let X be a set and let τ be a collection of subsets of X. We say τ is a* **topology on the set** X, *if the following four conditions are satisfied.*

(1) $X \in \tau$ *(The set X is itself a member of the collection τ.)*
(2) $\phi \in \tau$ *(ϕ, the empty set, is a member of τ.)*
(3) *If A and B are in τ, then $(A \cap B) \in \tau$.*
(4) *If J is an arbitrary index set, and $A_\beta \in \tau$, for all $\beta \in J$, then $\bigcup_{\beta \in J} A_\beta \in \tau$.*

A *topological space* is defined to be a pair (X, τ) consisting of a set X and a topology τ on set X. In order to facilitate this discussion, it is necessary to introduce some standard terminology. The subsets of X which are elements of τ are called *open sets* in the topological space (X, τ). There is a maximal topology on any set X, called the *discrete* topology, in which $\tau = 2^X$ is the set of all subsets of X (i.e., every set is an open set).

Two topological spaces, (X_1, τ_1) and (X_2, τ_2), are defined to be *homeomorphic* (or *topologically equivalent*), if there exists a one-to-one correspondence between the sets X_1 and X_2, which induces a one-to-one correspondence between the sets, τ_1 and τ_2. When we speak of a *topological property*, we mean by that any hypothesis which is preserved by topological equivalences.

Suppose $x \in X$ and N is a subset of X. We say N is a *neighborhood of x* in the topological space (X, τ), if and only if there exists an open set $U \in \tau$ such that $x \in U \subset N \subset X$. Thus, an open set (an element of τ) can be characterized by the property that it is a neighborhood of each of its elements.

A subset $B \subset X$ is called a *closed set* in the topological space (X, τ), if and only if the complement of B is open, $(X - B) \in \tau$. (It is worth noting that in a topological space, *finite intersections* and *arbitrary unions* of open sets are

[6] For example, Kelly (1955).

themselves open sets, while the reverse holds for closed sets: i.e., *finite unions* and *arbitrary intersections* of closed sets are again closed sets.)

If Y is an arbitrary subset of set X and (X, τ) is a topological space, then

$$\tau_Y = \{U \cap Y \mid U \in \tau\}$$

is a topology on the set Y. The resulting topological space, (Y, τ_Y) is often referred to as the *subspace Y* of the topological space X. It is a common abuse of language (employed here) to simply refer to '*the space X*' rather than the topological space (X, τ), when τ is understood.

The separation axioms for topological spaces have been thoroughly studied and are usually presented as the following logical hierarchy of properties ordered by implication:

$$T_4 \Rightarrow T_3 \Rightarrow T_2 \Rightarrow T_1 \Rightarrow T_0.$$

We introduce all five of these properties in order to give some perspective on the nature of topological separation axioms; however, we will finally focus on just the critical properties, T_1 and T_2.

In the following five definitions, we merely suppose that (X, τ) is a topological space.

Definition 2 (X, τ) is called a **T_0-space** *if the following condition is satisfied: $y, z \in X$ and $y \neq z \Rightarrow$ there exists an open set $U \in \tau$, such that either $y \in U$ and $z \notin U$, or $z \in U$ and $y \notin U$.*

Definition 3 (X, τ) is called a **T_1-space** *if the following condition is satisfied: $y, z \in X$ and $y \neq z \Rightarrow$ there exists an open set $U \in \tau$, such that $y \in U$ and $z \notin U$.*

The following is an attractive alternative to the description in definition 3. Of course, the definitions 3 and *3 are logically equivalent.

Definition *3 (X, τ) is called a **T_1-space** *if every one-element subset of X is a closed subset. (That is, $x \in X \Rightarrow (X - \{x\}) \in \tau$.)*

Definition 4 (X, τ) is called a **Hausdorff space** *(or a **T_2-space**) if the following condition is satisfied: $y, z \in X$ and $y \neq z \Rightarrow$ there exist open sets U_y and U_z in τ such that $y \in U_y$, $z \in U_z$, and $U_y \cap U_z = \phi$.*

There are other ways to describe the properties T_1 and T_2. For example, a space is T_1, if and only if every *constant sequence* (in the space) converges *uniquely*. (Of course, a constant sequence must converge to that constant; the question is, does it converge to anything else?)

In a Hausdorff space, sequences behave quite reasonably; if a sequence converges in a T_2-space, then it does so *uniquely*. There is a valid converse to this statement, and it allows us to characterize Hausdorff spaces in terms of convergence. However, in this generality, sequences do not suffice to describe this characterization of T_2. What we require is the concept of a *net converging* in a topological space, generalizing the concept of a sequence converging in a space (see Kelly[6] for details about nets). With this somewhat sharper technology, it is not difficult to prove that a space is Hausdorff, if and only if *every convergent net in the space converges uniquely*.

Definition 5 (X, τ) *is called a* **T_3-space** *if it is a Hausdorff space and the following condition is satisfied: $y \in X$, $B \subset X$, B is closed in X, and $y \notin B \Rightarrow$ there exist open sets U and W in τ, such that $y \in U$, $B \subset W$ and $U \cap W = \phi$.*

Definition 6 (X, τ) *is called a* **T_4-space** *if it is a Huusdorff spacc and the following condition is satisfied: $A, B \subset X$, A, B closed in X, and $A \cap B = \phi \Rightarrow$ there exist open sets U and W in τ, such that $A \subset U$, $B \subset W$, and $U \cap W = \phi$.*

Now it is necessary to clarify what is meant by the word '*local.*' In fact, this term can be understood to have (at least) two quite distinct meanings. We refer unambiguously to these by using the two terms '*locally*' and '*strongly locally,*' which are defined (rather abstractly) in the following way. Suppose Q is a topological property, so that in case two spaces are topologically equivalent (homeomorphic), either they both possess Q, or else, neither satisfies property Q.

Definition 7 *We say that the topological space (X, τ) is a* **locally Q space***, if for each element $x \in X$, there exists a neighborhood of x, N_x, which (as a subspace of X) has property Q.*

Definition 8 *We say that the topological space (X, τ) is a* **strongly locally Q space***, if for each element $x \in X$, and each neighborhood of x, say L_x, there exists a neighborhood of x, denoted N_x, with $N_x \subset L_x$, such that N_x (as a subspace of X) has property Q.*

Obviously, a strongly locally Q space must be a locally Q space. Also, it is clear that if the space (X, τ) has property Q (globally), then it must be a locally Q space. However, examples can be constructed to show that there exist no

6 Stochastically branching spacetime topology

other implications (at this generality) between the three types of properties: global, strongly local, and local.

Connectedness is one of the most elementary of topological properties; also, it yields both counter-examples required by the claim in the previous paragraph. Following the definition of connectedness two standard examples of topological spaces are presented. The first (rather trivial) space is strongly locally connected, but not (globally) connected, while the second example is (globally) connected, but not strongly locally connected.

Definition 9 *Let (X, τ) be a topological space. We say that (X, τ) is* **disconnected**, *if there exist two open sets, U, $V \in \tau$, such that $U \neq \phi$, $V \neq \phi$, $U \cup V = X$, and $U \cap V = \phi$. The topological space, (X, τ) is said to be* **connected**, *if it is not disconnected.*

The first example really is trivial, as promised. Choose any set X which has at least two elements, and let $\tau = 2^X$, the power set of X, which *is* the discrete topology on X. Then it is easy to see that (X, τ) is strongly locally connected, but is not (globally) connected.

The second example is a *subspace* of the Cartesian (x, y-coordinate) plane, $\Re^2$. Thus, it suffices to define the subset of the plane, $X \subset \Re^2$, which is the underlying set for this example. We define X as the closure of a graph,

$$X = \{(x, y) \mid x \neq 0 \text{ and } y = \cos(1/x)\} \cup \{(0, y) \mid -1 \leq y \leq 1\}.$$

This subspace of the plane is (globally) connected, but not strongly locally connected. Thus, the space X serves as a counter-example, which shows that the property of (global) connectedness does not imply the property of strong local connectedness.

Incidentally, this subspace of the plane is interesting from several other viewpoints. I will briefly sketch just one. Consider any continuous path[7] in this subspace, $X \subset \Re^2$. The *x-coordinates* of the points on this path are either always positive, or always negative, or always zero. Thus, we have an example of a connected space, which is not *path-connected*. However, path-connectedness does imply connectedness.

For some topological properties, Q, there is no distinction between *locally Q* and *strongly locally Q*. One class of such properties consists of those which are called *hereditary*.

---

[7] The concept of a continuous path is treated in detail in Kelly (1955), here simple intuitive understanding is sufficient.

Definition 10 *Q is an* **hereditary** *topological property, if every subspace of a Q space is also a Q space.*

Lemma 11 *If Q is an hereditary topological property, then* **locally Q** *and* **strongly locally Q** *are equivalent.*

Obviously, *connectedness* is not hereditary. Also, it is not hard to see that all five of the separation axioms are hereditary. Thus, locally T_n is equivalent to strongly locally T_n, for each $n = 0, 1, 2, 3, 4$.

Not only hereditary properties behave this way. If Q is the property of being *topologically equivalent to Euclidean space*, then it is easy to see that while Q is *not hereditary*, it is none the less true that *locally Q* and *strongly locally Q* are equivalent.

Now we are in a position to consider the separation axioms. We will focus on the critical properties T_1 and T_2. The properties (globally) T_1 and locally T_1 are logically equivalent (see appendix for proof) and this makes T_1 critical.

However, there exist topological spaces which are locally T_2, but are not (globally) T_2. Indeed, locally Euclidean spaces which are not Hausdorff are constructed in section 3. Spaces of this type may be considered to be the guiding theme of this chapter.

Definition 12 *If (X, τ) is a locally Euclidean, Hausdorff space, then we say that (X, τ) is a* **topological manifold**.

Manifolds are (relatively speaking) very well behaved mathematical objects. They have been extensively studied from the points of view of both algebraic topology and differential geometry; also, manifolds have been exploited vigorously in physics, at least since the discovery of general relativity.

The (global) Hausdorff property certainly simplifies the mathematics of our models; unfortunately, it is also an entirely inappropriate restriction for models of spacetime, precisely because (as we have seen) it is a strictly global constraint, in spite of any naive intuition to the contrary.

3 Stochastically branched spacetime: the many-world universe

There are many (including myself) who think that, at the most fundamental level, quantum mechanics requires a mathematical model of spacetime which can account for probabilistic events, without having to incorporate the idea

6 Stochastically branching spacetime topology

that *observers collapse amplitude waves to create apparent reality*. In this section, we will attempt to explicitly construct the simplest model of spacetime which has probability theory included as an intrinsic part of the topology of the model. This section concludes with some remarks about less simple mathematical models, which are immediately suggested by the attempt to 'up-grade' the simple model to reality.

Since our model will be locally Euclidean, we can locally isolate space from time (relative to a choice of inertial frame), and then combine them as a *product space*. We assume that both space and time are locally Euclidean, and we also assume that space satisfies the Hausdorff axiom (T_2) globally. However, we will not expect our model of time to be a Hausdorff space.

In order to illustrate the nature of the construction in its most elementary context, we consider a *one-experiment universe* and suppose that this *elementary experiment has only two possible results, which are equally likely to be observed*. In other words, we would like to consider a *single probabilistic event in complete isolation*.

To represent the regional time parameter in this rather unstimulating universe, we require the following topological space, denoted by (Υ, τ). In this model of time, the set, $\Upsilon \subset \Re^2$, can be described as the following subset of the Cartesian plane:

$$\Upsilon = \{(x,0) \,|\, x < 0\} \cup \{(x,1) \,|\, x \geq 0\} \cup \{(x,-1) \,|\, x \geq 0\}.$$

The set, Υ, is endowed with a topology, τ, which is *very different* from the subspace topology inherited from $\Re^2$.

Definition 13 *We define an **open interval** to be one of the five types of subsets of Υ listed below. In this list, α and β are arbitrary real numbers, which are constants for any one specific open interval.*

$$W_1(\alpha, \beta) = \{(x,0) \,|\, \alpha < x < \beta \leq 0\}$$
$$W_2(\alpha, \beta) = \{(x,1) \,|\, 0 \leq \alpha < x < \beta\}$$
$$W_3(\alpha, \beta) = \{(x,0) \,|\, \alpha < x < 0\} \cup \{(x,1) \,|\, 0 \leq x < \beta\}$$
$$W_4(\alpha, \beta) = \{(x,-1) \,|\, 0 \leq \alpha < x < \beta\}$$
$$W_5(\alpha, \beta) = \{(x,0) \,|\, \alpha < x < 0\} \cup \{(x,-1) \,|\, 0 \leq x < \beta\}.$$

Definition 14 *The **open sets** in (Υ, τ), the elements of τ, are precisely those subsets of Υ which are **unions of open intervals**.*

We begin with the observation that (Υ, τ) is *locally Euclidean*. Indeed, it is one dimensional, since each open interval in Υ is homeomorphic to the real line, $\Re$, and each element of Υ is contained in some open interval of Υ. Thus, locally, (Υ, τ) is exactly like the real line, $\Re$. Of course, that is just the way we experience time.

From the discussion in section 2, it is clear that (Υ, τ) satisfies the T_1 separation condition, but not the Hausdorff (T_2) separation condition. According to definition 4, the Hausdorff condition requires that each pair of distinct elements can be separated, using a pair of disjoint open neighborhoods. The Hausdorff condition fails in (Υ, τ) for *only one pair* of elements: $(0, 1)$ and $(0, -1)$ do not have disjoint neighborhoods, because they have a *common past*. If N_+ and N_- are neighborhoods of $(0, 1)$ and $(0, -1)$, respectively, then there exist real numbers, α and β, with $\alpha < 0 < \beta$, such that $W_3(\alpha, \beta) \subset N_+$ and $W_5(\alpha, \beta) \subset N_-$; of course, $W_3(\alpha, \beta) \cap W_5(\alpha, \beta) \neq \phi$. Removing either one of these two elements, $(0, 1)$ or $(0, -1)$, leaves a subspace of (Υ, τ) which is a Hausdorff space; in fact, such a subspace is homeomorphic to two disjoint copies of the real line, and every element has a *unique future*.

Notice that the inseparable points, $(0, 1)$ and $(0, -1)$ are only inseparable in (Υ, τ). Among themselves, they are easily separated. The set of all the inseparable points forms a subspace of (Υ, τ), and the subspace topology is **discrete**, which means every set is an open set. This behavior will persist in the more general model, since it is locally Euclidean.

Within this model of *time*, (Υ, τ), we have *embedded* two *time-lines* for this one event. They have their common past and distinct futures. The words embedded and time-line are defined as technical terms.

Definition 15 *We say one topological space, (X, τ) is **embedded** in another, (Y, σ), if (X, τ) is homeomorphic to a subspace of (Y, σ).*

Definition 16 *Suppose (Υ, τ) is a model of time (for instance the one constructed above). The term **time-line** refers to any connected, Hausdorff subspace of (Υ, τ). A maximal time-line will be called a **reality**.*

These definitions of a *time-line* and a *reality* will remain adequate in the context of the following more realistic model for time. Due to several severe limitations, modelling the entire universe (or any substantial part of it) goes far beyond the scope of this chapter, and so it is necessary to propose a more modest goal. Perhaps it is possible to construct a reasonable model for a

6 Stochastically branching spacetime topology

region of spacetime, which still contains only one elementary probabilistic event; but with as complicated an elementary event as we might wish to model.

The event in question has a set of possible distinguishable outcomes, and each of these has a specific probability of being observed, then and there, in the context of this model. If there are only a finite number of distinct outcomes and if they each have a *rational* probability, then the model of the time parameter can be constructed with only a finite number of branches. We limit our discussion to this special case, because it is the most easily understood, and it does not require an excursion through abstract measure theory. However, there is no technical problem in generalizing this construction to any number of outcomes with arbitrary probability *measures*.

Let N be the least common denominator for the finite number of rational fractions, which are the probabilities of the possible results of the event. Let Ω be the following subset of the plane:

$$\Omega = \{(0,1), (0,2), \ldots, (0,N)\} \subset \Re^2.$$

In our new model of time, the set $\Gamma \subset \Re^2$ can be described as the following subset of the Cartesian plane:

$$\Gamma = \{(x,0) \,|\, x < 0\} \cup \{(x,1) \,|\, x \geq 0\} \cup \{(x,2) \,|\, x \geq 0\} \cup \ldots \cup \{(x,N) \,|\, x \geq 0\}.$$

The set Γ will be given a topology, λ, which differs critically from the subspace topology (that may be obtained from the plane, $\Re^2$).

Definition 17 *We define an **open interval** to be one of the $2N + 1$ types of subsets of Γ listed below. (In this list, α and β are arbitrary real numbers, which are constants for any one specific open interval.)*

The first type is:

$$W_0(\alpha, \beta) = \{(x,0) \,|\, \alpha < x < \beta \leq 0\}$$

and there are two others:

$$W_{2k-1}(\alpha, \beta) = \{(x,k) \,|\, 0 \leq \alpha < x < \beta\}$$
$$W_{2k}(\alpha, \beta) = \{(x,0) \,|\, \alpha < x < 0\} \cup \{(x,k) \,|\, 0 \leq x < \beta\}$$

for each integer value of $k = 1, 2, \ldots, N$.

Definition 18 *The **open sets** in (Γ, λ), the elements of λ, are precisely those subsets of Γ which are unions of open intervals.*

Here again, we see that (Γ, λ) is a one-dimensional, *locally Euclidean* topological space, since each open interval in Γ is homeomorphic to the real line $\mathfrak{R}$, and each element of Γ is contained in some open interval of Γ. Thus, (Γ, λ) is said to be **locally homeomorphic** to the real line, $\mathfrak{R}$.

Also, it is clear that (Γ, λ) satisfies the T_1 separation condition, but not the Hausdorff (T_2) separation condition. The Hausdorff condition fails in (Γ, λ) only for pairs of distinct elements in Ω. Suppose that $1 \leq j < k \leq N$. To see that $(0, j) \in \Omega$ and $(0, k) \in \Omega$ do not have disjoint neighborhoods, recall that they have a *common past*. If L_j and L_k are neighborhoods of $(0, j)$ and $(0, k)$, respectively, then there exist real numbers, α and β, with $\alpha < 0 < \beta$, such that $W_{2j}(\alpha, \beta) \subset L_j$ and $W_{2k}(\alpha, \beta) \subset L_k$; of course, $W_{2j}(\alpha, \beta) \cap W_{2k}(\alpha, \beta) = W_0(\alpha, 0) \neq \phi$. Removing all but one of these N elements in the set Ω leaves a subspace of (Γ, λ) which is a Hausdorff space; specifically, such a subspace is homeomorphic to N disjoint copies of the real line, and every element has a *unique future*.

Notice that the N inseparable points, $(0, j) \in \Omega$ (for $j = 1, 2, \ldots, N$ are only inseparable in (Γ, λ). Among themselves, they are easily separated. The set of all the inseparable points, Ω, forms a subspace of (Γ, λ), and the subspace topology on Ω is *discrete*, which means every set is an open set. This is so, since (Γ, λ) is locally Euclidean.

Within this model of time, (Γ, λ), we have embedded N time-lines for this one event. They have their common past and distinct futures. Moreover, the number of these time-lines with a common outcome is proportional to the probability of that outcome.

We will require the *product space* construction, at least locally; thus, it will be necessary to discuss the relevant construction in topology.

Definition 19 *If (X, τ) and (Y, σ) are two topological spaces, then their* **product space***, denoted $(X \times Y, \tau \times \sigma)$, is a topological space, where the set,*

$$X \times Y = \{(x, y) \mid x \in X, y \in Y\}$$

is just the indicated set of all ordered pairs, and the topology on this set, denoted $\tau \times \sigma$, consists of all subsets of $X \times Y$ which are expressible as arbitrary unions of sets of the form:

$$U \times V = \{(x, y) \mid x \in U, y \in V\}$$

for all possible pairs (U, V) with $U \in \tau$ and $V \in \sigma$.

Now we can make use of our new model of time, (Γ, λ), combining it with some *manifold* model of physical space (perhaps three-dimensional, at the outset), by forming the product space of the two. Let (M, μ) be a *three-dimensional*, connected manifold representing physical space. The product space

$$(M \times \Gamma, \mu \times \lambda)$$

is a *four-dimensional, connected, locally Euclidean T_1-space*, but it is *not Hausdorff*. This is the model of spacetime that may be most appropriate for the *one-event universe* described above. However, we appear to be in a universe with a very high *density* of such probability events. Therefore, our spacetime model, $(M \times \Gamma, \mu \times \lambda)$, should really be viewed as a special kind of open neighborhood of the closed set:

$$M \times \Omega \subset M \times \Gamma \subset \langle universe \rangle$$

Such a constraint may be appropriate, since it is 'local' in some *weak* sense, which the reader is invited to explore. Of course, any phase space (or phase manifold) can be used to replace the physical space in similar constructions.

4 Conclusions: T_2 or not T_2? That is the question!

In contrast with the rest of this chapter, this section contains speculation on the nature of time (in the real world), a subject in which the most fundamental question seems to be experimentally undecidable.[8]

Consider our regional description of spacetime as represented by the model:

$$(M \times \Gamma, \mu \times \lambda)$$

Notice that *only* the *topology* of this model has been specified. Since we have taken the trouble to explicitly construct it, perhaps we should explain why, by listing some of the *attractive properties* of this model.

(1) This model attempts to adhere closely to the principle of *avoiding global topological constraints*, for which there can be no empirical data. As observed previously, the topology λ on the set Γ was designed with this in mind.

---

[8] Are we really discussing physics? My definition allows any hypothesis to be included within the purview of physics, provided one can imagine an experiment that could, in principle, be performed, and which, a priori, is capable of contradicting the hypothesis in question. By that standard, this work may fill a much needed gap in the literature of mathematical physics.

(2) This model branches in time, so that *no event requires* microscopic *choices* be made by elementary particles. After a probabilistic event, we might wish to ask, 'What happened?' Of course, the answer given by this branching-time model must be, 'Everything possible!' The model even counts how often each occurs. Moreover, the probability theory is indelibly inscribed in the topology λ on the set Γ.

(3) The local *direction* of the arrow of time is implicitly determined by the topology λ on the set Γ. This direction is easily detected by monotone convergent sequences (or more generally, *monotone convergent nets*). If they are monotone *decreasing*, then they converge to a *unique limit*. However, if $\{g_k \,|\, k = 1, 2, 3, \ldots\} \subset \Gamma$ is a convergent, monotone *increasing* sequence in (Γ, λ), then either it will converge to a unique limit, which is not in Ω, or we will find that, for all $\omega \in \Omega$:

$$\lim_{k \to \infty} g_k = \omega.$$

The *unattractive properties* and criticisms of this model also deserve examination. Thus far, these seem to fall into two broad classes of objections, denoted by (α) and (β), below. (I would be pleased to hear of any other, possibly new criticism of the model, which does not fall into either class.)

(α) If whole universes are being created, why does that not violate the principle of the conservation of energy? One could ask the same question regarding other conserved quantities, such as momentum, or angular momentum. Of course, there is no violation of a conservation law. The Noether Theorem (equating conservation with symmetry) requires *conservation* of these quantities *only along time-lines*. Thus, there is no problem with conserved quantities.

(β) With this model, the universe seems to go to a great deal of trouble, in a deterministic way, to create the illusion that the behavior of fundamental particles necessarily must be described statistically. Surely, there must be a less complicated way of *explaining* the peculiar phenomena confronting us in quantum mechanics.

The dissatisfaction expressed in (β) is a reasonable reaction, considering the unbelievably vast branching of time, as indicated by the model. It does seem a rather high price to pay for a return to determinism, and one might well wonder whether there is an easier way. Indeed, I find myself in near agreement with Bell's view that there is an easier way.

After I have rejected various models for their failure to provide (at least to me) a reasonable picture of what is really happening, then I am left with two models. One of these is the branching-time model that is developed here, while the other is the so-called 'Pilot Wave Theory.' This model was originally constructed by L. de Broglie.[9] Due to limitations of space and time, I will not try to explain the concept of a pilot wave, for fear of not doing justice to such a fine idea. I would urge more physicists to seriously consider it. Some explanation of the pilot wave concept and its meaning can be found in Bell.[5]

Finally, I should point out that this chapter differs fundamentally from the other contribution in this volume dealing with 'branched' models. In that contribution, Storrs McCall has constructed such a model, with a very different purpose. His model is intrinsically indeterministic; in contrast, the model described here is deterministic by motivation and design. This is certainly the major philosophical difference.

However, McCall's indeterministic model (as presented) has a serious problem, which must be addressed. The problem is quite explicitly presented in the first section of his chapter; more specifically, in the two paragraphs following the quotation of Paul Davies (on page 156). These paragraphs graphically indicate a dynamic of the spacetime model. Unfortunately, this 'losing of branches' requires us to postulate a second 'time' parameter, which can be used to parameterize the 'aging' of the universe-tree. Dismissing the problem, as McCall does, is simply not helpful or convincing. Of course, this problem does not arise for the deterministic model.

Another difficulty with McCall's model is the relationship between the process of 'losing branches' and consciousness. The Schrödinger equation seems to govern the evolution of quantum systems in an entirely deterministic way, when no observation is made (and the wave function is not made to 'collapse'). To dismiss the dynamic of 'losing branches' as some unspecified random process invites the criticism that the inevitable demand for an explanation has merely been postponed. The deterministic model attempts to simulate the physical universe. In contrast, the indeterministic model appears more suited to the psychological depiction of a single conscious awareness of the universe.

Appendix: The equivalence of locally T_1 and (globally) T_1

The observations following definition 8 and lemma 11, together with the following proposition, show that the three properties, *strongly locally T_1, locally T_1, and (globally) T_1* are equivalent. This fact justifies the intuition that the T_1 *property is intrinsically local.*

[9] de Broglie (1939).

Proposition 20 *If (X, τ) is a* **locally** *T_1-space, then it is* **(globally)** *T_1*.

Proof: Suppose (X, τ) is a locally T_1-space, and $y, z \in X$ are two arbitrary, but distinct elements of X. To show that (X, τ) is (globally) T_1, we must find an open set $U \in \tau$ such that $y \in U$ and $z \notin U$.

By hypothesis, there exists N_y, a neighborhood of y in the space X, such that (as a subspace of X), N_y is a T_1-space. Since N_y is a neighborhood of y, there exists an open set $W \in \tau$, so that $y \in W \subset N_y \subset X$.

It is easy to see that W receives the same topology, whether W is viewed as a subspace of N_y, or W is considered as a subspace of X. Since the T_1 property is hereditary, it follows that W is a T_1-space.

If $z \notin W$, then (letting $U = W$) we have accomplished our goal. Thus, we only need to consider the case in which $z \in W$.

Since W is a T_1-space and with $y, z \in W$, there exists an open set $U \in \tau_W$ (the subspace topology on W) such that $y \in U$ and $z \notin U$. However, since $U \in \tau_W$, it follows that there exists $V \in \tau$ so that $U = V \cap W$. Therefore, $U \in \tau$, by condition (3) in definition 1, since $W \in \tau$. QED.

Acknowledgements

This research is partly supported by NSERC of Canada.

PART 3 **Thermodynamics and time's arrow**

7 The elusive object of desire: in pursuit of the kinetic equations and the Second Law[1]

LAWRENCE SKLAR

1 Introduction

It is impossible to over-rate the important contribution which statistical mechanics makes to our contemporary understanding of the physical world. Cutting across the hierarchy of theories which describe the constitution of things and related in subtle and not yet fully understood ways to the fundamental dynamical theories, it provides the essential framework for describing the dynamical evolution of systems where large domains of initial conditions lead to a wide variety of possible outcomes distributed in a regular and predictable way. For the special case of the description of systems in equilibrium, the theory provides a systematic formalism which can be applied in any appropriate situation to derive the macroscopic equation of state. Here the usual Gibbsian ensembles, especially the microcanonical and canonical, function as a general schematism into which each particular case can be fit. In the more general case of non-equilibrium, the situation is less clear. While proposals have been made in the direction of a general form or schematism for the construction of non-equilibrium ensembles, most of the work which has been done relies upon specific tactical methods, of validity only in a limited area. BBGKY hierarchies of conditional probability functions work well for the molecular gas case. Master equation approaches work well for those cases where the system can be viewed as a large collection of nearly energetically independent subsystems coupled weakly to one another. But the theory still lacks much guidance in telling us what the macroscopic constraints ought to be which determine the appropriate phase space over

---

[1] This paper appeared originally in *PSA 1986: Proceedings of the 1986 Biennial Meeting of the Philosophy of Science Association*, volume 2, eds. A. Fine & P. Machamer, pp. 209–225. East Lansing, Michigan: Philosophy of Science Association. It is reprinted here, slightly revised, with the generous permission of Prof. Sklar and the Philosophy of Science Association.

which a probability distribution is to be assigned in order to specify an initial ensemble.[2]

Not surprisingly, most of the effort expended in statistical mechanics is devoted to applying its general principles to specific cases. Here one wishes to derive equations of state for equilibrium systems, dynamical equations of evolution for non-equilibrium systems, numerical values of such coefficients as the transfer coefficients which appear in the evolutionary equations, etc. For all but the very simplest systems the problem is one of great difficulty, with even moderately dense gases, plasmas, etc. requiring enormous ingenuity and a vast grab-bag of analytical tricks, prudent guesses, and so on.

From the philosopher's perspective, however, it is the alleged ability of the theory to provide insight into one of the two fundamental thermodynamical laws, the laws independent of the particular constitution of the system in question, which is the focus of attention. The First Law of Thermodynamics, being the energy conservation principle, presumably has its roots outside the domain explored by statistical mechanics. But the Second Law, the general principle of entropic non-decrease, of the thermal asymmetry of phenomena in time, has been, since the first developments of the field in the middle and late nineteenth century, the fundamental principle the explanation of which is the ultimate aim of the theory. Why is the world asymmetrical in time? Why does heat flow from hot to cold and not vice versa? Why do gases expand to fill volumes uniformly and never, statistical fluctuations to the side, show the reverse behaviour? In particular, why does the world show this irreversible behaviour when the underlying dynamical laws seem, except for subtle effects, perhaps, in weak interactions, effects which are demonstrably irrelevant in the cases in question, to be completely symmetrical in time in their description of the connection of states at one time to states at another?

The purport of this brief review is to argue that although enormous advances have been made in understanding the explanatory ground on which the familiar statistical posits of statistical mechanics rest, the final understanding of the basic principle of temporal asymmetry still eludes us. Even making the appropriate allowances for the weakened, statistical, sense in which we wish to hold the Second Law true, no amount of reliance upon the subtleties of dynamics or of the constitution of systems allows us

[2] For an outline of the schemes used in non-equilibrium statistical mechanics see chapter 6 of Sklar (1993).

7 The elusive object of desire

to extract coherently from the theory an explanatory understanding of the temporal asymmetry of the world. This remains true, I will claim, even if the famous cosmological asymmetry of the world with singular big-bang in the 'near' past is included in our explanatory resources. We can extract temporal asymmetry from the theory if we make certain basic assumptions either about the physical world or about our way of dealing with it, but, I will suggest, these additional factors will hardly resolve our sense of mystery about the temporal asymmetry of the world. If anything they make matters appear even more mysterious.

2 History of the problem

The first attempt at a derivation of time asymmetry from the kinetic theory of motion of the microscopic components of a system was, of course, Boltzmann's famous H-theorem. A kinetic equation alleged to describe the evolution in time of the function giving the distribution of the micro-components of a gas in space and momentum space is formulated. First it is shown that the equilibrium distribution of Maxwell and Boltzmann is a stationary solution of that equation. Next it is shown that no other distribution can be stationary. This is done by defining a kinetic surrogate for entropy and showing this must monotonically change toward its value at the equilibrium distribution so long as the system has not yet obtained equilibrium.

But this derivation of time asymmetry was immediately subjected to the famous refutations by recurrence and by reversibility. Given the nature of the underlying dynamical laws, it was easy to show that over time a properly bounded system would return arbitrarily closely to its initial state and that for each evolution from microstate to microstate corresponding to approach to equilibrium there was a reverse motion possible away from equilibrium.

The response was the move to the statistical version of the H-theorem, murkily presented by Boltzmann in a series of short pieces and presented in a superbly clear version by the Ehrenfests at the endpoint of a critical development at the hands of Burbery, Jeans, and others.[3] Here it is the evolution of a collection of systems which is the object of investigation. Any allegation of time asymmetric behaviour is the attribution of this asymmetry to the ensemble, not to every individual system. Over infinitely extended time each individual system is supposed to display the time symmetric behaviour of being almost

[3] Ehrenfest & Ehrenfest (1959).

always near equilibrium with occasional excursions from it. The curve of Boltzmann's H-theorem with its monotonic behaviour in time is now taken to be the so-called 'concentration curve' of a collection of systems. Here one looks at a collection of systems started in any one of the microstates compatible with the initial macroscopic constraints, and then one looks at the entropy assigned to each member of the collection at a specific later time, taking the entropy for the concentration curve as that dominating the collection at the time. There is nothing incompatible between the monotonic behaviour of this curve and time symmetric behaviour of each individual system.

But problems remain. The first is to go beyond an argument which establishes the consistency of the appropriate ensemble behaviour with recurrence and reversibility to offer some ground for thinking that the behaviour of the concentration curve not only *can* be as described but *will* be so. Here, as the Ehrenfests clearly explained, the results obtained relied upon an assumption that Boltzmann's originally posited molecular chaos, the total uncorrelation of particles prior to collision, would hold at each instant of evolution. But the specification of the initial ensemble of microstates fully determines all the future ensembles, and one has no right to arbitrarily impose such continual rerandomization as a posit. How can we derive from the underlying dynamics and, perhaps, from the specification of the ensemble at the initial time, the appropriate ongoing randomization necessary to get the concentration curve to evolve in the right way?

The second problem is to give an acceptable account of time asymmetry of physical systems in this new ensemble approach. This is a problem distinct from the one just noted above, and, as we shall see, one that remains in some obscurity. Basically the problem is this: the rerandomization which generates the monotonic behaviour of the concentration curve, if imposed by fiat, is matched by a temporally reversed rerandomization posit which generates exactly the opposite monotonic behaviour in time. What explains why we are to adopt one posit rather than the other? In the case where the rerandomization is generated out of the dynamics alone, or out of the dynamics plus the initial condition of the ensemble, we still meet perplexity. The kind of rerandomization we can generate from the dynamics alone is totally time symmetric. It generates a time reverse parallel for each consequence it generates about the ensemble in one direction of time. If we rely on the position of special initial conditions for the ensemble to break the symmetry, the question remains regarding the physical origin of these special initial conditions, for, once again, symmetrical conditions giving time reversed behaviour for the ensemble are easily generated.

Let me just very briefly sketch a couple of approaches to the first problem: generating the rerandomization out of dynamics instead of positing it in a way which seems illegitimate given the deterministic relation among states at different times. Then I will go on to discuss why that does not solve the time asymmetry problem and to explore some proposals to fill in that remaining lacuna in the theory.

3 Two formal approaches to irreversibility

Perhaps the closest thing we have to a successful 'derivation' of a kinetic equation in its ensemble interpretation is Lanford's derivation of the Boltzmann equation. Here appropriate assumptions are made as to the constitution of the system, a sufficiently rarified gas. As is frequently the case, the derivation is actually carried out not for a finite system but in an appropriate limit of systems with an infinite number of micro-constituents. Here the limit used is the Boltzmann–Grad limit, in which the number of particles, n, goes to infinity; the diameter of the particles, d, goes to 0; and nd^2 approaches a finite, non-zero limit. Next a quite strong non-correlation posit is made on the initial ensemble at time zero. It can then be shown that in an appropriate ensemble sense the evolution of the ensemble for some finite time into the future obeys the Boltzmann equation. Alas, the time is very short for which this can be shown, being of the order of one-fifth the mean free time between collisions of a particle. As Lanford says, the derivation 'does not justify its [the Boltzmann equation's] physically interesting applications.'[4]

The other 'derivation' of kinetic behaviour is more general than Lanford's derivation of the Boltzmann equation. But its generality is purchased at a price. Here we are dealing with a whole corpus of important mathematical results deriving originally from the attempt at putting ergodic theory on a firm mathematical basis.

In the 1930s Birkhoff and von Neumann provided the sufficient conditions for a system to be such that, except for a set of measure zero trajectories, the time average of any phase function along a trajectory would equal its ensemble or phase average. This result replaced the early attempt by Maxwell and Boltzmann to obtain the wanted result by the demonstrably false posit that

[4] Lanford (1976), 3, and Lebowitz (1983).

each trajectory would go through every point in the allowed phase space. In the 1960s Sinai, culminating a long research effort, showed that for interesting models (including hard spheres in a box) the appropriate conditions did in fact hold. The resulting demonstration of 'ergodicity' for a system is of some importance in rationalizing the standard ensemble method for equilibrium theory which identifies equilibrium thermal properties with averages over an ensemble of certain phase functions, where the averages are calculated using a standard probability measure.

But ergodicity is not sufficient to play a rationalizing role in dealing with the dynamic evolution of an ensemble which is supposed to represent in our theory the approach to equilibrium of a non-equilibrium system. For non-equilibrium theory we would like some reason to believe that an ensemble initially so constructed as to represent a system not in equilibrium will, over time, evolve to something like an equilibrium ensemble. It is possible to have ergodicity without anything like this happening. A system of phase points initially confined to a proper subset of the phase space allowed by the constraints, a subset which is intuitively not 'spread out' with any kind of uniformity over the allowed phase space, can evolve in such a way as to retain its 'coherence' while at the same time 'sweeping' over the allowed phase space in such a way as to have phase averages and infinite time averages be equal for almost all trajectories. Now it is easy to show that the initial set of phase points cannot evolve into the equilibrium ensemble for the newly allowed constraints in an exact sense. This follows trivially from the fundamental theorem of phase evolution, Liouville's theorem. But there are interesting senses in which results stronger than ergodicity can be obtained which show an ensemble behaviour plausibly related to the desired statistical surrogate for evolutionary approach to equilibrium.

The simplest (but not the weakest) such property of an ensemble is for it to be mixing. Pick any measurable region of the phase space at an initial time. Essentially the system is mixing if in the infinite time limit the overlap of the region into which the phase points of the original region evolves with any fixed measurable region of the space is proportional to the product of the size of the two regions. In an obvious and clear sense this condition guarantees that the initial region of phase points evolves until it is, in a measure theoretic sense, uniformly distributed over the entire allowed phase space.

Stronger results than mixing can be obtained as well. The mixing property is just one feature that follows from the radical instability of trajectories in the ensemble. This leads to a wild divergence of trajectories initially close to one another and a kind of 'randomness' in where one will be in the phase space at a later time relative to one's initial location. A condition strictly stronger than being mixing is for an ensemble to constitute a K-system. A coarse-graining is a partitioning of the phase space into a finite number of exclusive and exhaustive subspaces. One can think of a macroscopic measurement as locating the phase of a system as being in one of the boxes of a partition. A system is a K-system if the infinite list of boxes occupied by a trajectory in the past is not enough to determine (with probability one) which box it will next occupy (where determinations of a box occupied are made at discrete intervals). Crudely a K-system is one whose trajectories are sufficiently random that no macroscopic system of measurements on them is deterministic in the statistical sense.[5]

Even more random is a Bernoulli system. Here a system of phase points can be shown to have an evolution which, in a definite sense, duplicates that of a purely random process, despite the underlying micro-determinism of the trajectories following from an initial phase point. A partition of the phase space is 'generating' if the doubly infinite sequence of boxes occupied by a point (whose location in a box is determined at discrete time intervals) is (except for a set of measure zero of trajectories) enough to fully determine the trajectory. In the statistical sense the doubly infinite sequence of locations in boxes of a generating partition has the same informational content as fully specifying the microstate of the system. If the system is Bernoulli, however, there will be such a generating partition which is yet such that the infinite past history of box location of a phase point will have no informational value whatever in predicting its next box location. In other words the probability of being in a box at a time is the same absolutely (i.e., conditional on no information) as it is conditional on any information whatever about past box locations of the system.

One can indeed prove that interesting models (hard spheres in a box, for example) have the properties of being mixing, being K-systems, and being Bernoulli systems, as well as the property of ergodicity. Once again one must not over-rate what can be shown. To begin with, while idealizations like that of hard spheres in a box can be shown to meet the requirements necessary for mixing, etc., more realistic systems can be shown *not* to meet those requirements. An important stability theorem (the KAM theorem) at least indicates

[5] Arnold & Avez (1968), chapters 2 and 3.

that the case, for example, of molecules interacting by a smooth inter-molecular potential will give rise to systems having regions of stability in the phase space, regions whose trajectories will not show the proper chaotic and random motion of the whole available phase space necessary for the theorems above. One might hope that outside these regions mixing, etc., would hold, but actually little is known about how to define or prove the existence of appropriately randomizing qualities for these regions outside the stable regions. Nor is there much hope that one can show that the fully-fledged mixing properties reappear when one goes to the limit of infinite systems, although there certainly is reason to believe that in that 'thermodynamic limit' the stable regions shrink to insignificance in size.

Even when these theorems do apply, one must still be cautious in understanding what has been shown. They are far from rationalizing the application of the standard kinetic equations in their 'physical application'. Some of them give results only in the infinite time limit. Others tell us that for finite time intervals some partitions will display randomness when measurement consists in locating the system in a box of the partition, but these pure existence results do not give us a definitive idea of how precise these measurements can be and still retain their Bernoullian independence. More generally we can reason as follows: how rapidly a given ensemble will approach the equilibrium ensemble, in some appropriate sense of 'approach', to any degree of accuracy, will surely depend upon the structure of the initial ensemble. Initial ensembles which are large pieces of the available phase space with even distributions of points over them and smooth contours will get close to equilibrium faster than small ensembles or those with quirky probability concentrations in them or those cleverly contrived to evade, at least initially, the flow toward equilibrium. But the results noted above are all derivative strictly from the constitution of the system and the underlying dynamical laws and ignore the variability imposed by choice of initial ensemble.[6]

Crudely summarizing, the situation is something like this: for a limited number of cases we can, by imposing a posit of repeated rerandomization, generate for an ensemble of systems a kinetic equation. Again in a limited number of cases we can define a statistical surrogate of entropy for the non-equilibrium situation and demonstrate a monotonic approach to equilibrium for the solutions to the equation. We can even, sometimes, by making additional posits

[6] Ornstein (1975).

and assumptions generate actual classes of solutions of the equations, calculate relaxation times and transfer coefficients, etc.

In our attempts at deriving these results without the necessity of an independent posit of continual rerandomization, a posit conceivably inconsistent with the equations governing the dynamical evolution of the ensemble generated by the determined dynamical evolution of its points, we can make dramatic and interesting claims; although dramatic as they are, the results go only a short distance to our goal of a real rationalization of the theory as we actually need to apply it.

In one approach we can, by positing a sufficiently strong constraint upon the initial ensemble, derive, for an idealized limit system, an interval, however brief, in which the assumed kinetic equation holds for the ensemble evolution. In the other approach we can, from the system constitution and the laws governing the underlying micro-dynamic evolution, derive, again for idealized and, alas, not quite realistic, systems, evolutions of the ensemble toward the equilibrium ensemble, in appropriate senses of 'toward', in the infinite time limit, and also derive various results regarding the 'statistical indeterminism' or even 'complete statistical randomness' of macroscopic measurements carried out on the system.

None of these results comes close, however, to allowing us to 'derive' in even weak senses, the full kinetic evolution for realistic systems that we assume does, in fact, describe the real world and that we can, at least in simple cases, extract by imposing additional rerandomization posits onto our theory. We should also note here that careful inspection shows curious conflicts implicitly hidden in these rationales between the rationales of one kind or another.[7]

4 The 'parity of reasoning' problem

It should be clear that the results we have just been looking at, while serving to some degree to rationalize and explain the success of statistical mechanics in its description of the approach to equilibrium, do not, by themselves, resolve the fundamental problem of the origin of entropic asymmetry in the world.

This is clearest in the case of those theorems which generalize upon the results of ergodicity. Every result about the approach of an ensemble to the equilibrium ensemble as time goes to infinity is matched by a result telling us that projecting the ensemble backward by means of the dynamical laws

[7] See Earman (1987).

also leads to convergence to the equilibrium ensemble as time goes to negative infinity. And every result about macroscopic indeterminism or probabilistic independence in the forward time direction is matched by a parallel result in the negative time direction. Given that the results are derived solely from the constitution of the system and the underlying dynamical laws, both of which are fully time symmetric, how could we expect otherwise?

Even the Lanford result has its time reversed analogue. The constraint upon the initial ensemble imposed in the theorem is itself time symmetrical, and a result about 'anti-Boltzmannian' behaviour backward in time from that state is derivable from the constraint. Other versions of the theorem are time asymmetrical, but they result from imposing a constraint on the ensemble which is itself not time symmetrical and whose rationale itself therefore becomes problematic in that there exist time reversed analogues to the initial states generating thermodynamic behaviour which would generate anti-thermodynamic behaviour. (One might think that Lanford's result is paradoxical, since the time reverse of an ensemble slightly later than the initial ensemble would seem to have to evolve backward in time both to the initial ensemble, by reversibility of dynamics, and to an 'anti-Boltzmannian' alternative to Lanford's theorem. The paradox is resolved, as Lanford pointed out, by the fact that his time symmetric but very stringent no-correlation constraint imposed on the initial ensemble does not itself hold at any instant after the initial time.)

These facts leave us, then, with the fundamental perplexity concerning irreversibility in statistical mechanics. Given a system prepared in accordance with certain macroscopic constraints, statistical mechanics licenses us to make predictions about the future behaviour of the system. Naturally these predictions are statistical in nature. The detailed prediction can be generated in certain cases by positing a suitable rerandomization postulate and using it to derive a kinetic equation whose solution describes the evolution of an initial ensemble into the future, which fixes the statistical prediction for future times. In some cases we can offer rationalizations of the needed rerandomization postulate of the two kinds noted above. All well and good.

But why can we not use the theory to do exactly the same thing in the direction of time into the past from our particular instantaneous macroscopically constrained state? A symmetrical rerandomization postulate is easily constructed. And given the time symmetry of the rationalizations we

7 The elusive object of desire

have noted above, surely such a positing of kinetic behaviour into the past is just as reasonable as positing it into the future? Of course it is *not* reasonable. Given a gas non-uniformly distributed in a box and isolated, we ought to infer, with high probability, that it will be more uniform in the future. But we ought not to infer, with high probability, that it evolved from a more uniform state in the past. On the contrary we infer that it evolved from some even less uniform state, although which one we do not know since the convergence of states is into the future and many far from equilibrium states will approach the same equilibrium condition.

What we want to know is what reason we can offer why the world is such that parity of reasoning does not legitimate the inferences into the past symmetrical with inferences into the future. We can, of course, impose a time asymmetric restriction on the way in which we derive ensembles so as to guarantee time asymmetrical predictions. For example, O. Penrose suggests the 'principle of causality', that 'phase space density at any time is completely determined by what happened to the system before that time and is unaffected by what will happen in the future.'[8] But what we want to know is what it is about the world which explains why such an asymmetric theory works so well and why its symmetric or asymmetric but anti-parallel alternatives do not work well at all.

I think we can, to begin with, dismiss one 'way out' as not doing the job at all. It might be thought that taking the line that the future is simply the direction of entropic increase, in the sense in which down is simply the direction of the gradient of the local gravitational field, might resolve our difficulty. It will not, even if that claim about the 'reducibility' of time asymmetry to entropic asymmetry is in any sense correct. It would help us to understand why the time-asymmetric theory works and its asymmetric anti-parallel alternative does not, but it will not do the much more important and much more difficult job of explaining why we ought not to treat the past and future symmetrically in our statistical theory.

Once we have understood why in a vast collection of systems, when we explore their entropic history from one time to another, we find overwhelmingly that they have their entropic low points in the same direction from their entropic high points, we could at that stage try to make plausible the claim that what we call the future direction of time is simply the direction of time toward which entropy has increased in the overwhelming

[8] Penrose, O. (1979), 1941.

majority of systems. But the real puzzle is why there is such a temporal parallelism at all in the evolution of systems. Given the underlying time reversal invariance of the laws, why do we not find just as many systems with their entropic orientation in time in one direction as in the other? It is the *parallelism* that we need to account for, and the use of the thesis of the reducibility of future to direction of entropic increase can only help once we have already understood why there is a direction in which entropic increase is overwhelmingly likely.

Nor can we hope for a resolution to our problem by deriving a 'Second Law' in the manner suggested by Jaynes in his subjective probability approach to statistical mechanics. Taking thermodynamic values to be given correctly by the usual ensemble averages of phase functions, using the equality of thermodynamic entropy and information theoretic entropy at the time of system preparation, using the constancy of information theoretic entropy over time as demonstrable from Liouville's theorem, and using the result that thermodynamic entropy is in all cases greater than or equal to information theoretic entropy, he is able to show that the thermodynamic entropy at a later time is greater than or equal to its initial value. But the proof suggests the paradoxical result that thermodynamic entropy at earlier times would also be greater than or equal to the thermodynamic entropy at the designated time.

The resolution of the paradox rests in the assumption made that measured values will be identical to ensemble values calculated by applying the dynamical laws to the ensemble at the designated time. It is claimed that this can be true only if the observed change of measured values from time to time is 'experimentally reproducible'. But while macroscopic states do evolve in a predictable and at least quasi-lawlike way into the future, many different macroscopic states can *precede* a given macroscopic state. So the asymmetry of the result is broken by the fact that macroscopic behaviour is experimentally reproducible into the future but not into the past.

But from the point of view of trying to explain the parallelism of the entropic behaviour of systems this is clearly of no help. For what we want to know is *why* experimental reproducibility has the time asymmetric nature which it does. What Jaynes[9] provides is a familiar selection rule for the applicability of statistical mechanical reasoning, much like that of Penrose's causality

---

[9] Jaynes (1965).

principle, but not an explanation of why the world is such that our statistical mechanical principles of reasoning need be applied in the temporally asymmetric way.

5 Interventionism

A more promising approach to the origin of irreversibility might rely on the familiar allegation that treating systems as genuinely isolated is a mistake. From the early days of kinetic theory it has been observed that the combination of the instability of trajectories of the system in question with the irreducible causal interaction of the system with the external world, by gravity (which cannot be shielded at all) if by no other means, can lead to a breaking of the strictly lawlike connection of internally described microstates of the system at different times. This is sometimes alleged to be the very ground of the use of statistical methods. In other cases it has been suggested that the true origin of entropic increase into the future can only be found in the intervention of outside disturbing forces acting on the so-called 'isolated' system.[10]

The core of the interventionist approach is clear. Imagine a system prepared in such a way that a macroscopically controllable non-equilibrium feature is induced on it, say by confining a gas to the left hand side of a box and then removing the partition. As time goes on, the information that the molecules were all originally on the left hand side of the box has been distributed into subtle many-particle correlations among the states of the molecules at later times. Given the sensitivity of trajectories to initial conditions, even a slight 'shaking up' of the positions and velocities of the molecules at a later time by means of their even slight interaction with the outside world will destroy the information that the molecules were originally confined to the left hand side of the box. Allow the interactions from the outside to be completely random, in an appropriate sense, and we can construct a scenario in which nonequilibrium ensembles are driven to equilibrium ensembles by the interference from outside the component system.

But I can see two problems with this approach to rationalizing the time asymmetry of statistical mechanics. The first problem is an objection to the fundamental role played by intervention from the outside in 'rerandomizing' the evolution of the ensemble. Systems of nuclear spins in crystals can be constructed which display a fascinating combination of two properties:

[10] Blatt (1959).

(1) they show a kind of evolution toward equilibrium; but (2) they can be demonstrated not to have dissipated their original information about their non-equilibrium situation into the outside world by means of interaction with that world.

The nuclei are prepared with spins parallel. As time goes on, the system of nuclei evolve to systems with the spin axes 'randomly' distributed. But a clever trick which serves as a 'Loschmidt demon' and reverses the precessional motion of the spins (sort of) leads them to evolve back in time to their original state of parallel spin axes. This can be done even if the nuclei interact with one another by means of spin–spin coupling. The reproducibility of the initial state of the system indicates plainly that the original information has not been lost by dissipation into the outer world. Yet the initial evolution of the system of spins shows an at least 'apparent' continual randomization which makes the evolution suitable for description the usual statistical mechanical way.[11]

Even where it is true that the system is continually rerandomized, it is not clear that reference to this irreducible interaction with the outside world can really handle the problem of explaining the origin of temporal asymmetry. The problem is that standard one: how can you avoid the argument which, by parity of reasoning, allows you to infer backward in time by means of statistical mechanics the same way you use it to infer forward in time? Given the time symmetry of the dynamical laws and the time symmetric nature of the 'interfering' boundary condition usually imposed, i.e., random interference, one ought to be able to infer that an ensemble of systems found at a time in a non-equilibrium condition is such that its distribution at an earlier time is closer to equilibrium. But, of course, such an inference would be utterly wrong.

Perhaps one could make 'interference' play a role in breaking time symmetry by maintaining that interference is causation and that causation is itself asymmetrical, acting from past to future. But until someone can explain what this means and how such a primitive causal asymmetry is to do the job and how to reconcile it with the lawlike symmetry governing the systems in the ensemble, it seems as if this 'way out' is more a hope and, perhaps, a begging of the question, than a genuine explanatory account. In any case I will spend the rest of this paper focussing on some approaches to asymmetry which allow for the genuine isolation of the systems in the ensemble. This is

[11] Hahn (1953); Rhim, Pines, & Waugh (1971).

the more common, if not universally accepted, approach to the foundations of the theory, and, as we shall see, many have thought that they could generate time asymmetry without relying on continual re-intervention into the evolution of the systems in question.

6 Cosmology and branch systems

There are a number of proposals to found the asymmetry in time of physical systems on a combination of the observation of certain cosmological facts and the observation that the systems with which we deal in statistical mechanics are temporally transient, being separated off from the 'main system' and in energetic isolation only for short intervals of time. The tradition starts with Reichenbach, but it is supported in various ways by Grünbaum, Davies, and others.

The cosmological side of this approach would have us reflect upon the fact that the universe, or at least that vast portion of it available to us observationally, is (1) very far from equilibrium and (2) increasing its entropy in the future time direction. While technical matters of defining entropy for infinite systems and for spacetime itself (as opposed to the matter in it) remain, these assertions are certainly true.

Why this is so is more problematic. Reference merely to the existence of the big-bang singularity in the 'near' temporal past and to the expansion of the universe will not by themselves do the job. Tolman showed models of universes in which the entropy stayed constant despite expansion at a finite rate. Models of the universe in which a maximum point of expansion is reached and is then followed by a contraction, but in which entropy continues to increase in both the expansion and contraction phase, are common. Time reversal invariance of the underlying laws tells us that if this picture represents a physical possibility, so does its time reverse in which both expansion and contraction are accompanied by decreasing entropy.

While matters are by no means universally agreed upon, the most plausible view at the present time seems to be that in order to get a reasonable picture of the entropic increase accompanying expansion of our current phase of (at least the 'local') universe, we must impose a low entropy initial condition on the big-bang singularity. Now in most cosmological models we take matter as being in a high entropy, thermalized equilibrium state in the early stages of the universe. So where does the low entropy reside? There must be some if we are to have overall entropic increase despite the decreasing entropy of matter represented by the segregation of initially uniform matter into hot stars and cold space. It is the spacetime itself which is the reservoir of low entropy in this picture. It starts uniform (which for spacetime is low entropy), and the increasing variation in its curvature as matter clumps steadily increases its entropy and the entropy of the universe as well. At least one way of putting this is to demand that at the big-bang the Weyl curvature tensor (that part of the curvature tensor you get by subtracting from the Riemann curvature tensor the Ricci tensor, which is, in General Relativity, the part of curvature correlated with non-gravitational stress-energy) is zero. As time goes on, the entropy of matter decreases, paid by the randomizing of spacetime curvature. Eventually the matter begins to return to a high entropy, uniform, thermalized state.

Of course this picture gives rise to the perplexing question: why is the originating entropy of the universe so low? Here a common attitude is to imagine the collection of all possible initial states of the universe and then to point out that in fact the low entropy initial state is grotesquely 'improbable'. Then why did it occur? R. Penrose suggests a new law of nature restricting white holes to having a zero Weyl curvature. P. C. Davies suggests that the problem can be explained away by pointing out that in a cyclic universe going from singularity to singularity it is highly improbable that the entropy will be the same at both ends of the cycle, and he then uses the old 'the future *is* the direction of entropic increase' line to eliminate the need for an explanation as to why the low entropy singularity is earlier than the high entropy one. As Hugh Mellor has noted, the whole question is methodologically peculiar. What grounds do we really have, after all, for assigning low probability, or any probability for that matter, to initial states of the universe? From this point of view a modern Hume would deny the need for Penrose's restrictive law just as Hume methodologically critiqued the teleological argument with its posited need for a divine watchmaker to account for the otherwise 'improbable' order of the lawlike universe.[12]

But let us agree: the universe is in a very low entropy state and is showing entropic increase in the future time direction. How does that help us account for the parallel increase of entropy in the future time direction of small, energetically isolated systems?

[12] Penrose, R. (1979).

7 The elusive object of desire

It is at this point that the notion of a 'branch system' is invoked to complete the story. The systems with which we deal in statistical mechanics are, it is said, not eternal. Rather they are at a certain time constituted by a piece of the energetically inter-connected main system being isolated from its environment. Now suppose that the system is isolated in a state of non-equilibrium, easily done since the main system itself is so far from equilibrium that we can use its low entropy to allow us to prepare such non-equilibrium energetically isolated branch systems. Make a basic statistical assumption: the distribution of microstates which are the initial conditions of such newly isolated non-equilibrium systems will be according to the usual standard measures over the allowed set of phase space points. Then the usual mixing results, etc., will tell us that the evolution of the ensemble representing such systems will be toward equilibrium, justifying our time asymmetric Second Law. We need not be concerned with the fact that the negative time versions of the mixing results would also lead us to infer anti-thermodynamic evolution of the ensemble in the past time direction as well, since the systems just do not exist prior to the time of preparation, having been brought into existence by their isolation from the main system at this initial time.[13]

How well does this approach answer the riddle of the origin of time asymmetry? Not very well at all, I think. First note the essential use of a basic postulate about the initial condition for the ensemble. To be sure, it is a time symmetric postulate in nature, and so its introduction cannot be taken to beg the question of the origin of the time asymmetry. But it is still the introduction into the theory of a statistical postulate as primitive and underivable from either the micro-dynamics or the constitution of the system and disturbing for that reason.

But there is a much greater problem with this approach. Systems are not only born by being isolated, they die by ceasing to be isolated. Consider an ensemble of systems at the moment of their reabsorption into the main system, all of which are in a macroscopically non-equilibrium condition. If we once again assume this basic time symmetric postulate about the distribution of microstates among these systems, we ought then to infer that the ensemble of systems was closer to equilibrium in its recent past, which is, of course, exactly the wrong inference to make.

I can think of only one way to defeat this familiar parity of reasoning objection. It is to argue that the statistical assumption is not justified in this

[13] Reichenbach (1956), chapter 3; Grünbaum (1973), chapter 8; Davies (1974), chapter 3.

latter case because we did not 'prepare' the ensemble in its terminal state, as we did the initial ensemble. But what does this mean? 'Preparation' is used frequently in attempts to rationalize statistical mechanics, but it is a slippery and difficult notion. If it means just selecting for our consideration a collection of systems, it is hard to see how one can justify the claim that initial ensembles are prepared but final ones are not. If it is, instead, a reference to the physical process by which the systems are brought to their states, the initial ones by being 'prepared' and the final by 'evolving from prepared states', it is hard to see how the time asymmetry here is not being smuggled into the rationalization by a reliance on some notion of time asymmetric causation which allows the derivation of the idea that in the evolution of systems from one time to another the 'preparation' of the system is always in the past time direction from its 'measurement'. Perhaps sense can be made of this, but then this seems to be the crucial way in which time asymmetry gets into the picture, rather than by the invocation of the notion of branch systems.

But have I left a crucial element of the branch system approach out of my account? Is it the point of branch-system theorists that the branch systems evolve in their entropy not only parallel to one another in time, but parallel in the time direction of the entropic increase of the main system? Does the direction of entropic increase of the cosmos play the crucial role in introducing time asymmetry into their picture?

Well, they sometimes talk as if it is supposed to serve that function, but how it is supposed to do so is totally opaque to me. I see no way in which the direction of entropic increase of the cosmos (or the local cosmos, if you prefer that version of the account) functions in any of the branch-system arguments to derive the parallelism of entropic increase of the branch systems. Nor do I see any way in which it could so be invoked. All that is ever used, or usable, is the availability of low entropy from which low entropy initial state branch systems can be cut off. The time asymmetry is built in by means of the assumption that the initial ensemble is to be constructed by use of the usual statistical assumption and final ensembles are not, that preparation is to be distinguished from mere observation, and that preparations are always to the same side of measurements in time order. None of this is incompatible with a world in which all branch systems evolved in their entropic direction in the opposite time direction to that of the main system – assuming, of course, that at any time only a small fragment of the universe consisted in these energetically isolated branch systems to avoid the paradox that the parts might 'add up' to the whole.

7 Krylov's solution

A number of investigators have attempted to resolve the foundational issues in statistical mechanics by relying upon analogies with the solutions in quantum mechanics. Can a treatment of such things as the preparation of a system and its measurement, or of the very notion of the state of a system, borrow insights from the notions of preparation, measurement, and state in quantum mechanics to help us make progress on the crucial issues?

One positive answer comes from the brilliant work of N. Krylov, who, alas, died at a very early age and left only a small fragment of his intended program complete.[14] Krylov is one of the pioneers in the development of such concepts as mixing, and some of his work is designed to show us how mixing differs from the weaker concept of ergodicity and why mixing is essential in a foundational understanding of the approach to equilibrium. More important for us here is Krylov's correct insistence on the importance of the structure of the initial ensemble. Even if a system is mixing or possesses rerandomizing features of a more powerful sort, we cannot obtain the results we want without specifying the specific kinds of regions of phase space in which we are to take the microstates to be confined and the probability distribution over those microstates which is to constitute our ensemble and which, from the statistical point of view, characterizes the system at the initial time from which evolution is to be studied.

Initial regions of phase space must not be allowed to be of arbitrarily complex shape; for if we allow this, we can easily construct ensembles which 'go the wrong way' in time. Simply take the time reverse of the later 'fibrillated' stage of a legitimate initial ensemble which has coarsely spread out over the newly allowed phase space. One will then get an initial ensemble which behaves (at least early in its career) anti-thermodynamically, as is shown by a reversibility argument at the ensemble level.

Within the allowed region of phase space the probability distribution must be uniform and not allowed to concentrate probability in irregularly shaped subregions of the phase space. Were we to allow such perverse probability

[14] Krylov (1979).

distributions, we could, again, easily construct ensembles which were initially anti-thermodynamic. Further, the regularly shaped initial regions must not be too small, for a small enough region, even regularly shaped and with its probability distribution uniform, would 'spread' too slowly over the modified allowed region of phase space to represent the actual fast relaxation of the system in question from non-equilibrium to near equilibrium.

Krylov offers brilliant critical examinations of alternative ways of guaranteeing that these constraints will be met. Especially fascinating is his attack on '*de factoism*', the claim that it is the 'real distribution' of happenstance initial states in the world which is responsible for the nature of initial ensembles. He objects to any account which would deprive the Second Law of its genuine (statistical) lawlike nature and has other detailed objections to this account as well.

His own approach relies upon an analogy with state preparation in quantum mechanics and, in particular, with the espousal of a kind of 'uncertainty relation' which is to be understood in the old Heisenbergian manner. That is, preparation is an actual physical operation on a system, and the uncertainty (in statistical mechanics the regularity, sufficient size, and uniform probability distribution over the initial region of phase space) is the result of our inability to control the physical, causal disturbance of the system due to our attempt to prepare it in an initial state. Of course, in the statistical case it is the sensitivity of trajectories to initial conditions which combines with this ineluctable interference to generate our desired result.

Unfortunately, it is hard to give a detailed critique of Krylov's positive view, since only chapter I of his book, attempting to show that statistical mechanics cannot be understood in a purely classical basis without the introduction of a kind of uncertainty for initial states, and four sections of chapter II, designed to show that the same argument holds when the underlying micromechanics is quantum mechanics, were written. The later chapters outlining the theory of preparation and demonstrating its inbuilt 'complementarity' and 'uncertainty' were never written due to Krylov's untimely death. But I think we can make at least two observations here which will indicate some problematic aspects of this approach.

One problem is that Krylov might be proving too much. It would seem that he would like to convince us that ensembles which behave anti-thermodynamically are *impossible* to prepare. But, quirky and anomalous as it is, the spin-echo result (described in section 5) makes us hesitate to accept

7 The elusive object of desire

such a strong conclusion. Here we do seem to be able macroscopically to manipulate a system in such a way that macroscopic order evolves out of macroscopic disorder. Perhaps we should not make too much of this. The systems are prepared in a tricky way. First, they are macroscopically ordered in the usual way. Then we 'reflect' them to generate the initial macroscopic order once more. This is not the same thing as being able to prepare an anti-thermodynamic ensemble without first relying on the more usual thermodynamic evolution in the forward time direction. It is also true that there might be a nice, systematic means of separating off such quirky systems as those of nuclear spins from the more important systems we are considering, for instance interacting molecules, such that the latter kind of systems could be taken to be immune to preparation of an anti-thermodynamic ensemble of any kind, although I do not know how this could be done.

In any case, my main problem with Krylov's program as a rationale for time asymmetry is different. What is *preparation*? Until we know what that is, and in particular how to distinguish preparing a system from measuring it, observing it, or simply destroying it, we do not know how to use any results of a Krylovian sort to block the standard parity of reasoning objection. What leads us to take the physical intervention in a system to be a preparation of it rather than just an observation of it? And how can we be assured that each system whose lifetime from one moment to another we are concerned with will have its 'preparation' on the same side of its 'measurement' as any other system? Until this is resolved, we will be able to use any results Krylov obtains only to tell us that systems will be statistically predictable to evolve in time – in either time direction – from order at preparation to disorder at measurement. But, once again, the parallelism seems to be being plugged in from elsewhere. Again a resort to 'causation' and its time asymmetry suggests itself. We prepare a system at the time beginning of its evolution, but at the time end we cannot be preparing because preparation is a causal process which acts 'in the forward time direction'. But what account are we to offer that justifies such assertions which simply does not beg the question of the origin of time asymmetry? To get the 'interventionist' account to work we must allow intervention during the process to be in one time direction only. Here we seem to rely on a quite parallel hidden posit of unidirectional causation, not during the evolutionary process but in order to differentiate the interactions the system has with the outside world at the beginning and end of its life.

8 Prigogine's solution

A more radical view of statistical mechanics, but one also founded, at least in part, on certain analogies with features of quantum mechanics, is that due to Prigogine and others working in his program.

Prigogine *et al.* have provided some interesting approaches to the formalism of irreversible ensemble behaviour.[15] One can view the evolution of an ensemble by tracking the trajectories of systems started with certain initial microstates. Alternatively, one can view the evolution of the initial probability distribution over microstates by means of a time varying operator which maps the distribution at one time to distributions at later times. This technique (due originally to B. Koopman) has proven invaluable in ergodic theory and its generalizations, as various properties of the evolution (being ergodic, mixing, etc.), are intimately related to properties of the operator on the distribution function (such as its spectrum).

Prigogine is able to show that when a sufficient rerandomization condition (being a K-system will do) holds for the ensemble, we can define an operator which transforms the probability distribution in a way which is one-to-one, so that we can recover the original distribution function from its transformed version, and which is such that the transformed function can be shown to display an irreversible behaviour. One such transformation will transform the original probability distribution, whose fine-grained entropy, defined in the usual way, is provably invariant, into a distribution whose 'entropy' (defined the same way) will monotonically increase. The mathematics here tends to conceal what is going on, but I think one can get some understanding by noting that from the usual ensemble point of view there are partitions ('coarse-grainings') which are generating – that is, the doubly infinite set of numbers saying what box of the partition a point is in at a given discrete time determines, except for a set of measure zero of trajectories, the unique trajectory. For suitably rerandomizing systems such generating partitions can be found which display 'irreversibility' in the sense of monotonic spreading out of the distribution among the boxes. So we have a transformation from the original probability distribution function to a distribution which is

---

[15] Courbage & Prigogine (1983).

7 The elusive object of desire

statistically equivalent to it yet which manifests its one-directional spreading outness on its face.

Prigogine combines this mathematical result with the claim that the true states of the world are the ensemble distributions. Since exact, pointlike microstates cannot be determined by us to exist, or even fixed in on by approximation if we want to fix on a microstate whose future trajectory will follow some determinate pattern (due to the wild divergence of trajectories from points no matter how close), we ought to dismiss these pointlike microstates as misguided idealizations. There are no such determinate microstates of the world, only states properly described by regions of phase space and probability distributions over them. The existence of the new irreversible representations then displays in our formalism the irreversible behaviour of the genuine, basic states of the world, which are the real states of individual systems.

Next Prigogine faces the problem that for each such change or representation which gives rise to a new distribution whose 'entropy' increases monotonically in time, there is another change of representation which gives rise to states whose behaviour is monotonically anti-thermodynamic. This duality is intimately connected with the duality of instability of orbits (from arbitrarily close-by points there are trajectories which exponentially diverge in the future *and* trajectories which exponentially diverge in the past) and the duality of mixing results (the measures of the intersections of sets converge to the product of the measures of both in the infinite future *and* the infinite past). Indeed, it is the same time symmetry which goes back to Boltzmann's first 'probabilistic' version of his theory. How do we get rid of the undesirable anti-thermodynamic representation?

To select the proper representation and to dismiss the improper one Prigogine resorts, we are not surprised to find out, to facts about initial ensembles. Initial distributions which are absolutely continuous in the standard measure will not do, he believes. All of these, i.e., the distributions which assign probability zero to sets of Lebesgue measure zero, show a symmetry in time, at least in the infinite time limit, as demonstrated by the time symmetric results of mixing. But there are singular initial distributions whose representation in the new form spread out in the future to fill the available phase space, yet they do not spread out in the other time direction. Then what he claims is that only those singular initial states which show this asymmetric limiting behaviour in the new representation are admissible as representing genuine physical states of systems. This is backed up by a definition of entropy and a demonstration that with this sense of entropy the admissible states can be prepared using only a finite amount of 'information' whereas their time inverses (which, of course, cannot spread out to uniformity in the future time direction but do so in the past) can only be prepared utilizing an infinite amount of information. So now, allegedly, we have some understanding of why it is that we can prepare ensembles which behave thermodynamically but not those which behave anti-thermodynamically.

But there is much here that is puzzling. For one thing, as we noted in section 3, most realistic systems are provably not K-systems (or mixing for that matter, which might give Krylov a problem, depending on how he is to be interpreted). Another minor objection is, again, the fact that this approach seems to show the preparation of anti-thermodynamic ensembles impossible, but the spin-echo results seem, at least, to suggest that this need not always be so.

More important, perhaps, are doubts one might have about the 'admissible' states being the only ones we ought to consider using and about their time inverses being the only ones we wish to be able to exclude. A perfectly parallel beam of particles is a member of a singular admissible set which then thermalizes. We cannot produce an ensemble of thermal systems which evolve back into perfectly parallelized ones. Well, yes. But we are usually dealing with such cases as a gas confined to the left hand side of a box which spreads over the whole box. The spontaneous reverse process is 'impossible' (in the ensemble sense). But here it seems that we wish both admissible and inadmissible initial states to be non-singular in the Lebesgue measure. Our initial probability distribution is one which is uniformly distributed over half the allowed phase space, not concentrated on a subset of measure zero.

More important still is the lack of some physical account in this theory of why it is that systems are parallel in their entropic behaviour which is not question begging. The inadmissible states require an infinite amount of information to be prepared in the definition of entropy given. But one can define entropy so that the situation is reversed. What exactly is the connection between the structure of the world, which governs our abilities and inabilities to prepare ensembles, and this formal apparatus? It is far from clear. Why is it that the same initial states are admissible for all systems, and not some for one and their reverses for another? To be sure, we can force parallel time behaviour on all systems in our theory by demanding uniformity of admissible states across systems, but this is not an explanation as to why that account works as well as it does. Yes, it is true that we can prepare parallel beams which thermalize in the future time direction but not systems which are thermal and which, in the future direction, parallelize. Yes, we may be able to 'select' the proper representation for systems by means of ensembles and their transformations by uniformly allowing only that transformation and its associated admissible initial states which gives us the parallelism in time we wish across all systems. But it is the old story: we have introduced a selection principle which works. We have not explained why such a uniform selection principle correctly describes the world.

In a broader way the Prigogine approach is problematic as well. Prigogine's denial of the existence of pointlike microstates for systems rests upon positivistic arguments of such great radicality as to be dubious even to those (like me) who are quite sympathetic to a broader positivism in general. It just is not enough to offer the argument that we cannot exactly determine a microstate, or even determine a region around it of points whose trajectories behave like that of the point, to deny the reality of the microstate. Just what does the claim that ensemble distributions give the full, actual state of the individual system really amount to? Analogies are drawn with the 'completeness' of the quantum state in quantum mechanics, but that state is not an ensemble distribution, to begin with, and, importantly, the completeness argument in quantum mechanics is backed up by 'no hidden variables' proofs completely lacking in the statistical mechanical case. Nor, for that matter, am I at all clear just how this 'no microstate' metaphysics really is supposed to back up the alleged 'demonstration' of the origin of the Second Law.

9 Conclusion

These attempts to rationalize the foundations of statistical mechanics, and in particular to explain at least the parallelism of entropic behaviour of systems in time, are not the only theories which have been offered. I do think that all the others I have seen also fail to block the familiar 'parity of reasoning' problem except by, again, introducing some other unexplained and problematic temporally asymmetric ingredient in the derivation.

The results obtained at such laborious cost regarding the instability of motion of trajectories and the consequent rerandomizing results about ensembles are surely major achievements. Even if we cannot fully explain the success of the kinetic equations obtained by simply sticking rerandomization in as a posit, we do seem to be on the path of grasping what it is about the physical world which lies behind the success of the rerandomizing assumptions. But, as Krylov, Prigogine, and others have freely admitted and often emphasized, we can never complete the picture without some understanding of what initial conditions are allowed in the world with high likelihood and what are not. It is this that leads to the problem of the initial ensemble, and it is this which makes the problem of explaining the parallelism of the entropic behaviour of systems in time so difficult to resolve. Throwing up our hands in the counsel of despair which just says, 'Well, that happens to be the way initial conditions fall out in this world,' certainly seems unsatisfactory. But I do not think we yet know what to put in its place.

8 Time in experience and in theoretical description of the world

LAWRENCE SKLAR

The asymmetric nature of time, the radical difference in nature between past and future, has often been taken to be the core feature distinguishing time from that other manifold of experience and of nature, space. The past is fixed and has determinate reality. The future is a realm to which being can only be attributed, at best, in an 'indeterminate' mode of a realm of unactualized potentiality. We have memories and records of the past, but, at best, only inferential knowledge of a different sort of what is to come. Causation proceeds from past to future, what has been and what is determining what will be, but determination never occurs the other way around. Our attitudes of concern and our concepts of agency are likewise time asymmetric.

Since the late middle part of the nineteenth century there have been recurrent claims to the effect that all the intuitively asymmetric features of temporality are 'reducible to' or 'grounded in' the asymmetry of physical systems in the world that is captured by the Second Law of Thermodynamics. The suggestion is, as far as I know, first made by Mach. It is taken up by Boltzmann and used by him in his final attempt to show the statistical mechanical reduction of thermodynamics devoid of paradox.[1] The claim is treated in detail in the twentieth century by Reichenbach, and accepted as legitimate by a host of other philosophers, mostly philosophers of science.[2]

The physical asymmetry to which the asymmetry of time is to be reduced is that which tells us that the entropy of an isolated system can only increase in the future direction of time and can never spontaneously decrease. In the statistical mechanical version of the theory, it is the overwhelming probability of entropy increase in one time direction rather than the other that becomes the physical ground for the intuitive asymmetry of time. The future time direction is, according to this view, by 'definition' the direction of time in which the

---

[1] Boltzmann (1897). See also Boltzmann (1964), 446–7.
[2] Reichenbach (1956). See also Grünbaum (1973), chapter 8, and Horwich (1987).

overwhelmingly great majority of energetically isolated systems have their entropy increase.

While this doctrine seems to be so appealing as to be almost manifestly self-evident to many philosophers of science, to others concerned with time it seems manifestly absurd. What is it about time and its asymmetry that makes the entropic account seem so ludicrous as to be dismissed without argument by many perceptive philosophers?

One objection against the entropic theory of time asymmetry can be found in the following remarks of J. L. Mackie: 'Our concept of time is based on a pretty simple, immediate, experience of one event's following straight after another ... Our experience of earlier and later on which our concept of time direction is based, itself remains primitive, even if it has some unknown causal source.'[3]

The idea is that we have direct epistemic access to the relation of temporal succession, and that, consequently, any reductive account of the nature of that relation must be ruled out. But that sort of dismissal of the entropic theory rests, I think, at least in one interpretation of it, on a misunderstanding of the theory's claims. It is sometimes suggested that the entropic theory is established because if we view a film of a process it is only by means of entropic considerations that we can tell which is the beginning and which is the end. Without entropic features we cannot tell whether the film is being run forwards or backwards. But, of course, Mackie is quite right in claiming that it is not at all by awareness of entropic or other causal processes that we know of events in our immediate experience what their time order is. We know it 'directly and non-inferentially' insofar as it is possible to know anything directly and non-inferentially.

The claim of the entropic theorist is not that we know the time order of events by knowing something about their entropic order and inferring the time order from that basic knowledge. The entropic theorist is not claiming a kind of reduction of the sort claimed by the phenomenalist or by the behaviorist who base their reductive claims about material objects or about mental states on an analysis of our epistemic basis for making claims about matter or about the mental states of others. The reductive claim, then, cannot be refuted by any demonstration that the evidential access we have to the temporal order of events is 'direct and immediate.'

---

[3] J. L. Mackie, 'Causal asymmetry in concept and reality,' unpublished paper presented at the 1977 Oberlin Colloquium in Philosophy. Quoted in Sklar (1981).

8 Time in experience and in theory of the world

The kind of reduction being claimed is, rather, a scientific reduction. Its analogues are the reduction of pieces of macroscopic matter to arrays of molecules or atoms and the reduction of light waves to electromagnetic radiation. The claim is not that we determine temporal order by being first aware of entropic order, but that we discover, empirically, that temporal order is nothing but, is identical to, entropic order.

Now there are lots of puzzles in understanding what such a claim of identity amounts to when it is relations that are being identified in the reduction and not substances. But claims of this sort can sometimes be very plausible indeed, as is made clear by reference to a parallel case, one cited by Boltzmann himself.

We distinguish between the upward and the downward direction in space. We can identify the downward direction indirectly, say by seeing in which direction unsupported rocks move. But we can also identify the downward direction by a simple, immediate and non-inferential awareness. In a darkened room, without using any of our ordinary sense organs, we know which direction is down.

But we now have a perfectly adequate reductive account of what the downward direction is: it is, at any point, the local direction of the gradient of the gravitational field. It is the existence of this gravitational force that is constitutive of the very origin of the up–down distinction among directions. The gravitational force explains why things behave the way they do with respect to the up and down directions, why rocks fall down and helium balloons float up. It also accounts for the existence of our ability to tell, non-inferentially, which direction is up and which down. It is the action of gravity on the fluid in our semi-circular canals that gives us the direct awareness of which is the downward direction, although, of course, we do not tell which direction is down by first being aware of facts about our semi-circular canals and inferring facts about up and down from these.

It is this comparison case that provides Boltzmann with his famous claims to the effect that in regions of the universe in which entropic increase is in opposite time directions, if such regions exist, it will be the case that the future directions of time will be opposite directions of time. This is no more surprising, he thinks, than the fact that the downward directions in New York and in Australia are anti-parallel to one another. And, he claims, just as there is no downward direction in regions of space where there is no net gravitational force, there will be no past and no future direction of time in regions of the universe, if there are any, that are at equilibrium and that consequently show no parallel, statistical entropy increase of isolated systems.[4]

Can we then reduce the asymmetry of time to the asymmetric behavior in time of systems with respect to their entropic properties in just the same manner that we certainly can reduce the apparent asymmetry of space which is the asymmetry of up and down to the asymmetry inherent in the local direction of the gravitational force? While the gravitational theory of up and down provides a successful case of reduction, another example can show us a case where a reductivist argument would pretty clearly fail.

There is an intuitive distinction between right-handed and left-handed objects. Right-handed and left-handed members of a pair of gloves have many aspects in common. But they are also distinguished from one another by the different way in which their component parts are oriented with respect to one another. Is there some feature of the world that would serve as a reduction basis for right- and left-handedness in the same way that the direction of the gravitational force serves as a reduction basis for the up–down distinction?

It is sometimes suggested that the empirically discovered asymmetry with respect to orientation of weak interaction processes in physics provides such a reduction to the distinction between right- and left-handedness. But the claim seems manifestly implausible. To be sure, right- and left-oriented physical processes may not both be possible according to the theory of weak interactions that denies invariance under mirror-imaging for its laws. This does mean, for example, that we could tell someone what a right-handed object was without displaying such an oriented object but by utilizing the fact that some physical systems exist but their mirror images do not. This would be something like telling someone which way to run the film properly by relying upon the fact that some processes occur in a given time order but their time reversals do not since they are entropy decreasing. But that by itself is most assuredly not enough to allow us to claim that in any sense the right–left asymmetry 'reduces to' or is 'identical with' the asymmetry of processes revealed to us by contemporary physics. Even if there was no such physical asymmetry of processes, there would still be a distinction between a right- and a left-handed glove. But if there were no gravitational force, there simply would not be any up–down distinction to be made.

[4] See the items by Boltzmann cited in footnote 1.

What is the crucial difference in the two cases? Surely it is this: in the gravitational/up–down case, the existence and nature of the gravitational force provide a complete account of the existence and nature of the up–down distinction as normally understood. It is the existence and nature of gravity that explain why all the phenomena are as they are that constitute the normal external phenomena by which we distinguish up from down. Furthermore, it is gravity and its effect on matter that fully explain our ability to determine, non-inferentially, the downward direction at our location. The ability of our semi-circular apparatus to reveal to us, directly, the downward direction is explained in terms of the working of gravity on its contained fluid. But no such explanatory relation holds between the processes described by the theory of weak interactions and any of the features of the world that constitute for us the distinction between right- and left-handed objects. What makes an object right-handed and not left-handed has nothing to do with weak interactions. Nor do weak interactions tell us anything about the processes by which we come to identify the orientational structure of an object. In no way are weak interaction processes an explanatory basis for orientational features of things, as gravity is an explanatory basis for all that constitutes the distinction between up and down in space.

Similarly we can ask, is entropy related to the past–future distinction in the way that gravity is related to the up–down distinction, or is the connection between entropy and time asymmetry like that between weak interaction processes and spatial orientational features of the world? I do not think we have any definitive answer to that question available to us. What we would need to convince us that an entropic theory of time order was a possible contender for the true account would be an explanation, based on entropic theories of the world, of all those aspects of the world that we normally take as constitutive of the asymmetry of time.

We would need, for example, a demonstration that the existence of records and memories of the past, but not of the future, was a consequence of the asymmetric behavior in time of the entropic features of systems. We would need an explanation, framed in entropic terms, of why we take it that the direction of causal determination is always from past to future. We would want an entropic account of the basis of our intuition that the past is fixed and determinate, but the future merely a realm of indeterminate possibilities. And, paralleling the gravity/up–down case, we would want an entropic account of the nature of our ability to directly and non-inferentially determine of the members of pair of events in our experience which one is the later event of the pair.[5]

A number of attempts have been made to fill in the explanatory picture, of course. There is the program of trying to characterize the nature of records beginning with Reichenbach.[6] There are characterizations of the causal asymmetry that connect it with features not too distant from the things entropic theorists talk about, such as that of David Lewis.[7] But we can say with assurance that the project of convincingly showing us that all of the intuitively constitutive aspects of the time asymmetry can be given an explanatory basis in entropic features of the world is one that remains incomplete. But suppose the only important, grand, asymmetry in time of physical systems in the world is the asymmetry characterized by the Second Law and its consequences. Then surely a general naturalistic stance would tell us that there is no plausible alternative to the entropic account of time asymmetry?

Suppose the explanatory account could be carried out. Would it then be reasonable to say that the asymmetric relation in time that events have to one another is nothing but the asymmetric entropic relation they bear to one another, just as it does seem right to say that one object's being in the downward direction from another is just the first object's being in the direction of decrease of gravitational potential from the other? Several philosophers doubt that we would be entitled to do so. One such doubter is A. Eddington. The quote is from his Gifford Lectures printed as *The Nature of the Physical World*:

> In any attempt to bridge the domains of experience belonging to the spiritual and physical sides of our nature, Time occupies the key position. I have already referred to its dual entry into our consciousness – through the sense organs which relate it to the other entities of the physical world, and directly through a kind of private door into the mind ... While the physicist would generally say that the matter of this familiar table is *really* a curvature of space, and its color is really electromagnetic wavelength, I do not think he would say that the familiar moving on of time is *really* an entropy gradient ... Our trouble is that we have to associate two things, both of which we more or less understand, and, so as we understand them, they are utterly different. It is absurd to pretend that we are in ignorance of the nature of organization in the external world in the same way that we are ignorant of the true nature of potential. It is absurd to pretend that we have no justifiable conception of 'becoming' in the

---

[5] For a full discussion of these issues see Sklar (1981).
[6] Reichenbach (1956), chapter 4.
[7] Lewis (1979).

> external world. That dynamic quality – that significance that makes a development from future to past farcical – has to do much more than pull a trigger of a nerve. It is so welded into our consciousness that a moving on of time is a condition of consciousness. We have a direct insight into 'becoming' that sweeps aside all symbolic knowledge as on an inferior plane.[8]

Some of the ways Eddington has of putting things will probably make sophisticated philosophers squirm. Nonetheless, I think he makes an important point. There is something about time that makes a treatment of its relation to entropic asymmetry on a parallel with the treatment of the relation of up–down asymmetry to gravity implausible. Can we clarify what is making Eddington so uneasy?

For Eddington, most of the entities we posit in the physical world and most of the features of them are 'somethings' and 'some features' identified only by the role they play in a structure that we posit to account for our perceptual experience. While he does not offer a formal theory of the meaning of the terms occurring in physical theory, I have little doubt that he would be delighted with the theoretical holism of F. Ramsey. Ramsey believes that to say that there are electrons and that they have electric charge is to say that there are somethings with some features that behave the way electrons and their charge are posited to behave in our total physical theory of the world. For terms other than those characterizing our immediate perceptual experience, meaning is solely a matter of place in theoretical structure. There is nothing to say about what 'strangeness,' for example, is, other than what the theory of strange particles says about it.[9]

But such an account of meaning is not suitable for the terms characterizing items of immediate experience. We understand them by their association with the appropriate 'immediately known' feature of the world. There are features that we seem to want to attribute both to the realm of the immediately perceived and the realm of the theoretically inferred. In these cases can we say that both our perceptual images and the physical structures of things are characterized, for example, in spatial terms? The natural response is to claim that we are equivocating when we say that immediately perceived visual elements are in space and chairs and electrons are also in space. There is the visual space of the visual field and there is the physical space posited by our common sense theory of the physical world and posited with remarkable

---

[8] Eddington (1928), 91–7.
[9] For an outline of the approach to theories initiated by Ramsey see Sneed (1971).

changes (as quantized, curved spacetime, I suppose) by our best available physical theory. But it is just a confusion to think that the spatial relations visual percepts bear to one another are the same sort of relations that physical objects bear to one another. We know, directly and immediately, what the former relations are like – we experience them. All we know about the latter relations is what our theoretical structures say about them. Admittedly all of this is a lot to swallow, but it is one initially coherent approach to understanding the relation of perception to the world and the relation of perceptual language to the language of theory. And it is far from clear what alternative to put in its place if we reject it.

Now let us think a little about time and temporal language. The entropic theorists say that one event's being after another in time is just the second event having a higher entropy than the first (it is not that simple, but this will have to do here), are they talking about the perceived temporal relation that events in the realm of the perceived bear to one another? Apparently not, says Eddington. We know what the asymmetric relation of one event is being later than another is like. Given the statistical mechanical characterization of higher entropy as, essentially, more dispersed order structure, we also have some kind of immediate knowledge of the asymmetric relation one state of a system bears to another when one has a higher entropy than the other. And we know that these relations are not the same relation!

Compare this case with the reduction of 'downness' to the local direction of the gravitational force. We have no direct 'sensed downness' to deal with, although we do have the visual spatial relations visual percepts have to one another. Although we know which physical direction is down, immediately and without inference, we do not have a perceptual 'downness' the way we have a perceptual 'afterness' that is part and parcel of the intrinsic characterization of the realm of the directly perceived. There is nothing to stand in the way of our saying that downness is a something (or some-relation) in the physical world known to us by its causal effects (on rocks and on our semicircular canal fluids), and that we have discovered that this something is just directed gravitational force. But in the case of time we must examine perceived afterness and fathom its relation to afterness predicated of physical events.

We know from perception what it is for one state of a system to be temporally after some other state. And we know what it is for one state to have a more dispersed order structure than another state. We also know that these two relations are not the same. (This can be compared with our knowledge of the yellowness of a visual sensation and our assurance that whatever brain processes are like, they are not like *that*. This claim appears in many arguments rejecting mind–brain process identity claims.) So whatever the relation of temporal 'afterness' is, it is not an entropically defined relation, even if 'downness' is identifiable with a relation defined solely in terms of the nature of the local gravitational field.

There is a fairly obvious suggestion to make here. We have already allowed ourselves the distinction between the space of perception and the space of the physical world. Why not do the same thing for time? Given that the time of the physical world is spacetime, surely it is obvious that we would have to do so at some point? From this perspective there is some set of relations, 'we know not what' among events in the physical world that we think of as the temporal relations. But the temporal relations of our experience are nothing like these physical temporal relations. We need an asymmetric relation among physical events, and we call it the 'after' relation. This turns out to be a relation characterized in terms of the relative order dispersion, the relative entropy, among states of systems in the physical world. It is nothing like our perceived 'afterwardness,' of course, but why should it be? None of the temporal relations that events or stages bear to one another in the physical world are anything at all like the temporal relations items in our perceptual experience bear to one another.

The suggestion that we ought to distinguish the time of perception from the time of the physical world is, of course, not new. General Berkeleian considerations would tell us that if some perceived properties are only features of 'ideas in the mind,' all perceived features must be. And the idea that some features of perceived experience 'resemble' those of physical things also receives its first blows at Berkeley's hands. The more the time of physics is posited to be something radical and strange by contemporary physics, the more pressure there is to distinguish that 'time' from the time of perception. When relativity denies absolute simultaneity for distant events, that drives some to deny the identity of the time of perception and the time of physics. When K. Gödel discovered the possibility of closed timelike curves in his models for General Relativity, he took the very possibility of such closed causal loops, even if they were not actual in the real physical world, to be the final argument against the identity of perceptual and physical time. Taking 'real time' to be the time of experience, he denied that the 't' parameter of physics stood for time at all.[10]

[10] Gödel (1949b). For a thorough discussion of Gödel's ideas see Yourgrau (1991).

But Eddington is hesitant to adopt this solution to our problem about time order and entropic order, and he is wise to be so. Something seems to tell us that if we go the final step and take the time relation of physics to be one more posited relation, fixed in its meaning solely by the role played in the total theory by the second-order variable whose existential quantifier asserts the existence of the relation in the Ramsey sentence replacement for our total physical theory, we will have gone too far. I believe that this is the gist of Eddington's remarks to the effect that time is given to us twice, once in our direct experience and again in our theory of the physical world, and that it must be the same time that is given in both of these modes.

What goes wrong if we radically distinguish perceived time from the time of physics? What harm is done if we say that the 't' of physics stands not for time as understood by us from our experience of the temporal relation of our perceptual experiences to one another, but, instead, for some posited physical relation that is no more identical with time as we know it and mean it from our experience than 'strangeness' in the theory of particles has to do with the strangeness of things as meant in ordinary discourse?

Put most crudely, the problem is that at this point the veil of perception has become totally opaque, and we no longer have any grasp at all upon the nature of the physical world itself. We are left with merely the 'instrumental' understanding of theory in that posits about nature bring with them predicted structural constraints upon the known world of experience. Once we deny that the terms used to describe our direct experience have any meaning in common at all with any of the terms used to describe the physical world (deny, that is, that meaning can be projected by 'semantic analogy' from words whose meaning is given to us by ostention to items of experience to words not directly associated with the items of experience at all), it becomes hard to take the atomistic meaning of terms referring to the physical world or the atomistic truth of sentences in such a theory seriously at all.[11] If what we mean by 'time' when we talk of the time order of events of the physical world has nothing to do with the meaning of 'time' as meant when we talk about the order in time of our experiences, then why call it time at all? Why not give it an absurd name, deliberately chosen to be meaningless (like 'strangeness'), and so avoid the mistake of thinking that we know what we are talking about when we talk about the time order of events in the world? Instead let us freely admit that all we know about 'time' among the physical things is contained in the global theoretical structure in which the 't'

[11] For a defense of this claim, see Sklar (1980).

8 Time in experience and in theory of the world

parameter of physics appears. And all the understanding we have of that global structure is that when we posit it, it constrains the structure among those things presented to our experience in a number of ways. And that is its full cognitive content.

Another way of seeing why the move to Ramsify physical time along with all the other parameters of our total physical theory, including those characterizing physical space, leads to an insidious collapse of the possibility of a realist interpretation for our theories is as follows. Once one has accepted the Ramsey transcription of a theory, it is very hard to see how to block the interpretation of a theory that takes only the observational vocabulary as genuinely denoting elements of the natural world. If the full meaning of the theoretical terms is given by their replacements in terms of second-order variables bound by existential quantifiers, why not take these variables as ranging only over a domain of abstract objects? Why not, that is, read the theory in a 'representationalist' vein? To say that the theoretical entities exist and have the theoretical features attributed to them, is to say nothing more than that we can embed the regularities that hold among the real things and real features of the world into an abstract structure that neatly encodes all of the regularities. Such an interpretation is quite reasonable when it is features like 'strangeness' that we are positing. What does positing such a feature really do, other than summarize some lawlike regularities holding at a lower level of the theory?

One standard way to try to avoid this collapse of realism into mere representationalism, to try to keep some naturalistic and not merely abstract reference for the theoretical terms of our theories, is to argue that what the theory demands is that appropriate *causal* relations hold between the theoretical entities and their features and the observable things and their features. While numbers or sets can represent our perceptions, they cannot affect them. But electrons and electric charge can be causally related to what we perceive. So we must reserve a role for causal discourse in order to somehow provide a constraint on our theory that prevents it, even after all its other theoretical structure has been Ramsified, from becoming merely the posit of a representational matrix into which the observables can be embedded.

But can we reserve such a role for causation if we Ramsify physical time along with all the other physical features of the world? If we take it that time itself, when meant as a relation among the physical events of the world, is just a 'something we know not what' fixed by its place in the theoretical structure, we must view it as merely representational, in the same way as we view 'strangeness.' And it is not too implausible to claim that with time so goes causation. This would leave us utterly perplexed as to how we can reserve for causation some extra-theoretical role, a status it must have if it is to serve as the lever by which we pry apart the intended physical interpretation for the theory from the threatening anti-realist 'mere representationalist' interpretation.

I think it is in this sense that if we try to split the time of physics apart from the time of perception in a radical way we will lose entirely our grip on the possibility of a theoretically realist account of the world. I think this is what is moving Eddington to say the things about time that he does. Physical light waves might very well be 'somethings we know not what' that turn out to be electromagnetic waves. And physical chairs might be 'somethings we know not what' that turn out to be nothing but regions of curved spacetime. The perceived light and the perceived chair have been stripped away from their physical counterparts and made into 'secondary objects in the mind' before the identification of the physical object with the scientifically discovered entity is ever made. But time itself, time as meant in physics, cannot so easily be dissociated from the time order as we perceive it, the time that relates the items of immediate perceptual awareness to each other and that is itself knowable as a relation by us through some kind of ostensive presentation. If the 'time' of physics is not real, perceived time, then the 'causation' of physics is not real causation. It is then hard to see how we can have any access to the real physical world at all.

At this point alleged problems for Kant may spring to mind. If Kant shares with Hume the view that causation is a relation that holds only within the realm of the perceived, then it cannot be used to explain to us the relation of things-in-themselves to things-as-perceived. And if we cannot use causation in the explanation, what can we use? We are threatened with some version of idealism, or of its current representationalistic avatar.

If the only asymmetric features of physical events in time are their entropic features, then a broad naturalism would have to suggest that entropic features must be the only features to be invoked in an explanatory account of the intuitive features of the world that we take to mark out the asymmetry of time: the existence of records and memories of past and not future and the asymmetry of causation, for example. The claim that we know time order without knowing entropic order is not a good objection to a reductivist account of

8 Time in experience and in theory of the world

time order to entropic order that is meant as a 'scientific identification.' But a simple model of the reduction of time order to entropic order based on the plainly correct reduction of the up–down relation to a feature of the world framed entirely in terms of the theory of gravity is too naive. We cannot just *identify* time order with entropic order, at least if we mean by time order perceived time order. The perceived 'afterness' of events and the entropic relations among events are, as Eddington claimed, too unalike to be identified. Any attempt to escape from the dilemma by distinguishing time order as it relates elements of the perceived from 'time order' as it appears in physical theory, tempting as that might be for other reasons as well, will, as Eddington saw, leave us with little hope of a realistic interpretation of our physical theory.

9 When and why does entropy increase?

MARTIN BARRETT AND ELLIOTT SOBER

1 Introduction

Like the Sirens singing to Ulysses, the concept of entropy has tempted many a thinker to abandon the straight and narrow course. The concept is well-defined for chambers of gases. However, the temptation to extend the concept has been all but irresistible. As a result, entropy is used as a metaphor for uncertainty and disorder. We have nothing against such extensions of usage and, in fact, will indulge in it ourselves. However, we do believe that the price of metaphor is eternal vigilance.[1]

One example of the temptation can be found in R. A. Fisher's *Genetical Theory of Natural Selection*.[2] In the second chapter of that book, Fisher states a result that he dubs the *fundamental theorem of natural selection*. The theorem states that the average fitness of the organisms in a population increases under selection, and does so at a rate given by the additive genetic variance in fitness. Fisher then proposes the following analogy:

> It will be noticed that the fundamental theorem ... bears some remarkable resemblances to the second law of thermodynamics. Both are properties of populations, or aggregates, true irrespective of the nature of the units which compose them; both are statistical laws; each requires the constant increase of a measurable quantity, in the one case the entropy of a physical system and in the other the fitness ... of a biological population [p. 39].

Fisher then quotes Eddington's famous remark that 'the law that entropy always increases – the second law of thermodynamics – holds, I think, the supreme position among the laws of nature.' Fisher comments that 'it is not a little instructive that so similar a law should hold the supreme position among the biological sciences.'

---
[1] We borrow this phrase from R. C. Lewontin.
[2] Fisher (1930).

9 When and why does entropy increase?

Fisher's analogy between the second law and the fundamental theorem does not lead him to deny that they differ. Among the differences he notes is the idea that 'entropy changes lead to a progressive disorganization of the physical world, at least from the human standpoint of the utilization of energy, while evolutionary changes are generally recognized as producing progressively higher organization in the organic world' (p. 40). Here we see the familiar idea that entropy is not simply a technical concept applicable to chambers of gases, but has a larger meaning as a measure of organization and order.

If order declines in thermodynamic processes, but increases in the process that Fisher described, why is this so? One standard suggestion is that the second law of thermodynamics applies to closed systems, but populations are able to evolve new adaptations because they extract energy from their surroundings. Although it is true that the second law applies to closed systems and that organisms need energy to survive and reproduce, we will see in what follows that this does not answer the question we have posed. For the fact of the matter is that entropy can decline in closed systems and it can increase in open ones. The distinction between closed and open systems is an important one, but it is not the key to the problem we wish to explore.

In order to fix ideas, we begin with an informal and we hope intuitive description of what entropy is, one that we will refine in what follows. Entropy is a property of probability distributions.[3] If there are n kinds of events, each with its own probability $p_1, p_2, \ldots, p_n$, then the entropy of the distribution is given by $-\sum p_i \log p_i$. For n states, the highest entropy distribution occurs when the n possibilities are equiprobable; the lowest entropy distribution occurs when one of the states has probability 1 and the others have probability 0. Additional properties of entropy are given in appendix A.

To the extent that both gases and evolving populations can be characterized by probability distributions, each can also be described by the entropy concept. A population of organisms may be characterized by the n genes that can occur at a given locus. The frequency of a gene in the population defines what we mean by the probability that a randomly chosen organism will contain a copy of that gene. A population with the n genes in equal frequencies will have a higher entropy than a population in which all but one of those n genes has disappeared.

---

[3] The entropy concept in physics is a property of the particular probability distributions employed in statistical mechanics.

Consider, by analogy, a chamber divided down the middle by a wall. In the left hand side, there are some molecules of oxygen; in the right hand side, there are some molecules of helium. When the dividing wall is removed, the molecules will reach a mixed homogeneous state. If we associate probability distributions with the gases before and after the wall is removed, we can compare the entropy of the system at those two times.

When in the process just described is entropy well-defined? If each isolated subsystem is at equilibrium before the wall is removed, a probability distribution can be provided and the entropies of each may be calculated. However, immediately after the wall is removed, the system is not at equilibrium. At this point, the theory of equilibrium thermodynamics says nothing about the system's probability distribution, and so its entropy is not defined. However, if we wait a while, the coupled system will move to a new equilibrium, and entropy once again makes sense. Within equilibrium thermodynamics, one cannot talk about the change in entropy that occurs when a system moves from an out-of-equilibrium state to its equilibrium configuration.[4] What, then, does the second law of thermodynamics mean when it says that entropy must increase?

What we need to focus on is the coupling process.[5] Two or more isolated systems are each at equilibrium. Then they are coupled and the new conjoint system is allowed to reach its new equilibrium. We then look at the equilibrium entropy before coupling and the equilibrium entropy after. The second law does not say that entropy increases on the way to equilibrium; it says that equilibrium entropy increases under coupling.

This picture of what the thermodynamic concept means should guide how we frame the question about entropy in nonthermodynamic contexts. The gene frequency distribution in an evolving population changes in every generation until it reaches an equilibrium point. An equilibrium occurs when the population stops changing.[6] If we look for biological parallels to the thermodynamic idea, we must consider a coupling process and ask what is true of the probability distributions that characterize populations at equilibrium. So the appropriate analog is not a single population moving from an out-of-equilibrium state to

---

[4] We leave to others the investigation of how entropy should be understood in non-equilibrium situations.

[5] This is not the term customarily employed in thermodynamics, but it well describes the process, e.g., of bringing a system into thermal contact with a heat bath.

[6] The definition of the equilibrium concept in population biology would have to be refined to take account of random fluctuations around equilibrium values due to sampling error. This refinement is not worth pursuing here.

an equilibrium configuration. Rather, we should consider two or more populations, each at equilibrium, which then are coupled. The new conjoint population is then allowed to reach its equilibrium. We then compare the pre-coupling equilibrium entropy with the equilibrium entropy that obtains after coupling. Under what circumstances will the second have a higher value than the first?

In section 2, we describe with more care what entropy increase means in thermodynamics. Then, in section 3, we discuss several biological examples in which populations are coupled and their pre- and post-coupling equilibrium entropies are compared. In section 4, we address the question that is the title of this chapter.

2 Entropy increase in statistical mechanics

The textbook explanation[7] for the increase of entropy in statistical mechanics is appealingly simple. We start with two systems of particles (S_1 and S_2). These could be chambers of gases or blocks of metal or any other kind of homogeneous physical system. Each system has, in this idealization, a definite energy E_i; the total energy of the two systems is $E = E_1 + E_2$. The energy E_i is realized by summing the contributions from each of the N_i particles in S_i. The pair of numbers (N_i, E_i) fixes the *macrostate* of system S_i, since it describes macroscopic quantities which are (at least approximately) observable. There are many ways to assign individual energies to the N_i particles compatible with the system energy E_i. Each such way is a *microstate* of S_i, *accessible* to the macrostate (N_i, E_i).

A simple example illustrates these ideas. Let each S_i consist of a row of coins. Each coin can be in one of two states: H (energy 1) and T (energy 0). Each microstate of the system is just a particular configuration of heads and tails. The energy of the system in that microstate is just the number of heads in the configuration.[8] Table 1 shows S_1 and S_2 in microstates accessible to the macrostates $(N_1 = 2, E_1 = 1)$ and $(N_2 = 3, E_2 = 2)$ respectively. The total energy of the combined systems is $1 + 2 = 3$.

The *Fundamental Postulate of Statistical Mechanics* provides a way of assigning probabilities to the microstates. It states that in a closed system (one with no changing external influences) at equilibrium, all microstates accessible to the

---

[7] See, e.g., Kittel and Kroemer (1980) pp. 33–48.
[8] There are real physical systems, *magnetic spin systems*, whose statistical mechanics is virtually isomorphic to our example. We prefer the example for its vividness.

Table 1. *Accessible microstates of systems 1 and 2*

System 1 (energy 1)	System 2 (energy 2)
H T	H H T
T H	H T H
	T H H

macrostate of the system are equally probable. In S_1, there are two accessible microstates in the macrostate $(2,1)$, since there are two ways of assigning one head to two coins. Each microstate thus has probability 1/2. The system S_2 has three accessible microstates in the macrostate $(3,2)$, since there are three ways of assigning two heads to three coins; each microstate has probability 1/3.

Underlying this assignment is the so-called *ergodic assumption* that if we could somehow observe a closed system at equilibrium repeatedly (without disturbing the equilibrium) to pin down its microstate, the accessible microstates would appear equally often.

The probability space specifying accessible microstates of equal energy and equal probability is called in statistical mechanics the *microcanonical ensemble* (or the microcanonical distribution). The concept of entropy provides a way to measure the width of this probability distribution. If there are n accessible microstates with probabilities $1/n$, the entropy is given by $\log n$ (see appendix A). This is the quantity that the textbooks take as the *definition* of entropy, denoted σ, of a closed physical system. If we write $\#(N,E)$ for the number of accessible microstates in macrostate (N,E), then $\sigma = \log \#(N,E)$.

Suppose that S_1 and S_2 are initially isolated from each other; this means that no energy or other physical quantity can be exchanged between them. Probabilistically this appears as the assumption that the two systems are independent. The macrostate of the two isolated systems taken together is given by $N = N_1 + N_2$ and $E = E_1 + E_2$; an accessible microstate is just a choice of an accessible microstate of S_1 together with an accessible microstate of S_2. This means that there are $\#(N_1, E_1) \times \#(N_2, E_2)$ accessible microstates of the combined system. All $2 \times 3 = 6$ accessible microstates for the combined example systems are shown in table 2. Since the logarithm of a product is the sum of the logarithms, we have for the entropy of the combined system $\sigma = \sigma_1 + \sigma_2$.

9 When and why does entropy increase?

Table 2. *Accessible microstates of combined system before coupling*

Microstates of uncoupled system (energy 3)	
System 1	System 2
H T	H H T
T H	H H T
H T	H T H
T H	H T H
H T	T H H
T H	T H H

Entropy is additive for independent systems. In the example, $\sigma_1 = \log 2$, $\sigma_2 = \log 3$, and $\sigma = \sigma_1 + \sigma_2 = \log 6$.

Now suppose that S_1 and S_2 are coupled by placing them in thermal contact. In our example, remove the wall between the two systems so that energy may flow between them. The total energy of the combined system remains unchanged, but when equilibrium is reestablished, *any* configuration of the $N = N_1 + N_2$ heads and tails with that energy is an accessible microstate. In our example, the combined energy is 3, and there are $5!/3!2! = 10$ ways of assigning three heads to five coins. These are shown in table 3. According to the Fundamental Postulate, all 10 microstates are equally probable, and the new entropy is $\sigma = \log 10$. Coupling has increased the entropy from $\log 6$

Table 3. *Accessible microstates of combined system after coupling*

Microstates of coupled system (energy 3)				
H	H	H	T	T
H	H	T	H	T
H	H	T	T	H
H	T	H	H	T
H	T	H	T	H
H	T	T	H	H
T	H	H	H	T
T	H	H	T	H
T	H	T	H	H
T	T	H	H	H

to log 10. To summarize the steps that lead to the increase of entropy after coupling:

(1) Prior to coupling the system is composed of two isolated subsystems in a definite macrostate.
(2) The accessible microstates are *constrained* by the requirements that the energy of S_1 be E_1, that the energy of S_2 be E_2, and that the energy of the whole system be $E = E_1 + E_2$.
(3) Coupling the two subsystems results in the removal of the first two of these constraints.
(4) As a result, the space of accessible microstates is *enlarged*: formerly accessible microstates remain accessible, while new ones (possibly) are added.
(5) Hence, the entropy, which is a measure of the number of accessible microstates, increases.[9]

Call this the *classical explanation* of entropy increase. We believe that the simplicity of this explanation conceals two defects, one intrinsic to physics and the other extrinsic.

The first involves the justification of the Fundamental Postulate, on which the explanation intimately depends. The assumption that the systems in question, after being perturbed from equilibrium by coupling, eventually settle down to an equilibrium of equally likely microstates is far from obvious. What is obvious is that coupling must be implemented by some sort of physical operation, such as the insertion of a thermal (conducting) membrane between the systems where before there was an adiabatic (insulating) wall, or the removal altogether of the wall between two chambers of gases. But why this removal of a barrier in the physical sense should amount to a removal of constraints in the mathematical sense of (3) above is unclear.

The proper setting for the formulation of this difficulty is quantum mechanics, and we do not propose to detour into this theory here.[10] We mention in this connection only that some have found the solution to the difficulty to lie in the notion of *coarse-graining*, which involves defining probabilities not over individual microstates, but over groups of microstates lumped together. We discuss coarse-graining in the biological examples in section 3.

We can give a sense of the nonobviousness of the Fundamental Postulate with an extension of our simple example. Although at equilibrium there is assumed to be a probability distribution over the space of microstates, the system at any

---

[9] Actually, what we have established is that entropy does not decline. We will continue to use the word 'increases' when no confusion would result.
[10] A concise discussion may be found in Landsberg (1978).

9 When and why does entropy increase?

Table 4. *Time evolution before coupling*

	Microstate of uncoupled system	
Time	System 1	System 2
$t+0$	H T	H H T
$t+1$	T H	T H H
$t+2$	H T	H T H
$t+3$	T H	H H T
$t+4$	H T	T H H
$t+5$	T H	H T H
$t+6$	(repeats)	

particular instant is in fact in one or another particular microstate. From there it evolves (in time) according to a set of deterministic physical laws. Let us specify a (fictitious) dynamics for our system. At time t the system is in a particular microstate; at time $t + \Delta t$ each unit of energy (i.e., each occurrence of heads) moves one coin to the right, unless it reaches a wall, in which case it comes back in on the left. (In other words, the row of coins is really a circle, and the heads march around it.) The first six time-steps of the un-coupled systems' evolution are shown in table 4; after six steps the pattern repeats.

Under the ergodic assumption, we may derive the equilibrium probabilities from the long-term frequencies of the microstates; the result is that each of the two microstates of S_1 are equally probable (occur equally often), each of the three microstates of S_2 are equally probable, and the two subsystems are probabilistically independent. This agrees with the probability assignments demanded by the Fundamental Postulate. After coupling, however, the situation is as shown in table 5. The time evolution repeats after only five steps,

Table 5. *Time evolution after coupling*

Time	Microstate of coupled system
$t+0$	H T H H T
$t+1$	T H T H H
$t+2$	H T H T H
$t+3$	H H T H T
$t+4$	T H H T H
$t+5$	(repeats)

and this is true no matter what the initial configuration is at the time of coupling. The result using the ergodic assumption is that the probability is spread evenly over only five accessible microstates, not ten.[11]

Of course, the dynamics we have specified is not real physics (neither, for that matter, are our coins). But the problem of convincingly justifying the Fundamental Postulate for systems of real particles obeying real physical laws is well illustrated by this made-up example. To the extent to which this crucial assumption of the explanatory story remains ungrounded, the explanation itself is unsatisfying.

The extrinsic difficulty with the classical explanation of entropy increase concerns its applicability to problems outside of statistical mechanics. We wish to examine a variety of systems governed by probability distributions and which consequently allow entropy to be defined. We ask when and why entropy increases after coupling such systems. The formulation of a particular problem may reveal an analogy to statistical mechanics, leading us to hope that we can apply the classical explanation. But the classical explanation requires that the set of accessible states after coupling be an enlargement of the set before; accessible states before coupling remain accessible afterward. Unfortunately, this condition is not met in all real examples; a specific case is discussed in the next section.

Our conclusion, then, is that the classical explanation of entropy increase may be true (as far as it goes), but its assumptions are opaque and it offers little guidance as to whether we may properly apply the second law in applications outside of physics.

So to better illuminate the problem of entropy outside physics, we seek an alternative explanation of entropy increase within physics itself. Such an explanation, in its essentials, is provided by Khinchin.[12] His exposition applies strictly to systems with a continuous range of admissible energy values, but it can be adapted to systems with discrete admissible energies.

In order to understand the key points of Khinchin's discussion, we first reconsider our simple coin example when the number of coins is very large. We use the term *orbital* to refer to the states which a single particle may attain; this term comes from quantum mechanics. In the coin example, there are two orbitals: H and T. Under the microcanonical ensemble, the marginal distribution over the orbitals is the same for every particle. Consider table 3,

---

[11] Notice that with the stipulated dynamics, coupling leads the entropy to *decline*.
[12] Khinchin (1949).

9 When and why does entropy increase?

in which there are five particles (i.e., coins) and 10 equally probable microstates. Coin 1 has a marginal distribution of $P(H) = 0.6$ and $P(T) = 0.4$. A perusal of the table confirms that the other coins have the same marginal distribution.

Now compute the probability that coin 1 is H and coin 2 is H, which we write $P(HH)$. From the table we see that this probability is 0.30, and we have

$$P(HH) = 0.30 \approx 0.36 = (0.6)(0.6) = P(H)P(H).$$

Similar approximate equalities hold for $P(HT)$, $P(TH)$, and $P(TT)$. This looks suspiciously like probabilistic independence. In fact, as the number of particles gets very large, the approximate equality in this equation approaches equality, and the marginal distributions of the particles become independent of one another. A derivation is given in appendix B. (In applications outside of physics, it is often appropriate, or at least not incorrect, to stipulate this independence at the outset.)

The probability distribution in which the marginal distributions of the particles are strictly independent is called in physics the *canonical ensemble*, or *canonical distribution*. It is of practical importance that even in a system distributed according to the microcanonical ensemble (a closed system), if the system is a large one, small chunks of it will be distributed approximately according to the canonical ensemble. This is discussed in appendix B. To put it another way, in the microcanonical ensemble there are correlations among the particles; but if the system is large, the correlations are weak, and small chunks of the system behave as though the particles were truly independent.

Now we can summarize the main postulates of Khinchin's demonstration of entropy increase for a homogeneous system. His development assumes that the particles are distinguishable.[13]

(1) The system consists of N identical particles whose (discrete) states are called orbitals. A microstate for the system is given by specifying the orbital that each particle occupies.

(2) At equilibrium, there is a probability distribution over the microstates, and consequently marginal distributions for each of the particles. Each particle has the same marginal distribution, and (in the limit of large N) the marginal distributions are independent.

(3) Associated with each orbital is an energy. Energy is additive: the energy of the system in any microstate is the sum of the energies of the particles in their

---

[13] However, Khinchin's treatment can be adapted to apply, for example, to Bose and Fermi gases, in which the particles are not distinguishable.

orbitals. (This implies that interactions between the particles are weak or absent.)

(4) When two isolated systems are coupled, total energy is conserved. (If the probability distribution is spread over a range of energies, the *mean* total energy is conserved.)

(5) The probability distribution over the microstates is governed by one or more underlying parameters, such as the number of particles and the total energy. Another such parameter is the *temperature*, denoted τ, which in many systems is proportional to the average energy per particle. The parameters determine the distribution *uniformly*; i.e., they apply in the same way both before and after coupling.

(6) The probability that a given particle will be found in a specific orbital of energy ϵ when the system is at temperature τ is proportional to an exponential: $P \propto \exp(-\epsilon/\tau) = [\exp(-1/\tau)]^\epsilon$. (See the remarks following.)

(7) The value of τ for the system is determined uniquely by the requirement that the mean energy for each particle, multiplied by N, must give the (known) observable energy for the whole system. (According to assumption (5), this must hold before and after coupling.)

(8) When the conditions above are fulfilled, and in particular when τ is constrained to give the correct value for the total energy, then Khinchin's theorem states that the entropy of the coupled system must be at least as great as that of the system before coupling.

With reference to condition (6): when τ is positive, as it is for most systems in nature (including gases), the quantity $\exp(-1/\tau)$ is a positive constant $c < 1$. (The smaller τ is, the smaller c is.) Suppose for simplicity that the energy values $\epsilon_0, \epsilon_1, \epsilon_2, \ldots$ are equally spaced integers $0, 1, 2, \ldots$ Then the probabilities are proportional to $c^0, c^1, c^2, \ldots$ This means that the probability of a state of a particular energy is always a constant factor times the probability of a state of the *next lower* energy. This gives us a picture of gradually declining probability as the energy of the orbital gets higher. In fact, whether the temperature is positive or negative, (6) means that the probabilities decrease or increase monotonically with increasing energy.

When N (the number of particles) is large, the system is governed by the microcanonical distribution, and the temperature is not disturbed by small changes in the total energy, it can be *proved* that conditions (6) and (7) hold (see appendix B). We list them as independent postulates because in examples outside of physics (such as those in the next section), the microcanonical distribution may not obtain, or energy and temperature may not be definable in such a way as to have the listed properties.

9 When and why does entropy increase?

In this development, much of the simplicity of the classical explanation has been lost. In particular, it is not obvious from the abbreviated discussion just given *why* Khinchin's Theorem (item (8)) is true. (See Khinchin[12] for the mathematics.) What is important, however, is that the conditions preceding item (8) collectively constitute a true and applicable sufficient condition for entropy increase. When statistical mechanical principles are applied outside of physics, all or some of the conditions above are often met. When some of them fail to be met, it is possible for entropy to *decline*. We will see from the biological examples how this can occur.

3 Some biological examples

Natural selection is a process that is set in motion by the existence of variation in fitness. The result of the process is that variation in fitness is eliminated. Variation in fitness is like the oxygen that fuels a fire; it is a precondition for the fire to exist, but the fire uses up oxygen and so will bring itself to a halt.

Although it is not standard to think of natural selection as initiated by the coupling of two populations, it does no violence to the biological concepts to impose this formulation. We will think of two populations that are each at equilibrium and then are coupled; after selection has run its course, acting on the variation created by the coupling process, we can compute the entropies before coupling and the entropy after.

The simplest case involves two populations and two traits, A and B. Each population has some equilibrium frequency of the two traits. Then the populations are coupled and the conjoint population moves to an equilibrium frequency. What will happen to the entropy in this case?

Let's imagine that A and B are the two alleles that can occur at a given diploid locus. There are then three genotypes AA, AB, and BB; their frequency independent fitnesses are ω_{AA}, ω_{AB}, and ω_{BB}. Table 6 describes what happens when two genetically different populations are coupled. Each is at equilibrium before coupling; after coupling, the fitness ordering determines the new equilibrium that the conjoint population attains.

When a population is monomorphic, its entropy is zero. So, in the first and last processes described, the summed entropy of the uncoupled populations is zero, and the entropy of the coupled population is zero as well.

Table 6. *The effect of coupling two populations*

Fitness ordering	Before coupling		After coupling
	Pop. 1	Pop. 2	
$\omega_{AA} > \omega_{AB} > \omega_{BB}$	100% A	100% B	100% A
$\omega_{AB} > \omega_{AA} > \omega_{BB}$	100% A	100% B	AB polymorph.
	AB polymorph.	100% B	AB polymorph.
$\omega_{AA} > \omega_{BB} > \omega_{AB}$	100% A	100% B	100% A or 100% B

The two middle cases, which involve heterozygote superiority, are more interesting. When each of the populations before coupling is monomorphic, coupling makes the entropy increase. And when one of the populations before coupling is polymorphic, the entropies before and after coupling are identical and nonzero.

So in the one-locus setting we have considered, entropy never declines under coupling. However, it is not hard to find selection processes in which this pattern is violated. Let us simply add a third allele C to our consideration. The fitnesses of the genotypes are given in table 7.

Now consider two populations. In the first, A and B are the only alleles present; they are stably maintained because there is heterozygote superiority. In the second population, C is monomorphic. If these two populations are coupled, C will sweep to fixation. In this case, the sum of the entropies of the pre-coupled populations is nonzero, but the entropy of the coupled population after it reaches equilibrium is zero. Entropy has declined.

The situations described thus far bear some analogy to the statistical mechanical setup described in the previous section. The populations are homogeneous systems, whose macrostates are defined by the number of organisms and the gene frequencies involved. Microstates corresponding to

Table 7. *Hypothetical fitnesses for genotypes with three alleles*

	A	B	C
A	1	2	3
B		1	3
C			4

the macrostates are assignments of allowable genotypes to the individuals in the populations. We may define probabilities over the microstates by treating as equally probable all (and only those) microstates which have individual genotypes in the proportions prescribed by the macrostate. There are one or more underlying parameters which determine the macrostate, and hence the probability distribution over the microstates. Here the parameters are the trait fitnesses and frequencies, and the recipe for determining from them the equilibrium macrostate is that the gene frequencies adjust until the overall fitness of the population attains a maximum (Fisher's Fundamental Theorem). This recipe applies uniformly to the populations both before and after coupling.

There are important differences as well. There is no clear analog in a selection process to the energy, a quantity which is additive and which is conserved when two systems are coupled. (Fitness could be regarded as additive – take the fitness of each organism as its expected number of offspring, and the fitness of the population as the sum of the fitnesses of the individuals – but it is certainly not conserved when two systems are coupled.) Also, if we order the genotypes *AA, AB, BB*, the genotype frequencies will not necessarily be monotonically increasing or decreasing (both before and after coupling), as condition (6) in section 2 would require.

The result of the failure of these examples to correspond to important parts of the statistical mechanical picture is that entropy may increase or decrease, depending on the specific circumstances. An artifact of the selection process is that something or other will happen to entropy. But there is nothing intrinsic to the process of natural selection that allows us to say anything general about the direction of entropy change.

Matters are quite different when we turn away from selection and consider another example in population genetics. Wahlund's Principle describes the consequences of coupling populations that were previously isolated.[14] Suppose that an allele A occurs in each of k populations with frequencies $p_1, p_2, \ldots, p_k$ and that the alternative allele a occurs with frequencies $q_1, q_2, \ldots, q_k$, where $p_i + q_i = 1$. Within each population there is random mating. Equilibrium in each population is reached after at most one generation, with the genotypes *AA, Aa*, and *aa* in the Hardy–Weinberg proportions of p_i^2, $2p_iq_i$, and q_i^2. If the populations sizes are $n_1, n_2, \ldots, n_k$, the average frequency of the A allele

[14] Crow and Kimura (1970) pp. 54–5.

across these separate populations is

$$\frac{n_1 p_1 + n_2 p_2 + \cdots + n_k p_k}{n_1 + n_2 + \cdots + n_k} = \bar{p}.$$

Here we may view p as a random variable with possible values $p_1, \ldots, p_k$ defined on the space $\{1, \ldots, k\}$ with weights $n_1/\sum n_i, \ldots, n_k/\sum n_i$; so $\bar{p}$ is simply the mean of p. Similarly, the average frequency of the AA homozygote across the separate populations is

$$\frac{n_1 p_1^2 + n_2 p_2^2 + \cdots + n_k p_k^2}{n_1 + n_2 + \cdots + n_k} = \overline{p^2}.$$

If these populations are combined, the allele frequency of A in the new pooled population remains $\bar{p}$, so the frequency of AA homozygosity after random mating is $\bar{p}^2$.

The variance in the random variable p is by definition $V_p = \overline{p^2} - \bar{p}^2$. Since variance is always nonnegative, this means that there is less (or at least, no more) AA homozygosity in the pooled population than there was, on average, before the populations were pooled. The same holds, *mutatis mutandis*, for aa homozygosity. Or to put the point in terms of heterozygosity (i.e., the frequency of the Aa genotype), pooling increases heterozygosity.

Heterozygosity reflects the degree of mixing in the population of the alleles A and a. In thermodynamical situations, mixing of disparate species of particles is accompanied by an increase in entropy; so we were naturally led to inquire what happens to the entropy of populations of organisms coupled in this manner.

When we look at this problem carefully, a close correspondence emerges between the biology and statistical mechanics. The organisms are 'particles'. The possible genotypes of the organisms are 'orbitals'. The assignment of a genotype to each organism is a 'microstate'. The number of organisms and the total number of A (or a) alleles in the population together comprise the macrostate. We initially employ the artifice that the genotypes Aa and aA are distinct; we will correct this later. We may assign 'energies' to the orbitals as follows: the allele A has energy 1 and the allele a has energy 0. Then the orbitals have energy 0, 1, or 2; this is just the number of A alleles in the genotype. Energy is additive and sums to give the number of A alleles in the whole population. At Hardy–Weinberg equilibrium, the orbitals have probabilities as given in table 8.

9 When and why does entropy increase?

Table 8. *Genotype probabilities at Hardy–Weinberg equilibrium*

Orbital	Probability	
AA	p_i^2	$= q_i^2 (p_i/q_i)^2$
Aa	$p_i q_i$	$= q_i^2 (p_i/q_i)^1$
aA	$p_i q_i$	$= q_i^2 (p_i/q_i)^1$
aa	q_i^2	$= q_i^2 (p_i/q_i)^0$

The last column rewrites the probabilities to show how they are in geometric proportion with the energy as the exponent, as in clause (6) of Khinchin's development.[15] For random mating we may validly employ the *beanbag* model, in which each organism determines its genotype by 'drawing' randomly-with-replacement twice from the gamete pool; this entails that the orbital distributions of the 'particles' are independent. This means that the entropy of the ith population is just n_i times the orbital entropy:

$$\sigma_i = n_i (p_i^2 \log p_i^2 + 2 p_i q_i \log p_i q_i + q_i^2 \log q_i^2).$$

The isolation of the populations means that the population distributions are independent of each other, so the entropy of the populations taken all together is just the sum of the population entropies. (See appendix A.)

When the populations are coupled (pooled), the total 'energy' (i.e., the number of A alleles) is conserved. The pooled population reaches an equilibrium in which each organism has the same (new) orbital distribution. Thus all the conditions of Khinchin's theorem are fulfilled, and we may conclude that entropy in the combined population increases.

We now must attend to the detail referred to above: in population genetics the genotype Aa is not in general distinguishable from aA. So there are really three (not four) orbitals, corresponding to AA, Aa, and aa. The probabilities and entropy have to be adjusted to take this into account. It turns out (we omit the proof) that this small change in the setup makes no difference to the conclusion; entropy increases anyway.

Notice that in this example we could not derive the conclusion that entropy increases by employing the model of equally probable microstates in which

---

[15] This means that temperature may be defined so that $\exp(-1/\tau) = q_i^2(p_i/q_i)$, and then clause (6) is satisfied by this example.

coupling entails removal of constraints on the admissible microstates. This model requires that microstates of the (combined) system which are admissible before coupling remain admissible after coupling. But this condition is *not* fulfilled here. Take for example two populations of equal size: one with 80% A and the other with 20% A. In the first population the genotype frequencies for AA, Aa, and aa are 64%, 32%, and 4%; in the second population the genotype frequencies are 4%, 32%, and 64%. In the first population, take as equally probable microstates all genotype-to-organism assignments which have exactly the correct proportions of AA, Aa, and aa; do the same in the second population. In the combined (but uncoupled) population, all admissible microstates have frequency of AA equal to $(64\% + 4\%)/2 = 34\%$. After coupling, however, all admissible microstates have frequency of AA equal to $(50\%)^2$, or 25%; so none of the previously admissible microstates are admissible. The classical explanation of entropy increase cannot be applied here; but Khinchin's alternative viewpoint, as we have seen, proves fruitful.

When we try to extend the result to more general situations in population biology, we begin to see ways in which the statistical mechanical analogy must be weakened. The most obvious extension is to consider loci possessing more than two alleles, with random mating as before. It is no longer possible to assign in a consistent way a single energy to the genotypes, so that the total energy is conserved when the populations are coupled and random mating produces a new equilibrium. However, the gene frequency of each of the alleles $A_1, \ldots, A_m$ *is* preserved after random mating, so we could imagine m different species of energy, each representing the total number of alleles $A_1, \ldots, A_m$ in the population, and each separately conserved upon coupling of the populations. With no single quantity to play the role of energy, we must abandon assumption (6) but the other assumptions remain in force. In this situation, it turns out that entropy increases.

The question then arises as to whether the list of conditions would suffice for entropy increase if assumption (6) were deleted. To address this question, consider a situation in which this condition is violated. Suppose that we have a single locus with two or more alleles, and a set of *phenotypes* which supervene[16] on the genotypes at this locus. The simplest example is furnished by a recessive trait and two alleles A and a. Suppose a is recessive. The organism exhibits the

---

[16] A set of properties P is said to *supervene* on a set of properties Q precisely when the Q properties of an object determine what its P properties will be. See Kim (1978) for discussion.

9 When and why does entropy increase?

recessive trait t when its genotype is aa; otherwise it exhibits the dominant trait $\neg t$. The probabilities of t and $\neg t$ are q^2 and $p^2 + 2pq$ respectively. This lumping together of (genotypic) states and their probabilities is a kind of *coarse-graining*. Underlying the phenotypic distribution is another distribution, the genotypic distribution, in which an energy can be defined and all the other conditions of Khinchin's theorem are fulfilled.

With coupling by pooling and random mating, we know from the results above that the entropy of the underlying genotypic distribution increases. It is natural to suppose that the entropy of any distribution supervening on the underlying distribution will also increase, so long as the supervenience is defined in a uniform manner before and after coupling. But the entropy of the recessive/dominant distribution does not! (Or rather, it *need not*; it increases for some values of the gene frequencies and decreases for other values.) For example, take two n-organism populations, the first with the a allele at 50% and the second with the a allele at 0%. The corresponding phenotype frequencies are 25% and 0%. The entropy of the first population is $-n(0.25 \log 0.25 + 0.75 \log 0.75) = 0.562n$. The entropy of the second population is $-n(0 \log 0 + 1 \log 1) = 0$. The entropy of the two populations before coupling is therefore the sum $0.562n$. After coupling there are $2n$ organisms and the a allele has a frequency of 25%. The phenotype frequency is 6.25% and the entropy is $-2n(0.0625 \log 0.0625 + 0.9375 \log 0.9375) = 0.466n$. Entropy after coupling has declined.

A similar example is supplied by the ABO blood-type system. Here we have three alleles, A, B, and O, and four phenotypes which we denote a, b, ab, and o. The phenotypic supervenience is as follows: phenotype a corresponds to genotypes AA and AO, b to BB and BO, ab to AB, and o to OO. We know from the results already described that the entropy of the genotypic distribution (over the six genotypes) must increase after coupling with random mating. However, it is by no means obvious what happens to the entropy of the phenotypic distribution. We employed a computer program to calculate the phenotypic entropies before and after coupling with various values for the gene frequencies. It turns out that for most values of the gene frequencies, the entropy after coupling increases, but that for a small range of gene frequencies, entropy decreases. For example, a population of 10% A, 87% B, and 3% O will produce an entropy decrease when coupled to an equal sized population of 10% A, 60% B, and 30% O. It is an empirical accident that the values which produce entropy decrease happen to be far removed from the actual values of the gene frequencies in naturally occurring human populations.[17]

4 Conclusion

In the biological examples we have surveyed, the entropy of a population is determined by the frequency distribution of traits of organisms. It is important to realize that the entropy of a population, defined in this way, has nothing to do with the orderliness of the organisms in the population. Under Wahlund coupling, entropy goes up. Yet, the organisms in the post-coupling population may be no less ordered than the organisms were when the populations were isolated. Although heterozygosity increases under coupling, there is no clear sense in which heterozygotes are less ordered than homozygotes. What has changed here is the entropy *of* the population, not the orderliness of the organisms *in* the population.[18]

Although the entropy of the genotype distribution increases under Wahlund coupling, the entropy of the phenotype distribution need not do so. It is of some interest that supervening distributions can move in one direction while the distributions on which they supervene go in another. This point applies to chambers of gases no less than it does to Mendelian populations. Coupling two closed systems to produce a third can reduce entropy. The second law of thermodynamics says what will happen to *some* probability distributions, not to *all* of them. It does not take an energy input to get entropy to go down. Merely reconceptualizing the traits on which distributions are defined can do the trick.

[17] Because our discussion of thermodynamic entropy was restricted to equilibrium situations, we have discussed entropy in population genetics also from the point of view of pre- and post-coupling equilibria. However, it is quite easy to think about entropy in population genetics for nonequilibrium situations. An evolving population on its way to equilibrium will move through a sequence of gene frequencies; since probability distributions are thereby well-defined in every generation; so too are the entropies. What then happens to the entropy of evolving populations? For many evolutionary processes, entropy declines. A novel variant X that sweeps to fixation by selection will lead entropy to rise and then fall. Imagine that it is introduced into a population of 100% Y individuals. The frequency of X increases and the entropy reaches a maximum when the frequencies are 50% X and 50% Y. When X reaches 100%, the population's entropy is again zero. Drift also tends to destroy variation and so reduce entropy. Mutation and recombination, however, tend to have the opposite effect, since they increase the range of variation.

For a sampling of opinion concerning how the entropy concept might be applied to evolutionary processes, see Brooks & Wiley (1986), Collier (1986), Morowitz (1986), Wicken (1987), and Layzer (1990).

[18] Barrett & Sober (1992).

9 When and why does entropy increase?

Table 9. *Properties of the examples*

	Statistical mechanics	Selection	Genotypic Wahlund	Supervening Wahlund
Particle distributions identical and (almost) independent	Yes	Yes	Yes	Yes
Same process generates probabilities, before and after coupling	Yes	Yes	Yes	Yes
Additive energy analog, conserved by coupling	Yes	No	Yes	Yes
Exponential probabilities	Yes	No	Yes	No
Must entropy increase?	Yes	No	Yes	No

In the classical explanation of entropy increase in thermodynamics, entropy goes up because the removal of a physical barrier is taken to imply that constraints on admissible microstates are removed. The post-coupling system thus can exhibit all the microstates associated with the pre-coupling systems, and some more besides. The Khinchin explanation of entropy increase is less transparent than this simple account, which is why physicists rarely advert to it. Yet, a more general treatment of the problem of entropy increase suggests that the classical account may be limited in important ways. The deep affinity of Wahlund coupling to thermodynamic processes becomes clear when we use the Khinchin account, a parallelism that the classical account fails to describe.

The Khinchin account has a further virtue. It provides a template for analyzing cases in which entropy fails to increase. The Khinchin account lists a set of jointly sufficient conditions; entropy decline must violate at least one of them, as table 9 shows.

Discussions of entropy have tended to vacillate between two extremes. On the one hand there are unstructured and metaphorical explorations of the issues, which allow one to see analogs to thermodynamic processes everywhere. On the other hand, there is the tendency to restrict the concept to its strict physical meaning; though this reduces the scope (and, dare we say, the glamour) of the issue, at least it can be said that the problem of entropy increase remains in such treatments a precise and intelligible one. We have tried to steer a middle course. We have tried to isolate abstract features of the physical process in which isolated equilibrium systems are coupled and allowed to find their new equilibria. The question we have explored is what it is about this abstract formulation that forces entropy to increase. Our results are summarized in table 9.

In order to give the problem some structure, we have assumed that two populations of objects attain an equilibrium frequency distribution by some definite process. Then the populations are coupled, and the same process at work before coupling then takes the coupled system to a new equilibrium. As the example of selection illustrates, this minimum way of framing the problem does not guarantee an increase in entropy. The reason entropy need not increase under selection is that there is no additive energy analog, which is conserved under coupling. In the Wahlund process as well as in thermodynamics, there is a measure on each population that is the sum of the measures of all the individuals. To be sure, fitness is a measure of this sort – the productivity of a population is just the sum of the productivities of its member organisms. However, the point is that the fitnesses at equilibrium of the coupled population need not be the same as the sum of the fitnesses before coupling. There is no principle of fitness conservation. This is because the coupling operation triggers a process whose dynamics is controlled by the principle that fitter traits increase in frequency while less fit traits decline.

However, even if we add the constraint that the process must possess an additive energy analog that is conserved under coupling, this still is not enough to guarantee an increase in entropy. The genotypic Wahlund process generates an entropy increase, but the phenotypic Wahlund process does not. The difference here is due to the fact that genotypes in the Wahlund coupling process can be represented as obeying an exponential probability distribution with energy as a parameter, whereas the phenotypes (in our examples of the recessive trait and the ABO blood groups) cannot. The Khinchin proof does not apply to entropies defined on arbitrary probability distributions.

One of the most interesting features of the second law is its description of a process that is temporally asymmetric. It is therefore worth considering how this asymmetry is represented in our account of the second law as a proposition concerning the coupling processes. If entropy increases under coupling, what happens under *de*coupling? That is, if a population is subdivided, what will happen to its entropy? Here we must recognize that the actual energy (or actual gene frequency, etc.) of populations at equilibrium fluctuates about the mean energy (or mean gene frequency, etc.) derived from the governing distribution. If the population size is extremely large (on the order of 10^{23}), these fluctuations almost never depart measurably from the mean value. When such populations are subdivided and separated, we may therefore expect that the temperatures (or the temperature analogs) in the two parts are equal and equal to that of the system before decoupling; entropy therefore will be unchanged. Coupling and decoupling have qualitatively different effects.

In the case of populations that may be large but not 'virtually infinite', we must modify the contrast a bit. Fluctuations about the mean will be more significant, and there is a small chance that subdividing a population will create subpopulations that differ from each other appreciably. But, in expectation, the subdivided populations will differ from each other only a little. Decoupling leads to very minor entropy declines. It is a contingent fact about the large populations found in the world that they often differ from each other appreciably. It follows that coupling two such populations often leads to entropy increases that are much larger than the entropy decline that would result from subdividing any one of them. This fact, we emphasize, depends on both the 'laws of motion' of the coupling process and on the initial conditions that happen to obtain.

For the sake of completeness, we must consider the difference between coupling and decoupling for populations that are small. Here fluctuations around the equilibrium values can be large – indeed, large enough to obliterate the asymmetries noted for the previous two cases. For small populations, everything will depend on contingent facts concerning the actual sizes and frequencies at the moment of decoupling. One can imagine a universe in which populations are small enough that subdividing makes entropy go down more than coupling makes entropy go up. For example, consider a universe in which there are 10 populations of nearly identical small size and temperature. Coupling any two of them will leave entropy virtually unchanged; but, since the populations are small, subdividing and separating any one of them is likely to reduce entropy more.

In summary, the asymmetry described by the second law owes something to the 'laws of motion' governing the processes of coupling and subdividing, but also depends on contingent facts about the populations found in nature. The world we live in – whether we are talking about thermodynamic systems or Mendelian populations – consists of populations that are large enough and different enough that coupling and subdividing yield qualitatively different results.

Our strategy in exploring the general question of entropy increase has been to forget the physical meaning of concepts like 'orbital' and 'energy' and to focus on their abstract mathematical representation. In Wahlund coupling, the number of alleles of each type is conserved; it is quite irrelevant that the physical energy of a population (whatever that might mean) is unrelated to the number of alleles the population contains. Although entropy increases in thermodynamic processes because of the *physical* processes that take place, we have tried to understand why entropy increases as a *mathematical* problem. This strategy conflicts with one of the contrasts that Fisher[2] draws between entropy and fitness:

> Fitness, although measured by a uniform method, is qualitatively different for every different organism, whereas entropy, like temperature, is taken to have the same meaning for all physical systems [pp. 39–40].

Fisher's point about fitness is that it is multiply realizable. The fruitfulness of this abstract representation of the organic world is beyond dispute; it is astonishing that a single framework can encompass the evolution of life forms that are so physically diverse. In a sense, we have attempted in this paper to follow Fisher's example – to describe entropy as a mathematical property that also is multiply realizable. It remains to be seen to what extent this abstract treatment will bear fruit.

Appendix A: Some simple properties of entropy

The entropy of a finite partition $\{X_i\}_{i=1,\ldots,n}$ of a probability space is defined by the formula $\sigma = -\sum_{i=1}^{n} P(X_i) \log P(X_i)$. (We take $0 \log 0 = 0$.) Entropy is a measure of the spread of the distribution: roughly, the more possibilities over which the probability is spread, the greater the entropy.

(1) $\sigma \geq 0$. Proof: If $0 < x < 1$ then $\log x < 0$.
(2) If the distribution consists of n equiprobable alternatives, that is, $P(X_i) = 1/n$ for $i = 1, \ldots, n$, then $\sigma = \log n$. Proof:

$$\sigma = -\sum_{i=1}^{m} 1/n \log 1/n = -n(1/n \log 1/n) = \log n.$$

(3) If the distribution is spread over n alternatives, the maximum σ is attained when the alternatives are all equiprobable. Proof: omitted.
Recall that partitions $\{X_i\}_{i=1,\ldots,n}$ and $\{Y_j\}_{j=1,\ldots,m}$ of a probability space are said to be *independent* if $P(X_i \wedge Y_j) = P(X_i)P(Y_j)$ for all $i = 1, \ldots, n$ and

$j = 1, \ldots, m$. When this holds, we also say that the probability space (partition) $\{X_i \wedge Y_j\}_{i=1,\ldots,n, j=1,\ldots,m}$ is the *product* of $\{X_i\}_{i=1,\ldots,n}$ and $\{Y_j\}_{j=1,\ldots,m}$.

(4) If the partitions $\{X_i\}_{i=1,\ldots,n}$ and $\{Y_j\}_{j=1,\ldots,m}$ are independent, then the entropy of the product partition is the sum of the entropies of $\{X_i\}_{i=1,\ldots,n}$ and $\{Y_j\}_{j=1,\ldots,m}$. Proof:

$$\sigma = -\sum_{i,j} P(X_i \wedge Y_j) \log P(X_i \wedge Y_j)$$

$$= -\sum_{i,j} P(X_i)P(Y_j)[\log P(X_i) + \log P(Y_j)]$$

$$= -\sum_{i,j} P(Y_j)P(X_i) \log P(X_i) - \sum_{j,i} P(X_i)P(Y_j) \log P(Y_j)$$

$$= -\sum_{i} P(X_i) \log P(X_i) - \sum_{j} P(Y_j) \log P(Y_j).$$

(5) Corollary: if the partition $\{X_i\}$ with entropy σ is the product of N independent copies of a partition $\{Y_j\}$ with entropy σ', then $\sigma = N\sigma'$.

Appendix B: Some statistical mechanics

In this appendix we derive the exponential form of the marginal distributions for a single particle from the Fundamental Postulate of Statistical Mechanics, and show that single-particle distributions are identical and approximately independent. We consider a system S with N distinguishable particles, where N is very large. Each of the particles may occupy one of m possible states (orbitals) of varying energy. (Different orbitals may have the same energy.) We also assume that these orbital energies, $\epsilon_1, \ldots, \epsilon_m$, remain fixed throughout. A *microstate* of S is given by specifying an orbital for each particle. Paramagnetic spin systems (with a fixed external magnetic field) are one example of a system satisfying these conditions with $m = 2$.

We suppose that S is a *closed* system with (unchanging) total internal energy U. Then according to the Fundamental Postulate of Statistical Mechanics, each of the accessible microstates of energy U is equally probable. We denote the number of such states by $\#(N, U)$ to indicate the dependence on N and U. The entropy of a closed system with n particles and energy u is $\sigma(n, u) = \log \#(n, u)$ (see appendix A). We write $\sigma_S = \sigma(N, U)$ for the entropy of S. The *temperature* τ of S is defined in terms of N and U by

$$\frac{1}{\tau} = \left[\frac{\partial \sigma(n, u)}{\partial u}\right]_{n=N, u=U}.$$

Now let us focus on a small chunk of S with k particles; $k \ll N$. The remainder of S we can call the reservoir and denote it R. The reservoir is so large and contains so much internal energy that we can assume that small amounts of energy added to it or taken from it or small numbers of particles removed from it make no difference to its temperature. In other words, the derivative $\partial \sigma(N - k, u)/\partial u$ is constant for values of u near U and small k. This means that near U, the graph of σ versus u is a straight line with slope $[\partial \sigma(n, u)/\partial u]_{n=N-k, u=U} = 1/\tau$, and we can write

$$\sigma_R(U - \Delta u) = \sigma(N - k, U - \Delta u) = \sigma(N - k, U) - \frac{1}{\tau} \Delta u. \tag{1}$$

Now a state s of the chunk is simply a k-tuple of orbitals: $s = (i_1, \ldots, i_k)$, with energy $U_s = \epsilon_{i_1} + \cdots + \epsilon_{i_k}$. (This assumes, as we usually do in statistical mechanics, that there is no interaction energy between the particles.) We seek the probability P_s that the chunk is in state s. Since all microstates of S are equiprobable, P_s is proportional to the number of states of S in which the chunk is in state s. Since each such state is simply a state in which the reservoir R has energy $U - U_s$, P_s is proportional to the number of states of R with energy $U - U_s$, i.e., $\#(N - k, U - U_s)$. Using (1) we compute the logarithm of this number:

$$\log \#(N - k, U - U_s) = \sigma(N - k, U - U_s) = \sigma(N - k, U) - \frac{1}{\tau} U_s.$$

So we can write $P_s \propto \exp[\sigma(N - k, U) - U_s/\tau]$, or $P_s = c[\exp(-U_s/\tau)]$, where we have absorbed the factor $\exp[\sigma(N - k, U)]$ into the constant of proportionality c. Probabilities must be normalized to sum to 1, so we have

$$1 = \sum_s P_s = \sum_s c\, e^{-U_s/\tau} = c \sum_s e^{-U_s/\tau}.$$

The quantity $Z_k = Z_k(\tau) = \sum_s \exp(-U_s/\tau)$ is called the *partition function* (for the k-particle chunk) and is ubiquitous in statistical mechanics. We obtain, finally, $c = 1/Z_k$ and

$$P_s = \frac{e^{-U_s/\tau}}{Z_k}. \tag{2}$$

When $k = 1$, the chunk is composed of a single particle, and the energies U_s are simply the orbital energies $\epsilon_1, \ldots, \epsilon_m$. The result (2) then becomes:

$$P_i = e^{-\epsilon_i/\tau}/Z_1, \tag{3}$$

where $Z_1 = \sum_{i=1}^{m} \exp(-\epsilon_i/\tau)$ is the single-particle partition function, and P_i is the probability of being in orbital i. Equation (3) holds no matter which particle we consider, so we have shown that the single-particle distributions are identical and that the single-particle probabilities have exponential form.

We next want to show that within a small k-particle chunk, the single particle distributions are independent. (This is how we formulate the notion of approximate independence.) First we relate the partition functions Z_k and Z_1. We have

$$Z_k = \sum_s e^{-U_s/\tau} = \sum_{(i_1,\ldots,i_k)} e^{-(\epsilon_{i_1}+\cdots+\epsilon_{i_k})/\tau}$$

$$= \sum_{i_1} e^{-\epsilon_{i_1}/\tau} \ldots \sum_{i_k} e^{-\epsilon_{i_k}/\tau}$$

$$= Z_1 \ldots Z_1 = Z_1^k.$$

Using this result we can write

$$P(i_1,\ldots,i_k) = P_s = \frac{e^{-U_s/\tau}}{Z_k} = \frac{e^{-(\epsilon_{i_1}+\cdots+\epsilon_{i_k})/\tau}}{Z_1^k}$$

$$= \left(\frac{e^{-\epsilon_{i_1}/\tau}}{Z_1}\right) \ldots \left(\frac{e^{-\epsilon_{i_k}/\tau}}{Z_1}\right)$$

$$= P_{i_1} \ldots P_{i_k}$$

and the independence property is established.

PART 4 **Time travel and time's arrow**

10 Closed causal chains

PAUL HORWICH

By the expression 'closed causal chain' I mean a process (a causally connected sequence of events) that loops back in such a way that each event is indirectly a cause of itself. In this chapter I will explore whether such processes might actually occur and whether certain scientific theories that suggest them might be correct. Since these are questions about what *might* exist, they are highly ambiguous. For there are *various* notions of possibility – including, for example, logical consistency, compatibility with the laws of physics, feasibility given current technology, and consonance with known facts – and the status of closed causal chains may be examined with respect to any of these notions. However, since our interest is not confined to just one kind of possibility, there is no need to choose between them, and so we can put our problem as follows. In what senses (if any) are closed causal chains possible, and in what senses (if any) are they impossible?

Another way in which the issue before us is somewhat unclear is with respect to what is to qualify as a closed causal chain – whether the often cited, potential examples really involve *causation*, strictly speaking. For one might well wonder if any relation that an event can bear to *itself* could really count as the *causal* relation, the relation of *making something happen*. I shall try to mitigate (or, rather, avoid) this problem by raising the questions of possibility, not primarily with respect to the abstract notion, 'closed causal chain', but rather with respect to three specific theories, and by construing the expression, 'closed causal chain', by reference to the processes they postulate. My first example is the theoretical spacetime discovered and investigated by Gödel. He says that 'by making a round trip in a rocket ship it is possible in these worlds to travel into any region of the past, present, and future and back again, exactly as it is possible in other worlds to travel to different parts of space'.[1] Therefore someone might travel back and introduce his parents to one another; and

[1] Gödel (1949b).

this would constitute what I shall regard as a closed causal chain. A second physical theory apparently implying closed causal chains is Feynman's idea that positrons are nothing but electrons going backwards in time for a while.² If this is so then things could be arranged so that the event at which a certain electron starts again to move forwards in time exerts some causal influence on the trajectory of the electron at a point before it began to move backwards in time. Again we would have what I shall take to be a closed causal chain. My third example is relatively unrealistic. It is the prospect of perfect precognition, whereby future events somehow cause the conviction that those events will occur. Given such an ability someone might foresee himself winning the lottery and consequently go out and buy a ticket – again resulting in a closed causal chain. These three examples show what I shall mean by this expression, and I will not be concerned with whether they involve what we, strictly speaking, call 'causation'. Let me elaborate this point.

It is sometimes said that closed causal chains are conceptually or semantically impossible. The rationale for this claim is that allegedly the very meaning of the word 'cause' requires us to accept the principle, 'if E causes F, then F does not cause E'. According to some advocates of this view it registers a basic fact about the meaning of 'cause'; according to others it arises, indirectly, from the essential time-asymmetry of causation – that is, from the fact that the meaning of 'cause' requires us to accept, 'if E causes F, then E is earlier than F'. Either way, it follows that closed causal chains are, by definition, impossible.

A common critical response to this sort of semantic argument is to dismiss it as presupposing the analytic–synthetic distinction. The response, in other words, is to follow Quine³ in denying the existence of any distinction between truths of meaning and matters of fact, and inferring that there is no such thing as conceptual or semantic impossibility.

Another possible response – one that I prefer – is to concede that perhaps there *is* an analytic–synthetic distinction, perhaps it *is* analytically false that E causes F and F causes E, and perhaps, strictly speaking, closed causal chains *are* conceptually impossible. However, one can argue, this would all be of no importance, since the existence of what we, strictly speaking, call 'closed causal chains' is not something with which we are, or should be, concerned. For what

² Feynman (1949).
³ Quine (1951).

10 Closed causal chains

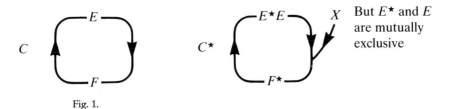

Fig. 1.

we really care about is what our response should be to the proposal of particular theories like the three mentioned above. It may or may not be the case that the adoption of such a theory would, if it is to be coherent, entail some *change* in the meaning of the word 'cause'. But even if it would, this linguistic fact has no bearing on whether or not the theory should be accepted. To put it another way, what concerns us is the truth of specific theories such as the ones that I have described. So we should be interested in any arguments tending to show that these theories are incoherent, or implausible for any other reason. But the issue of whether the word 'cause' would no longer mean precisely what it means now has no bearing on the question of plausibility. Therefore we should not be concerned with the semantic argument against closed causal chains.

The genuinely interesting (and most prominent) criticism of theories that would seem to permit closed causal chains is known as *the bilking argument*.[4] A familiar example of it is that if time travel were possible then it would be possible to go back and murder one's infant self (autofanticide!) – which is a contradiction – therefore time travel is impossible. Similarly it might be argued that if there really were electrons moving backwards in time we could arrange things so that an electron sent back in time would trigger a device designed to prevent the electron from being sent back in the first place – again a contradiction – therefore Feynman's theory cannot be true. The general form of the bilking argument is that, given a closed causal chain that regularly occurs in a certain sort of environment, it would be possible to *modify* the environment so that some element of the chain would cause something different from what it causes in the original environment, and where this difference would work its way around the chain in such a way that the 'first' event would be precluded. In other words, as shown in figure 1, if there were a closed causal chain of type C, involving E causing F causing E, then there could be a causal chain of type C^*, with the same spatio-temporal structure

---

[4] See Tolman (1917), Flew (1954), Black (1956), Pears (1957), Dummett (1964), Earman (1972), Mellor (1981), and Horwich (1987).

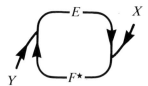

Fig. 2.

as C, involving E causing F^* causing E^* – but where E^* and E are logically incompatible. Therefore, since C^* evidently cannot occur, C must also be impossible.

This argument presupposes that the background environment of the closed process, C, is in fact susceptible to modification into conditions in which E would cause E^*. For if not, then the existence of C would not imply the existence of C^*; so bilking could not occur. But this assumption about the nature and modifiability of C's background environment is an empirical assumption – not necessarily true. Consequently the bilking argument shows, at best, that closed causal chains are empirically implausible – not that they are conceptually impossible.

Let me put the bilking argument in a slightly different way – a way designed to show how the presupposed assumption, although empirical, is very plausible indeed. Ask yourself what would happen if chains of type C occurred and if we *tried repeatedly* to construct self-defeating chains of type C^*. Of course, we would fail. We know this because C^* is contradictory. But the question is: what would account for these failures? How could E continue to cause itself even in those circumstances in which we disrupt the process by means of some new input X? Evidently the addition of our input X to the environment of E would not allow E to ultimately engender itself unless the environment of the process co-operated with some *further* change – call it Y (as shown in figure 2).

That is to say, in the modified background environment, containing X, E would cause itself only given a specific additional input Y from the environment that 'compensates' for X by enabling F^* to cause E. But, the bilking argument continues, there would be no explanation for the fact that every time we add X to the environment of C, the appropriate correction Y occurs. Since neither of these changes in the environment of C causes the other and since they have no common cause, the correlation between X and Y would be a sheer coincidence. And it is an empirical fact that sheer coincidences do not occur. Thus it is indeed the case (a matter of empirical fact) that the conditions surrounding

C would be modifiable – they are not constrained in the way needed to block the inference from C to C^*. Therefore, the empirical premise of the bilking argument is indeed plausible, and its conclusion is probably correct.

Suppose, for example, people try over and over again to commit autofanticide. On every attempt something happens to frustrate their plans – guns are constantly jamming, poisons spilling, etc. But there is no causal connection between decisions to commit autofanticide and the guns' jamming so often. Neither one causes the other, nor is there any common cause. Thus Gödelian time travel would imply massive (indeed, limitlessly extendable) coincidence: a phenomenon of the sort we know from experience to be absent from our world. We can infer, therefore, that Gödelian time travel will not take place. It is epistemologically impossible.[5]

The core of the bilking argument, as we have just seen, is that the existence of closed causal chains implies the existence of inexplicable coincidences. On this basis it is then natural to reject the prospect of closed causal chains, together with any theories that would countenance them. However, I now wish to suggest that this natural move is not generally legitimate. Let us explore its domain of legitimacy by looking at the argument in the context of our three examples.

Firstly let us consider the case of perfect precognition. Suppose I adopt the policy of trying to prevent the events that I have foreseen. In each case I will be thwarted. Therefore there will be a correlation between *occasions on which I decide to embark on some such bilking attempt* and *the occurrence of circumstances that prevent me from doing things that I can ordinarily do more or less at will.* For example, normally, if I decide that I am not going to have dinner with a certain irritating person tomorrow night, I am able to achieve this goal. Now, suppose that whenever I foresee a dinner party involving me and someone I don't like, I try to prevent it from happening by changing my plans. But suppose these manoeuvres always fail; sometimes I simply forget the change of plan, sometimes the person I am avoiding unexpectedly shows up at another place – in one way or another, things invariably go wrong. Thus my ability to do something that is normally fairly easy to do is thwarted every time by unfavourable events, and such events are highly correlated with my having a bilking intention. Given what we know of the world, this could happen only if there were some causal connexion between these things. But my bilking decision does not cause the unfavourable events, nor do they

---

[5] For an extended discussion of this line of thought, see Horwich (1987).

cause my bilking decision, and nor is there some third thing that causes them both. Thus we know that such a correlation would never arise. So we can conclude that perfect precognition of the sort imagined would never occur. This is a case in which the bilking argument is quite effective. It shows that a certain sort of perfect precognition is epistemologically impossible.

Secondly there is the case of Gödelian time travel. Here again we can argue, given our knowledge of human nature, that the practice of time travel into the immediate past would bring with it the occurrence of repeated bilking attempts (attempts, for example, to assassinate monstrous dictators while they were still children), implying the occurrence of circumstances that will cause failure – and so there would be an unexplained coincidence in the correlation between such frustrating circumstances and bilking attempts. But such coincidences are ruled out by what we know of the world. Consequently, closed causal chains deriving from Gödelian time travel are epistemologically impossible.

So far so good. It is a further step, however, to infer from this result that our spacetime does not have the kind of Gödelian structure that would permit such chains. One might think that we must acknowledge the non-Gödelian nature of spacetime in order to account for the conclusion of the bilking argument – namely, the non-existence of closed causal chains. However, there are alternative ways of accommodating this conclusion. One such alternative was urged by Gödel himself. He showed that even if spacetime permitted time travel to the immediate past, such trips would always be technologically impossible because the energy required would amount to the mass of several galaxies. Thus the bilking argument shows, in this case, that Gödelian closed causal chains are epistemologically impossible. But since this is explained by the prohibitive fuel requirements, we cannot infer the impossibility of Gödelian spacetimes. And therefore the door remains open for time travel into those regions of the past that would not provide the opportunity for closed causal chains and for bilking attempts.

Finally let us consider the import of the empirical bilking argument for Feynman's theory of positrons. We begin with a closed chain in which the event ZIG, at which the electron begins to move back in time, causes an event ZAG, at which the electron encounters a gamma ray and begins to move forward in time again, which causes a signal S that perturbs the trajectory of the incoming electron (as shown in figure 3(a)).

In order to develop a bilking modification of this set-up it must be assumed that the causal chain is repeatable at will – which implies that we have some

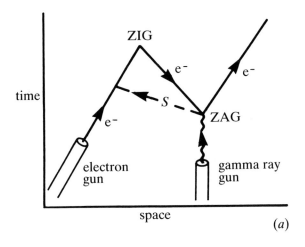

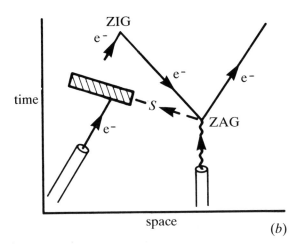

Fig. 3.

control over when and where the ZIG and ZAG events occur. If this assumption is false then there can be no bilking argument against Feynman's theory. If however the repeatability assumption is true, then a bilking modification could be made by the inclusion of a device whereby the signal S would trigger a screen that would prevent the electron from entering the region (figure 3(b)). In that situation any incoming electron must have come from beyond the screen. Thus, given the addition of the screening device (which is finally triggered only by ZAG, the arrival of a positron), there is a correlation between the electron and gamma guns being fired, and the invariable arrival of a positron

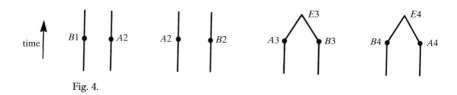

Fig. 4.

to trigger the screen. In other words there is, in the bilking set-up, a coincidence of *the firing of the electron and gamma guns* and *an occurrence of ZAG*.

It is tempting to infer, as in the previous examples, that such a coincidence is ludicrously improbable, and thereby to conclude that the Feynman theory is false. But this would be an error, I believe. For it is an error to suppose that *all* coincidences are improbable. What we know on the basis of observation is, at most, the following. There do not occur cases in which events of type A are correlated with events of type B, and where each A and each B is the product of entirely separate causal chains.[6] In other words, we never observe the event patterns represented in figure 4. Rather, whenever A and B are correlated they are embedded in structures of spatio-temporally continuous, nomological determination that are either *linear*, as in figure 5, or V-*shaped*, as in figure 6.

Notice that I have not yet given a *causal* characterization of these structures. In a context in which unusual causal orientations are under consideration there is room for a causal interpretation of some of the linear or V-shaped patterns of determination according to which A and B are not causally connected. One might conceivably maintain, in the linear case, that the direction of 'causation' *changes* at some point between A and B – that it goes from past to future before that point and from future to past afterwards. Similarly, regarding certain V-shaped structures of determination, one might say that the direction of 'causation' is *toward* the point of the V. Thus we would have cases of correlation without causal connexion – but cases that could not be regarded as empirically improbable.

Turning back to the positron example we can see that the uncaused correlation there is precisely of this kind. More specifically, it provides an uncaused correlation whose structure is represented in the *linear* pattern – where A represents the firing of the electron and gamma guns and B represents the

---

[6] This general phenomenon was noted in Reichenbach (1956). It has recently come under critical scrutiny, for example in Sober (1988) and in Arntzenius (1990). The moral of this work, it seems to me, is not that there is no such phenomenon, but rather that the phenomenon is difficult to characterize accurately.

10 Closed causal chains

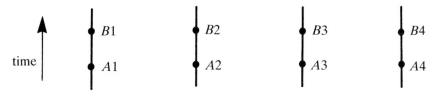

Fig. 5.

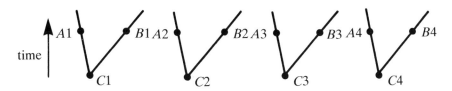

Fig. 6.

ZAG event. Thus the empirical bilking argument does not suggest that Feynman's theory of positrons is epistemologically impossible.

My conclusions are as follows. Theories that countenance what are naturally termed 'closed causal chains' cannot be dismissed on *a priori* semantic grounds, and cannot invariably be ruled out by the *a posteriori* bilking argument. The latter argument shows that closed causal chains imply uncaused correlations. But the import of this result is not uniform. In some cases (for example, perfect precognition) the proper conclusion is that such a phenomenon is incompatible with what we know. In others (for example, Gödelian time travel) we may infer that closed causal chains do not and will not occur – but we cannot conclude that a spacetime structure permitting them is not actual. And in a third class of cases (including, for example, Feynman's theory of positrons) the bilking argument has minimal force; for the uncaused correlations to which it draws our attention are not improbable. The overall moral is that the bilking argument, even in its strongest form, should not be regarded as a *general* argument against backwards causation and closed causal chains. Its significance can be assessed only with respect to specific hypotheses.

11 Recent work on time travel

JOHN EARMAN

Introduction

Over the last few years leading scientific journals have been publishing articles dealing with time travel and time machines. (An unsystematic survey produced the following count for 1990–1992. *Physical Review D*: 11; *Physical Review Letters*: 5; *Classical and Quantum Gravity*: 3; *Annals of the New York Academy of Sciences*: 2; *Journal of Mathematical Physics*: 1. A total of 22 articles involving 22 authors.[1]) Why? Have physicists decided to set up in competition with science fiction writers and Hollywood producers? More seriously, does this research cast any light on the sorts of problems and puzzles that have featured in the philosophical literature on time travel?

The last question is not easy to answer. The philosophical literature on time travel is full of sound and fury, but the significance remains opaque. Most of the literature focuses on two matters, backward causation and the paradoxes of time travel.[2] Properly understood, the first is irrelevant to the type of time travel most deserving of serious attention; and the latter, while always good for a chuckle, are a crude and unilluminating means of approaching some delicate and deep issues about the nature of physical possibility. The overarching goal of this chapter is to refocus attention on what I take to be the

---

[1] See Carroll, Farhi, & Guth (1992); Charlton & Clarke (1990); Cutler (1992); Deser, Jakiw, & 't Hooft (1992); Deutsch (1991); Echeverria, Klinkhammer, & Thorne (1991); Friedman & Morris (1991a, 1991b); Friedman, Morris, Novikov, Echeverria, Klinkhammer, Thorne, & Yurtsever (1990); Frolov (1991); Frolov & Novikov (1990); Gibbons & Hawking (1992); Gott (1991); Hawking (1992); Kim & Thorne (1991); Klinkhammer (1992); Novikov (1992); Ori (1991); 't Hooft (1992); Thorne (1991); Yurtsever (1990, 1991). See also Deser (1993); Goldwirth, Perry, & Piran (1993); Mikheeva & Novikov (1992); Morris, Thorne, & Yurtsever (1988); and Novikov (1989).

[2] A representative but by no means complete sample of this literature is given by Brown (1992); Chapman (1982); Dummett (1986); Dwyer (1975, 1977, 1978); Ehring (1987); Harrison (1971); Horwich (1987); Lewis (1976); MacBeath (1982); Mellor (1981); Smith (1986); Thom (1975); and Weir (1988).

important unresolved problems about time travel and to use the recent work in physics to sharpen the formulation of these issues.[3]

The plan of the chapter is as follows. Section 1 distinguishes two main types of time travel – Wellsian and Gödelian. The Wellsian type is inextricably bound up with backward causation. By contrast, the Gödelian type does not involve backward causation, at least not in the form that arises in Wellsian stories of time travel. This is not to say, however, that Gödelian time travel is unproblematic. The bulk of the chapter is devoted to attempts, first, to get a more accurate fix on what the problems are and, second, to provide an assessment of the different means of dealing with these problems. Section 2 provides a brief excursion into the hierarchy of causality conditions on relativistic spacetimes and introduces the concepts needed to assess the problems and prospects of Gödelian time travel. Section 3 reviews the known examples of general relativistic cosmological models allowing Gödelian time travel. Since Gödel's discovery,[4] it has been found that closed timelike curves (CTCs) exist in a wide variety of solutions to Einstein's field equations (EFE). This suggests that insofar as classical general relativity theory is to be taken seriously, so must the possibility of Gödelian time travel. Section 4 introduces the infamous grandfather paradox of time travel. It is argued that such paradoxes involve both less and more than initially meets the eye. Such paradoxes cannot possibly show that time travel is conceptually or physically impossible. Rather the parading of the paradoxes is a rather ham-fisted way of making the point that local data in spacetimes with CTCs are constrained in unfamiliar ways. The shape and status of these constraints has to be discerned by other means. Section 5 poses the problem of the status of the consistency constraints in terms of an apparent incongruence between two concepts of physical possibility that diverge when CTCs are present. Section 6 considers various therapies for the time travel malaise caused by this incongruence. My preferred therapy would provide an account of laws of nature on which the consistency constraints entailed by CTCs are themselves laws. I offer an account of laws that holds out the hope of implementing the preferred therapy. This approach is investigated by looking at recent work in physics concerning the nature of consistency constraints for both non-self-interacting systems (section 7) and self-interacting systems (section 8) in

[3] Although it is a truism, it needs repeating that philosophy of science quickly becomes sterile when it loses contact with what is going on in science.
[4] Gödel (1949a).

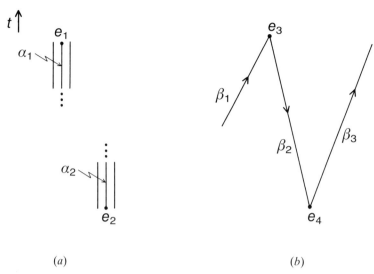

Fig. 1. Two forms of Wellsian time travel

spacetimes with CTCs. Section 9 investigates a question that is related to but different from the question of whether time travel is possible; namely, is it possible to build a time machine that will produce CTCs where none existed before? Some concluding remarks are given in section 10.

1 Types of time travel; backward causation

Two quite different types of time travel feature in the science fiction and the philosophical literature, though the stories are often vague enough that it is hard to tell which is intended (or whether some altogether different mechanism is supposed to be operating). In what I will call the *Wellsian type*[5] the time travel takes place in a garden variety spacetime – say, Newtonian spacetime of classical physics or Minkowski spacetime of special relativistic physics. So the funny business in this kind of time travel does not enter in terms of spatiotemporal structure but in two other places: the structure of the world lines of the time travellers and the causal relations among the events on these world lines. Figure 1 illustrates two variants of the Wellsian theme. Figure 1(*a*) shows the time traveller α_1 cruising along in his time machine. At e_1 he sets the time travel dial to 'minus 200 years', throws the

[5] This appellation is suggested by some passages in H. G. Wells' *Time Machine* (1895), but I do not claim to have captured what Wells meant by time travel. For a review of and references to some of the science fiction literature on time travel, see Gardner (1988).

switch, and *presto* he and the machine disappear. Two hundred years prior to e_1 (as measured in Newtonian absolute time or the inertial time (t) of the frame in which α_1 is at rest) a person (α_2) exactly resembling the time traveller both in terms of physical appearance and mental state pops into existence at e_2. Even if we swallow these extraordinary occurrences, the description given so far does not justify the appellation of 'time travel'. That appellation requires that although α_1 is discontinuous with α_2, α_2 is in some appropriate sense a continuation of α_1. Whatever else that sense involves, it seems to require that events on α_1 cause the events on α_2. Thus enters backward causation.

Figure 1(*b*) also involves funny world line structure, but now instead of being discontinuous, the world line 'bends backwards' on itself, the arrows on the various segments indicating increasing biological time. Of course, as with the previous case, this one also admits an alternative interpretation that involves no time travel. As described in external time, the sequence of events is as follows. At e_4 a pair of middle aged twins are spontaneously created; the β_3 twin ages in the normal way while his β_2 brother gets progressively younger; meanwhile, a third person, β_1, who undergoes normal biological aging and who is the temporal mirror image of β_2, is cruising for a fateful meeting with β_2; when β_1 and β_2 meet at e_3 they annihilate one another. Once again, the preference for the time travel description seems to require a causal significance to the arrows on the world line segments so that, for example, later events on β_1 (as measured in external time) cause earlier events on β_2 (again as measured in external time).

Much of the philosophical literature on Wellsian time travel revolves around the question of whether backward causation is conceptually or physically possible, with the discussion of this question often focusing on the 'paradoxes' to which backward causation would give rise. I will not treat these paradoxes here except to say that they have various similarities to the paradoxes of Gödelian time travel that will receive detailed treatment in section 4. But aside from such paradoxes, there is the prior matter of whether the phenomena represented in figure 1 are physically possible, even when shorn of their time travel/backward causation interpretations. In figure 1(*a*), for example, the creation *ex nihilo* at e_2 and the extinction *ad nihilo* at e_1 are at odds with well entrenched conservation principles. Of course, the scenario can be modified so that conservation of mass–energy is respected: at e_1 the time traveller and the time machine dematerialize as before but now their mass is replaced by an equivalent amount of energy, while at e_2

a non-material form of energy is converted into an equivalent amount of ponderable matter. But this emended scenario is much less receptive to a time travel/backward causation reading. For the causal resultants from e_1 can be traced forwards in time in the usual way while the causal antecedents to e_2 can be traced backwards in time, thus weakening the motivation for seeing a causal link going from e_1 to e_2.

At first blush Gödelian time travel would seem to have three advantages over Wellsian time travel. First, on the most straightforward reading of physical possibility – compatibility with accepted laws of physics – Gödelian time travel would seem to count as physically possible, at least as regards the laws of the general theory of relativity (GTR). Second, unlike stories of Wellsian time travel, Gödelian stories are not open to a rereading on which no time travel takes place. And third, no backward causation is involved. On further analysis, however, the first advantage turns out to be something of a mirage since (as discussed below in sections 5–8) Gödelian time travel produces a tension in the naive conception of physical possibility. And the second and third advantages are gained in a manner that could lead one to object that so-called Gödelian time travel is not time travel after all.

To begin the explanation of the claims, I need to say in some detail what is meant by Gödelian time travel. This type of time travel does not involve any funny business with discontinuous world lines or world lines that are 'bent backwards' on themselves. Rather the funny business all derives from the structure of the spacetime which, of course, cannot be Newtonian or Minkowskian. The funny spacetimes contain continuous and even infinitely differentiable timelike curves such that if one traces along such a curve, always moving in the future direction as defined by the globally defined external time orientation, one eventually returns to the very same spacetime location from whence one began. There is no room here for equivocation or alternative descriptions; hence the second advantage. (More cautiously and more precisely, there are some spacetimes admitting Gödelian time travel in the form of closed, future directed timelike curves, and the curves cannot be unrolled into open curves on which events are repeated over and over *ad infinitum* – at least such a reinterpretation cannot be made without doing damage to the local topological features of the spacetime; see section 3.) As for the third advantage, consider a CTC γ that is instantiated by, say, a massive particle. Pick two nearby events $x, y \in \gamma$ such that x chronologically precedes y ($x \ll y$ in the notation defined in section 2). One might be tempted to say that backward causation is involved since although y is chronologically later than x, y causally affects x. But the situation here is quite different from that in Wellsian time travel. In universes with Gödelian time travel it is consistent to assume – and, in fact, is implicitly assumed in standard relativistic treatments – that all causal influences in the form of energy–momentum transfers propagate forward in time with a speed less than or equal to that of light. So in the case at issue, y is a cause of x because $y \ll x$ and because there is a continuous causal process linking y to x and involving *always future directed causal propagation*. Of course, one could posit that there is another kind of causal influence, not involving energy–momentum transfer, by which y affects x backwards in time, so that even if the future directed segment of γ from y to x were to disappear, y would still be a cause of x. But the point is that Gödelian time travel need not implicate such a backward causal influence.

We are now in a position to see why the second and third advantages have been purchased at a price. One can object that Gödelian time travel does not deliver time travel in the sense wanted since so-called Gödelian time travel implies that there is no time in the usual sense in which to 'go back'. In Gödel's universe,[4] for example, there is no serial time order for events since for every spacetime point x, $x \ll x$; nor is there a single time slice which would permit one to speak of the Gödel universe at-a-given-time. I feel that there is a good deal of justice to this complaint. But I also feel that the phenomenon of 'time travel' in the Gödel universe and in other general relativistic cosmologies is a worthy object of investigation, whether under the label of 'time travel' or not. The bulk of this chapter is devoted to that investigation.

Before turning to that investigation, it is worth mentioning for sake of completeness other senses of time travel that appear in the literature. For example, Chapman and Zemach[6] devise various scenarios built around the notion of 'two times'. One interpretation of such schemes would involve the replacement of the usual relativistic conception of spacetime as a four-dimensional manifold equipped with a pseudo-Riemannian metric of signature $(+ + + -)$ (three space dimensions plus one time dimension) with a five-dimensional manifold equipped with a metric of signature $(+ + + - -)$ (three space dimensions and two time dimensions). This scheme is worthy of investigation in its own right, but I will confine my attention here to standard relativistic spacetimes.

[6] Chapman (1982) and Zemach (1968).

2 The causal structure of relativistic spacetimes

There is an infinite hierarchy of causality conditions that can be imposed on relativistic spacetimes.[7] I will mention only enough of these conditions to give some flavor of what the hierarchy is like. The review also serves the purpose of introducing the concepts needed for an assessment of Gödelian time travel.

A *relativistic spacetime* consists of a manifold M equipped with an everywhere defined Lorentz signature metric g_{ab}. For the real world, of course, we are interested in the case $\dim(M) = 4$ and metric signature $(+++-)$, but for ease of illustration examples will often be given for $\dim(M) = 2$ and a metric of signature $(+-)$. The basic presupposition of the causality hierarchy is that of temporal orientability.

(C0) M, g_{ab} is *temporally orientable* iff the null cones of g_{ab} admit of a continuous division into two sets, 'past' and 'future'.[8]

Which set is which is part of the problem of the direction of time, a problem that for present purposes we may assume to have been resolved.

With a choice of temporal orientation in place, we can say that for $x, y \in M$, x *chronologically precedes* y (symbolically, $x \ll y$) iff there is a smooth future-directed timelike curve from x to y. Similarly, x *causally precedes* y ($x < y$) iff there is a smooth future-directed non-spacelike curve from x to y. It follows without any further restrictions on M, g_{ab} that $\ll$ and $<$ are transitive relations. The first condition of the causality hierarchy says that $\ll$ has the other property we expect of an order relation, viz. irreflexivity.

(C1) M, g_{ab} exhibits *chronology* iff $\neg \exists x \in M$ such that $x \ll x$.

Chronology is, of course, equivalent to saying that the spacetime does not permit Gödelian time travel. The next condition up the hierarchy is simple causality.

(C2) M, g_{ab} exhibits *simple causality* iff $\neg \exists x \in M$ such that $x < x$.

To go further up the hierarchy we need the definitions of the *chronological future* $I^+(x)$ and *chronological past* $I^-(x)$ of a point $x \in M$: $I^+(x) \equiv \{y \in M : x \ll y\}$ and $I^-(x) \equiv \{y \in M : y \ll x\}$.

---

[7] Carter (1971).
[8] One way to make (C0) precise is to require that there exists on M a continuous, everywhere defined, timelike vector field.

(C3) M, g_{ab} is *future* (respectively, *past*) *distinguishing* iff $\forall x, y \in M, I^+(x) = I^+(y) \Rightarrow x = y$ (respectively, $I^-(x) = I^-(y) \Rightarrow x = y$).

Stronger than both simple causality and past and future distinguishing is the condition of strong causality.

(C4) M, g_{ab} is *strongly causal* iff $\forall p \in M$ and $\forall$ open neighborhoods $N(p) \subseteq M$, $\exists$ open $N'(p) \subseteq N(p)$ such that once a future directed causal curve leaves N' it never returns.

Intuitively, (C4) not only rules out closed causal curves but also 'almost closed' causal curves. The reader can now envision how requiring no 'almost almost closed' or no 'almost almost almost closed' etc. causal curves can produce a countable hierarchy of causality conditions.

(C4) is still not strong enough to guarantee the existence of a time structure similar to that of familiar Newtonian or Minkowski spacetime, both of which possess a time function. The spacetime M, g_{ab} is said to possess a *global time function* iff there is a differentiable map $t: M \to \mathbb{R}$ such that the gradient of t is a past directed timelike vector field. This implies that $t(x) < t(y)$ whenever $x \ll y$. The necessary and sufficient condition for such a function is given in the next condition in the hierarchy.

(C5) M, g_{ab} is *stably causal* iff $\exists$ on M a smooth non-vanishing timelike vector field t^a such that M, g'_{ab} satisfies chronology, where $g'_{ab} \equiv g_{ab} - t_a t_b$ and $t_a \equiv g_{ab} t^b$.

Intuitively, (C5) says that it is possible to widen out the null cones of g_{ab} without allowing CTCs to appear.

None of the conditions given so far are enough to guarantee that causality in the sense of determinism has a fighting chance on the global scale. That guarantee is provided for in the next condition.

(C6) M, g_{ab} possess a *Cauchy surface* iff $\exists$ a spacelike hypersurface $S \subset M$ such that S is intersected exactly once by every timelike curve without endpoint.

On any spacelike hypersurface S of M, g_{ab} one can specify initial data and attempt to use the relevant laws to project the data into the future and into the past. If S is not a Cauchy surface, the attempt is liable to break down in some region of M. If M, g_{ab} admits one Cauchy surface, then it can be partitioned by them. In fact, a global time function t can be chosen so that each of the level surfaces $t =$ constant is Cauchy.

There are even stronger conditions above (C6), but they will play no role in what follows. For future reference, a *time slice* of M, g_{ab} is defined to be a spacelike hypersurface without edges. (This is the generalization of a $t =$ constant surface of Minkowski spacetime.) A spacelike hypersurface that is not intersected more than once by any timelike curve is said to be *achronal*. A partial *Cauchy surface* is an achronal time slice. For the familiar hyperbolic partial differential equations that govern fields in relativistic physics, initial conditions on a partial Cauchy surface S of M, g_{ab} will determine the state of the field in the *domain of dependence* $D(S)$ of S, which consists of the union of the future domain $D^+(S) \equiv \{x \in M:$ every past endless causal curve through x meets $S\}$ and the past domain $D^-(S) \equiv \{x \in M:$ every future endless causal curve through x meets $S\}$. Of course, if S is a Cauchy surface for M, g_{ab} then $D(S) = M$. The future boundary of $D^+(S)$, called the *future Cauchy horizon* of S, is denoted by $H^+(S)$.

The philosophical literature has devoted most of its attention to the ends of the hierarchy, principally to (C0), (C1), and (C6), and has largely neglected (C2)–(C5) and the infinity of other conditions that have not been enumerated. There are both good and dubious reasons for this selective attention. The intimate connection of (C0) and (C6) respectively to the longstanding philosophical problems of the direction of time and determinism is enough to explain and justify the attention lavished on these conditions. Focusing on (C1) to the exclusion of (C2) and (C5) and all that lies between can be motivated by two considerations. First, if one takes seriously the possibility that chronology can be violated, then one must *a fortiori* take seriously the possibility that all the higher conditions above can fail. Second, Joshi[9] showed that a good bit of the hierarchy above (C1) collapses under a natural continuity condition. Define the *causal future* $J^+(x)$ and the *causal past* $J^-(x)$ of a point $x \in M$ analogously to $I^+(x)$ and $I^-(x)$ respectively with $<$ in place of $\ll$. The continuity condition in question says that $J^\pm(x)$ are closed sets for all x. For this condition to fail there would have to be a situation where $x < y_n$, $n = 1, 2, 3, \ldots$, with $y_n \to y$ but $\neg(x < y)$, i.e. a causal signal can be sent from x to each of the points y_n but not to the limit point y.

Despite these good reasons for the selective focus, I suspect that most of the philosophical attention lavished on (C1) derives from the fascination with the

[9] Joshi (1985).

paradoxes of time travel, and that I take to be a dubious motivation. But before taking up this matter in section 4, I turn to reasons for taking seriously the possibility of chronology violation.

3 Why take Gödelian time travel seriously?

Any relativistic spacetime M, g_{ab} based on a compact M contains closed CTCs.[10] Stronger results are derivable for cosmological models of GTR. A *cosmological model* consists of a triple M, g_{ab}, T^{ab} where M, g_{ab} is a relativistic spacetime and T^{ab} is a tensor field describing the mass–energy distribution on M. In addition to requiring the satisfaction of Einstein's field equations (EFE), general relativists typically demand that any candidate for the source T^{ab} of the gravitational field satisfies various energy conditions. For example, the *weak energy condition* requires that $T_{ab}V^aV^b \geq 0$ for any timelike V^a; intuitively, the demand is that the energy density of the source, as measured by any observer, is non-negative. By continuity, the weak energy condition entails the *null energy condition* which requires that $T_{ab}K^aK^b \geq 0$ for every null vector K^a. The model is said to be *generic* iff every timelike or null geodesic feels a tidal force at some point in its history (see Hawking and Ellis[11] for the technical details and physical motivation). Tipler[12] established that if the cosmological model M, g_{ab}, T^{ab} satisfies EFE with zero cosmological constant, the weak energy condition, and the generic condition, then compactness of M entails that the spacetime is *totally vicious* in that $x \ll x$ for every $x \in M$.

CTCs are not confined to compact spacetimes. In Gödel's cosmological model, $M = \mathbb{R}^4$.[13] This example is also important in that it illustrates that the failure of chronology can be *intrinsic* in that chronology cannot be restored by 'unwinding' the CTCs. More precisely, an intrinsic violation of chronology occurs when the CTCs do not result (as in figure 2(b)) by making identifications in a chronology respecting covering spacetime (figure 2(a)).

[10] See Geroch (1967).
[11] Hawking & Ellis (1973).
[12] Tipler (1977).
[13] Good treatments of Gödel's model are to be found in Malament (1984), Pfarr (1981), and Stein (1970). Gödel's original (1949a) cosmological model is a dust filled universe, i.e. $T^{ab} = \rho U^a U^b$ where ρ is the density of the dust and U^a is the normed four velocity of the dust. This model is a solution to Einstein's field equations only for a non-vanishing cosmological constant $\Lambda = -\omega^2$, where ω is the magnitude of the vorticity of matter. Alternatively, the model can be taken to be a solution to EFE with $\Lambda = 0$ and $T^{ab} = \rho U^a U^b + (\omega^2/8\pi)g^{ab}$; however, this energy–momentum tensor does not seem to have any plausible physical interpretation. Gödel's original model gives no cosmological redshift. Later Gödel (1952) generalized the model to allow for a redshift.

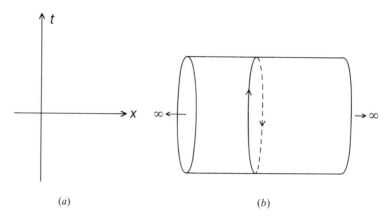

Fig. 2. (a) Two-dimensional Minkowski spacetime. (b) Rolled up Minkowski spacetime

Gödel spacetime is totally vicious. But there are other cosmological models satisfying EFE and the energy conditions where chronology is violated but not viciously. This raises the question of whether it is possible to have the chronology violating set V so small that it is unnoticeable in the sense of being measure zero. The answer is negative since V is always an open set.[11]

In the Gödel universe all CTCs are non-geodesic, necessitating the use of a rocket ship to accomplish the time travel journey. Malament[14] provided estimates of the acceleration and the fuel/payload ratio needed to make such a journey. These quantities are so large as to make the journey a practical impossibility. It was on this basis that Gödel[15] himself felt justified in ignoring the paradoxes of time travel. Such complacency, however, is not justified. Oszvath[16] produced a generalization of Gödel's model that accommodates electromagnetism. De[17] showed that in such a universe time travellers do not need to use a rocket ship; if they are electrically charged they can use the Lorentz force to travel along CTCs. Even better from the point of view of a lazy would-be time traveller would be a cosmological model with intrinsic chronology violation where some timelike geodesics are closed. An example is provided by the Taub–NUT model which is a vacuum solution to EFE (energy conditions trivially satisfied).

[14] Malament (1984).
[15] Gödel (1949b).
[16] Oszvath (1967).
[17] De (1969).

As a result of being totally vicious and simply connected, Gödel spacetime does not contain a single time slice so that one cannot speak of the Gödel universe at-a-given-time. But there are solutions to EFE which are intrinsically chronology violating but which do contain time slices. Indeed, the time slices can themselves be achronal, and thus partial Cauchy surfaces, even though CTCs develop to the future. This raises the question of whether GTR allows for the possibility of building a time machine whose operation in some sense causes the development of CTCs where none existed before. This matter will be taken up in section 9.

None of the chronology violating models discussed so far (with the exception of the trivial example of figure 2(b)) are asymptotically flat. But CTCs can occur in such a setting. The Kerr solutions to EFE form a two-parameter family described by the value of the mass m and the angular momentum a. The case where $m^2 > a^2$ is thought to describe the unique final exterior state of a non-charged stationary black hole. In this case chronology is satisfied. When $m^2 < a^2$ the violation of chronology is totally vicious. Charlton and Clarke[18] suggest that the latter case could arise if a collapsing rotating star does not dissipate enough angular momentum to form a black hole.

The solution to EFE for an infinite rotating cylinder source contains CTCs.[19] Tipler suggests that chronology violation may also take place for a finite cylinder source if the rotation rate is great enough.

The Gödel, Oszvath, Kerr, and Tipler models all involve rotating matter. But this is not an essential condition for the appearance of CTCs in models of GTR – recall that the Taub–NUT model is a vacuum solution. Also Morris, Thorne, and Yurtsever[20] found that generic relative motions of the mouths of traversable 'wormholes' (multiply connected surfaces) can produce CTCs, as can generic gravitational redshifts at the wormhole mouths.[21] However, the maintenance of the traversable wormholes implies the use of exotic matter that violates standard energy conditions. Such violations may or may not be allowed by quantum field theory (see section 9 below).

Gott[22] found that the relative motion of two infinitely long cosmic strings can give rise to CTCs. This discovery has generated considerable interest and

[18] Charlton & Clarke (1990).
[19] Tipler (1974).
[20] Morris, Thorne, & Yurtsever (1988).
[21] Frolov & Novikov (1990).
[22] Gott (1991).

controversy.²³ Part of the interest lies in the fact that Gott's solution, unlike the wormhole solutions, does not violate the standard energy conditions of classical GTR. The global structure of Gott's solution has been elucidated by Cutler.²⁴

The upshot of this discussion is that since the pioneering work of Gödel over forty years ago, it has been found that CTCs can appear in a wide variety of circumstances described by classical GTR and semi-classical quantum gravity. And more broadly, there are many other known examples of violations of causality principles higher up in the hierarchy. One reaction, which is shared by a vocal if not large segment of the physics community, holds that insofar as these theories are to be taken seriously, the possibility of violations of various conditions in the causality hierarchy, including chronology, must also be taken seriously. (This is the attitude of the 'Consortium' led by Kip Thorne.²⁵) Another vocal and influential minority conjectures that GTR has within itself the resources to show that chronology violations can be ignored because, for example, it can be proved that if chronology violations are not present to begin with, they cannot arise from physically reasonable initial data, or because such violations are of 'measure zero' in the space of solutions to EFE. (This position is championed by Hawking.²⁶) If such conjectures turn out to be false, one can still take the attitude that in the short run classical GTR needs to be supplemented by principles that rule out violations of the causality hierarchy, and one can hope that in the long run the quantization of gravity will relieve the need for such *ad hoc* supplementation. Which of these attitudes it is reasonable to adopt will depend in large measure on whether it is possible to achieve a peaceful coexistence with CTCs. It is to that matter I now turn.

4 The paradoxes of time travel

The darling of the philosophical literature on Gödelian time travel is the 'grandfather paradox' and its variants. For example: Kurt travels into the past and shoots his grandfather at a time before grandpa became a father, thus preventing Kurt from being born, with the upshot that there is no Kurt

---

[23] See Carroll, Farhi, & Guth (1992); Deser, Jakiw, & 't Hooft (1992); Deser (1993); Ori (1991); and 't Hooft (1992).
[24] Cutler (1992).
[25] Friedman, Morris, Novikov, Echeverria, Klinkhammer, Thorne, & Yurtsever (1990).
[26] Hawking (1992).

to travel into the past to kill his grandfather so that Kurt is born after all and travels into the past ... (Though the point is obvious, it is nevertheless worth emphasizing that killing one's grandfather is overkill. If initially Kurt was not present in the vicinity of some early segment of his grandfather's world line, then travelling along a trajectory that will take him into that vicinity, even if done with a heart innocent of any murderous intention, is enough to produce an antinomy. This remark will be important to the eventual unravelling of the real significance of the grandfather paradox.)

On one level it is easy to understand the fascination that such paradoxes have exercised – they are cute and their formal elucidation calls for the sorts of apparatus that are the stock in trade of philosophy. But at a deeper level there is a meta-puzzle connected with the amount of attention lavished on them. For what could such paradoxes possibly show? Could the grandfather paradox show that Gödelian time travel is not logically or mathematically possible?[27] Certainly not, for we have mathematically consistent models in which CTCs are instantiated by physical processes. Could the grandfather paradox show that Gödelian time travel is not conceptually possible? Perhaps so, but it is not evident what interest such a demonstration would have. The grandfather paradox does bring out a clash between Gödelian time travel and what might be held to be conceptual truths about spatiotemporal/causal order. But in the same way the twin paradox of special relativity theory reveals a clash between the structure of relativistic spacetimes and what were held to be conceptual truths about time lapse. The special and general theories of relativity have both produced conceptual revolutions. The twin paradox and the grandfather paradox help to emphasize how radical these revolutions are, but they do not show that these revolutions are not sustainable or contain inherent contradictions. Could the grandfather paradox show that Gödelian time travel is not physically possible? No, at least not if physically possible means compatible with EFE and the energy conditions, for we have models which satisfy these laws and which contain CTCs. Could the paradox show that although Gödelian time travel is physically possible it is not physically realistic? This is not even a definite claim until the relevant sense of 'physically realistic' is specified. And in the abstract it is not easy to see how the grandfather paradox would support that claim as opposed to the claim that time travel is flatly impossible.

---

[27] Some philosophers apparently think that time travel is logically or conceptually impossible; see Hospers (1967), 177, and Swinburne (1968), 169.

Does the grandfather paradox at least demonstrate that there is a tension between time travel and free will? Of course Kurt cannot succeed in killing his grandfather. But one might demand an explanation of why Kurt does not succeed. He has the ability, the opportunity, and (let us assume) the desire. What then *prevents* him from succeeding? Some authors pose this question in the rhetorical mode, suggesting that there is no satisfactory answer so that either time travel or free will must give way. But if the question is intended non-rhetorically, it has an answer of exactly the same form as the answer to analogous questions that arise when no CTCs exist and no time travel is in the offing. Suppose, for instance, that in the time travel scenario Kurt had his young grandfather in the sights of a 30–30 rifle but did not pull the trigger. The reason that the trigger was not pulled is that laws of physics and the relevant circumstances make pulling the trigger impossible at the relevant spacetime location. With CTCs present, *global* Laplacian determinism (which requires a Cauchy surface – see section 2) is inoperable. But *local* determinism makes perfectly good sense. In any spacetime M, g_{ab}, chronology violating or not, and any $x \in M$ one can always choose a small enough neighborhood $N(x)$ such that N, $g_{ab}|_N$ possesses a Cauchy surface S (where $g_{ab}|_N$ denotes the restriction of the metric g_{ab} to $N \subset M$). And the relevant initial data on S together with the coupled Einstein–matter equations will uniquely determine the state at x. Taking x to be the location of the fateful event of Kurt's pulling/not pulling the trigger and carrying through the details of the deterministic physics for the case in question shows why Kurt did not pull the trigger. Of course, one can go on to raise the usual puzzles about free will; viz. granting the validity of what was just said, is there not a way of making room for Kurt to have exercised free will in the sense that he could have done otherwise? At this point all of the well choreographed moves come into play. There are those (the *incompatibilists*) who will respond with arguments intended to show that determinism implies that Kurt could not have done otherwise, and there are others (the *compatibilists*) waiting to respond with equally well rehearsed counterarguments to show that determinism and free will can coexist in harmony. But all of this has to do with the classic puzzles of determinism and free will and nothing to do with CTCs and time travel *per se*.

Perhaps we have missed something. Suppose that Kurt tries over and over again to kill his grandfather. Of course, each time Kurt fails – sometimes because his desire to pull the trigger evaporates before the opportune moment, sometimes because although his murderous desire remains unabated his hand cramps before he can pull the trigger, sometimes because although he pulls the trigger the gun misfires, sometimes because although the gun fires the bullet is deflected, etc. In each instance we can give a deterministic explanation of the failure. But the obtaining of all the initial conditions that result in the accumulated failures may seem to involve a coincidence that is monstrously improbable.[28] Here we have reached a real issue but one which is not easy to tackle.

A first step towards clarification can be taken by recognizing that the improbability issue can be formulated using inanimate objects. (Consider, for example, the behavior of the macroscopic objects in my study as I write: a radiator is radiating heat, a light bulb is radiating electromagnetic waves, etc. If the world lines of these objects are CTCs, it would seem to require an improbable conspiracy to return these objects to their current states, as required by the completion of the time loop.) Since free will is a murky and controversial concept, it is best to set it aside in initial efforts at divining the implications of the grandfather paradox. After some progress has been made it may then be possible to draw some consequences for free will. As a second step we need to formalize the intuition of improbability. One method would be to define a measure on the space of solutions to EFE and to try to show that the solutions corresponding to some kinds of time travel (those involving the functional equivalent of Kurt trying over and over again to kill his grandfather) have negligible or zero measure. Even if such a demonstration is forthcoming, we still have to face the question: so what? (After all, some types of space travel will be measure zero, but this hardly shows that the concept of space travel is suspect.) The answer will depend crucially on the justification for and significance of the measure. This matter will receive some attention in section 9. But for the moment I want to note that the impression of improbability in connection with time travel stories may not be self-reenforcing. In the above example the judgment of the improbability of the failure of Kurt's repeated attempts to kill his grandfather was made relative to our (presumably chronology respecting) world; but perhaps from the perspective of the time travel world itself there is no improbability. By way of analogy, suppose that the actual world is governed by all of the familiar laws of classical relativistic physics *save for* Maxwell's laws of electromagnetism. If we peered into another world which is nomologically accessible

[28] Horwich (1987).

from our world but which is governed by Maxwell's laws we would see things that from our perspective are improbable ('measure zero') coincidences. We would find, for example, that the electric and magnetic fields on a time slice cannot be freely specified but must satisfy a set of constraints; and we would find that once these constraints are satisfied at any moment they are thereafter maintained for all time. Amazing! But, of course, from the perspective of the new world there is no improbability at all; indeed, just the opposite is true since the 'amazing coincidences' are consequences of the laws of that world. That this analogy may be apt to the case of time travel will be taken up in sections 5 and 6.

What then remains of the grandfather paradox? The paradox does point to a seemingly awkward feature of spacetimes that contain CTCs: local data are constrained in a way that they are not in spacetimes with a more normal causal structure. But the forms in which the paradox has been considered in the philosophical literature are of little help in getting an accurate gauge of the shape and extent of the constraints. And by itself the paradox is of no help at all in assessing the status of these consistency constraints.

5 Consistency constraints

The laws of special and general relativistic physics that will be considered here are all local in the following two-fold sense. First, they deal with physical situations that are characterized by local geometric object fields O (e.g. scalar, vector, and tensor fields) on a manifold M. Second, the laws governing these fields are in the form of local ordinary or local partial differential equations. The result is a *global-to-local property*: if M, g_{ab}, O satisfies the laws and $U \subseteq M$ is an open neighborhood, then U, $g_{ab}|_U$, $O|_U$ also satisfies the laws. (This property holds whether or not CTCs are present.) Thus, it would seem at first blush that the question of whether some local state of affairs is physically possible can be answered by focusing exclusively on what is happening locally and ignoring what is happening elsewhere.

In Minkowski spacetime and in general relativistic spacetimes with nice causality properties we typically have the reverse *local-to-global property*: any local solution can be extended to a global solution.[29] Consider, for example,

---

[29] Sometimes the local-to-global property may fail in causally nice spacetimes because singularities develop in solutions to some field equation. But one may regard such a failure as indicating that the field is not a fundamental one.

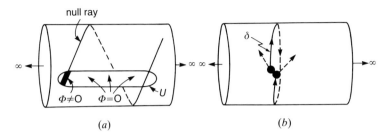

Fig. 3. Light rays and billiard balls on rolled up Minkowski spacetime

the source-free wave equation for a massless scalar field Φ: $g^{ab}\nabla_a\nabla_b\Phi \equiv \Box\Phi = 0$, where ∇_a is the derivative operator associated with g_{ab}. On Minkowski spacetime ($M = \mathbb{R}^4$ and $g_{ab} = \eta_{ab}$ (Minkowski metric)), any C^∞ solution on an open $U \subset \mathbb{R}^4$ can be extended to a full solution on $\mathbb{R}^4$. But obviously this local-to-global property fails for the chronology violating spacetime of figure 2(b). Figure 3(a) shows a local solution with a single pencil of rays traversing U. This solution is obviously globally inconsistent since the light rays from U will trip around the spacetime and reintersect U, as shown in figure 3(a).

The point is straightforward, but some attempts to elaborate it make it sound mysterious. Thus, consider the presentation of the Consortium:

> The only type of causality violation the authors would find unacceptable is that embodied in the science-fiction concept of going backward in time and killing one's younger self ('changing the past' [grandfather paradox]). Some years ago one of us (Novikov) briefly considered the possibility that CTC's might exist and argued that they cannot entail this type of causality violation: Events on a CTC are already guaranteed to be self-consistent, Novikov argued; they influence each other around the closed curve in a self-adjusted, cyclical, self-consistent way. The other authors have recently arrived at the same viewpoint.
>
> We shall embody this viewpoint in a *principle of self-consistency*, which states that *the only solutions to the laws of physics that can occur locally in the real universe are those which are globally self-consistent*. This principle allows one to build a local solution to the equations of physics only if that local solution can be extended to be part of a (not necessarily) unique global solution, which is well defined throughout the nonsingular regions of spacetime. (Friedman et al.[25] pp. 1916–17)

The first part of the quotation seems to invoke either a notion of pre-established harmony or else a guiding hand that prevents the formation of an inconsistent scenario. But once such connotations are removed, the 'principle of self-consistency' (PSC) threatens to deflate into a truism. Here is the Consortium's comment:

> That the principle of self-consistency is not totally tautological becomes clear when one considers the following alternative: The laws of physics might permit CTC's; and when CTC's occur, they might trigger new kinds of local physics, which we have not previously met ... The principle of self-consistency is intended to rule out such behavior. It insists that local physics is governed by the same types of physical laws as we deal with in the absence of CTC's ... If one is inclined from the outset to ignore or discount the possibility of new physics, then one will regard self-consistency as a trivial principle. (Friedman et al.[25] p. 1917)

What the Consortium means by discounting the possibility of 'new physics' is, for example, ignoring the possibility that propagation around a CTC can lead to multi-valued fields, calling for new types of laws that tolerate such multi-valuedness. I too will ignore this possibility. But I will argue shortly that CTCs may call for 'new physics' in another sense. For the moment, however, all I want to insist upon is that taking the PSC at face value seems to force a distinction between two senses of physical possibility.

In keeping with the global-to-local property introduced above, we can repeat, more pedantically, what was said informally: a local situation is *physically possible$_1$* iff it is a local solution of the laws. But the PSC – which says that the only solutions which can occur locally are those which are *globally* self-consistent – seems to require a more demanding and relativized sense of physical possibility; viz. a local situation is *physically possible$_2$* in a spacetime M, g_{ab} iff it can be extended to a solution of the laws on all of M, g_{ab}. Calling the conditions a local solution must fulfil in order to be extendible to a global solution the *consistency constraints*, one can roughly paraphrase physical possibility$_2$ as physical possibility$_1$ plus satisfaction of the consistency constraints.[30]

This distinction might be regarded as desirable for its ability to let one have one's cake and eat it too. On one hand, we have the intuition that it is physically possible to construct and launch a rocket probe in any direction we like with any velocity less than that of light. This intuition is captured by physical possibility$_1$. But on the other hand it is not possible to realize all of the physically possible$_1$ initial conditions for such a device in spacetimes with certain

---

[30] The need for such a distinction has been previously noted by Bryson Brown (1992).

kinds of CTCs since the traverse of a CTC may lead the probe to interfere with itself in an inconsistent way. (Or so it would seem; but see the discussion of section 8 below.) This impossibility is captured by the failure of physical possibility$_2$.

But on reflection, however, having one's cake and eating it too is, as usual, too good to be true. Thus, one might maintain with some justice that to be physically impossible just is to be incompatible with the laws of physics – as is codified in the definition of physical possibility$_1$. So as it stands the notion of physical impossibility$_2$ seems misnamed when it does not reduce to physical impossibility$_1$, and it appears not to when CTCs are present. To come at this point from a slightly different angle, let us reconsider the grandfather paradox. It was suggested in the preceding section that Kurt's failure to carry out his murderous intentions could be explained in the usual way – by reference to conditions that obtained before the crucial event and the (locally) deterministic evolution of these conditions. But while not incorrect, such an explanation deflects attention from a doubly puzzling aspect of spacetimes with CTCs. First, it may not even be possible for Kurt to set out on his murderous journey, much less to carry out his intentions. And second, the ultimate root of this impossibility does not lie in prior contingent conditions since there are no such conditions that can be realized in the spacetime at issue and which result in the commencement of the journey. The ultimate root of this impossibility taps the fact that (as we are supposing) there is no consistent way to continue Kurt's journey in the spacetime. But, one might complain, to call this impossibility 'physical impossibility$_2$' is to give it a label that is not backed by any explanatory power since the way the story has been told so far the local conditions corresponding to the commencement of Kurt's journey are compatible with all of the postulated laws. In such a complaint the reader will detect a way of trying to scratch the residual itch of the grandfather paradox.

The itch must be dealt with once and for all. I see three main treatments, the first two of which promise permanent cures while the third denies the existence of the ailment.

6 Therapies for time travel malaise

(T1) This treatment aims at resolving the tension between physical possibility$_1$ and physical possibility$_2$ by getting rid of the latter. The leading idea is to argue that GTR shows that, strictly speaking, the notion of consistency constraints is incoherent. For example, looking for consistency constraints in a spacetime M, g_{ab} for a scalar field obeying $\Box\Phi = 0$ makes sense if Φ is treated as a test field on a fixed spacetime background. But (the argument continues) this is contrary to both the letter and spirit of GTR. For Φ will contribute to the total energy–momentum – the usual prescription being that $T_{ab}(\Phi) = \nabla_a \Phi \nabla_b \Phi - 1/2 g_{ab} \nabla_c \Phi \nabla^c \Phi$ – that generates the gravitational field cum metric. And (one could conjecture) if Φ and Φ' are interestingly different (say, they differ by more than an additive constant), then the metrics g_{ab} and g'_{ab} solving EFE for the corresponding $T_{ab}(\Phi)$ and $T'_{ab}(\Phi')$ will be different (i.e. non-isometric). This therapy is radical in that if it succeeds it succeeds by the draconian measure of equating the physically possible$_1$ local states with the actual states. This is intuitively unsatisfying. If we restrict attention to Φ's such that $T_{ab}(\Phi)$ is small in comparison with the total T_{ab} and the spacetime is stable under small perturbations of T_{ab}, then Φ can to a good approximation be treated as a test field. Questions of stability will be examined in section 9, but meanwhile I will assume that they can be set aside. One could also object that (T1) is inapplicable in cases where there are CTCs and where the laws entail that the spacetime structure is non-dynamical and that a variety of physically possible$_1$ states can be realized on a given local region of the fixed spacetime. However, the strength of this objection is hard to grasp since the laws in question would have to be rather different from those of our world, at least if something akin to GTR is true. And recall that the success of GTR is the main reason for taking Gödelian time travel seriously.

(T2) The second treatment strategy is to naturalize physical possibility$_2$. The idea is, first, to insist that physical possibility$_2$ (relative to a world) just is the compatibility with the laws (of that world) and, second, to go on to argue that physical possibility$_2$ can be brought into the fold by showing that in chronology violating environments the consistency constraints of physical possibility$_2$ have law status. Thus, (T2) insists that, contrary to the Consortium's explanation of the PSC, there is a sense in which CTCs do call forth 'new physics'.

(T2) can take two forms. (a) The naturalization of physical possibility$_2$ would amount to a reduction to physical possibility$_1$, understood as consistency with the local laws of physics, if the consistency constraints/new laws were purely local so that, even in the chronology violating environments, what is physically possible locally is exactly what is compatible with the (now augmented) local laws of physics. (b) Unfortunately, the reduction of (a) can be expected in only very special cases. In general the consistency constraints may have to refer to the global structure of spacetime. In these latter cases, insofar as (T2) is correct, the concept of physical possibility$_1$ must be understood to mean consistency of the local situation with all the laws, local and non-local. The patient who demands a purely local explanation of the difference between local conditions that are physically possible and those which are not will continue to itch.

(T3) If the first two therapies fail, the discomfort the patient feels can be classified as psychosomatic. The therapist can urge that the patient is getting over excited about nothing or at least about nothing to do specifically with time travel; for global features of spacetime other than CTCs can also impose constraints on initial data. For example, particle horizons in standard big-bang cosmologies prevent the implementation of the Sommerfeld radiation condition which says that no source-free electromagnetic radiation comes in from infinity.[31] Here the patient may brighten for a moment only to relapse into melancholy upon further reflection. For the constraints entailed by the particle horizons are of quite a different character than those entailed by the typical chronology violating environment; the former, unlike the latter, do not conflict with the local-to-global property and thus do not drive a wedge between physical possibility, and physical possibility$_2$. And in any case the choice of the particle horizons example is not apt for therapeutic purposes since these horizons are widely thought to be so problematic as to call for new physics involving cosmic inflation or other non-standard scenarios.[32] Clearly this line of therapy opens up a number of issues that require careful investigation; but such an investigation is beyond the scope of this chapter.

My working hypothesis favors (T2). This is not because I think that (T2) will succeed; indeed, I am somewhat pessimistic about the prospects of success. Nevertheless, making (T2) the focus of attention seems justified on several grounds: the success of (T2) would provide the most satisfying resolution of the nagging worries about time travel, while its failure would have significant negative implications for time travel; whether it succeeds or fails, (T2) provides an illuminating perspective from which to read recent work on the physics of

---

[31] See Penrose (1964).
[32] For a review of the horizon problem and attempted solutions, see Earman (1994).

time travel; and finally, (T2) forces us to confront issues about the nature of the concept of physical law in chronology violating spacetimes, issues which most of the literature on time travel conveniently manages to avoid.

It is well to note that my working hypothesis is incompatible with some analyses of laws. For example, Carroll[33] rejects the idea that laws supervene on occurrent facts, and adopts two principles which have the effect that laws of the actual world $W_@$ are transportable *as laws* to other possible worlds which are nomologically accessible from $W_@$. The first principle says that 'if P is physically possible and Q is a law, then Q would (still) be a law if P were the case'. The second says that 'if P is physically possible and Q is not a law, then Q would (still) not be a law if P were the case'. Let P say that spacetime has the structure of Gödel spacetime or some other spacetime with CTCs. And let us agree that P is physically possible because it is compatible with the laws of $W_@$ (which for the sake of discussion we may take to be the laws of classical general relativistic physics). It follows from Carroll's principles that if spacetime were Gödelian, the laws of $W_@$ would still be laws and also that these would be the only laws that would obtain.

I will not attempt to argue here for the supervenience of laws on occurrent facts but will simply assume it. In exploring my working hypothesis, I will rely on an account of laws that can be traced back to John Stuart Mill;[34] its modern form is due to Frank Ramsey and David Lewis.[35] The gist of the M-R-L account is that a law for a logically possible world W is an axiom or theorem of the best overall deductive system for W (or what is common to the systems that tie for the best). A deductive system for W is a deductively closed, axiomatizable, set of (non-modal) sentences, each of which is true in W. Deductive systems are ranked by how well they achieve a compromise between strength or information content on one hand and simplicity on the other. Simplicity is a notoriously vague and slippery notion, but the hope is that, regardless of how the details are settled, there will be for the actual world a clearly best system or at least a non-trivial common core to the systems that tie for best. If not, the M-R-L theorist is prepared to admit that there are no laws for our world.

The M-R-L account of laws is naturalistic (all that exists is spacetime and its contents); actualistic (there is only one actual world); and empiricistic (a world is

---

[33] Carroll (1994).
[34] Mill (1904).
[35] Ramsey (1978) and Lewis (1973).

a totality of occurrent facts; there are no irreducible modal facts). In addition I would claim that this account fits nicely with the actual methodology used by scientists in search of laws. The reader should be warned, however, that it is far from being universally accepted among philosophers of science.[36] I will not attempt any defense of the M-R-L account here. If you like, the ability to illuminate the problems of time travel can be regarded as a test case for the M-R-L account.

Suppose for sake of discussion that the actual world has a spacetime without CTCs; perhaps, for example, its global features are described more or less by one of the Robertson–Walker big-bang models. And suppose that the M-R-L laws of this world are just the things dubbed laws in textbooks on relativistic physics, no more, no less. Now consider some other logically possible world whose spacetime contains CTCs. But so as not to waste time on possibilities that are too far removed from actuality, let us agree to restrict our attention to worlds that are nomologically accessible from the actual world in that the laws of the actual world, taken as non-modal propositions, are all true of these worlds. Nevertheless, we cannot safely assume that the M-R-L laws of our world 'govern' these time travel worlds in the sense that the set of laws of our world coincides with the set of laws of time travel worlds.

One possibility is that the M-R-L laws of a time travel world W consist of the M-R-L laws of this world *plus* the consistency constraints on the test fields in question. If so, we have a naturalization of physical possibility$_2$, though it would remain to be seen whether the naturalization takes the preferred (a) form or the less desirable (b) form. Additionally, time travel would have implications for free will. In cases where an action is determined by the laws plus contingent initial conditions, compatibilists and incompatibilists split on whether the actor could be said to have the power to do otherwise and whether the action is free. But all parties to the free will debate agree that if an action is precluded by the laws alone, then the action is not in any interesting sense open to the agent. Thus, under the scenario we are discussing, there are various actions that, from a compatibilist perspective at least, we are free to perform in the actual world that we are not free to perform in the time travel world.

Other possibilities also beg for consideration. For instance, it could turn out that although (by construction) the M-R-L laws of this world are all *true* of a time travel world W, they are not all *laws* of W, except in a very tenuous

[36] See, for example, van Fraassen (1989) and my response in Earman (1993a).

sense. I will argue that this possibility is realized in cases where the consistency constraints are so severe as to supplant the laws of this world. In such cases the time travel involved is arguably such a remote possibility that it loses much of its interest. But note that since the consistency constraints are still subsumed under the laws of the time travel world, we retain the desirable feature that physical possibility$_2$ is naturalized.

Finally, these remarks point to the intriguing possibility that purely local observations can give clues to the global structure of spacetime without the help of a supplementary 'cosmological principle'; namely, local observations may reveal the absence of consistency constraints that would have to obtain if we inhabited certain kinds of chronology violating spacetimes.

What is needed as a first step in coming to grips with these matters is a study of the nature of consistency conditions on test fields that arise for various chronology violating spacetimes. The recent physics literature has made some progress on the project. In the next two sections I will report on some of the results for self-interacting and non-self-interacting fields. On the basis of these results I will advance some tentative conclusions about the nature of physical possibility and laws in chronology violating spacetimes.

7 Test fields on chronology violating spacetimes: non-self-interacting case

The simplest regime to study mathematically is the case of a non-self-interacting field, e.g. solutions to the source free scalar wave equation $\Box \Phi = 0$. Of course, the grandfather paradox and the related paradoxes of time travel that have been discussed in the philosophical literature typically rely on self-interacting systems. Even so we shall see that non-trivial consistency conditions can emerge in the non-self-interacting regime. But on the way to illustrating that point it is worth emphasizing the complementary point that in small enough regions of some chronology violating spacetimes the consistency constraints for non-self-interacting fields do not make themselves felt so that local observations in such regions will not reveal the presence of CTCs.

Following Yurtsever,[37] call an open $U \subseteq M$ *causally regular* for the spacetime M, g_{ab} with respect to the scalar wave equation iff for every C^∞ solution Φ of $\Box \Phi = 0$ on U, there is a C^∞ extension to all of M, i.e. there is a $C^\infty \tilde{\Phi}$ on M

[37] Yurtsever (1990).

such that $\Box\tilde{\Phi} = 0$ and $\tilde{\Phi}|_U = \Phi$. In addition M, g_{ab} can be said to be *causally benign* with respect to the scalar wave equation just in case for every $p \in M$ and every open neighborhood U of point p there is subneighborhood $U' \subset U$ which is causally regular.

The two-dimensional cylinder of figure 2(b) is causally benign with respect to the scalar wave equation. The following remarks, while not constituting a proof of this fact, give an indication of why it holds in the optical limit. In that limit Φ waves propagate at the speed of light (i.e. along null trajectories). At any point on the cylinder a small enough neighborhood can be chosen such that any null geodesic leaving this neighborhood in either the future or past direction never returns. Consider then any solution on this neighborhood. To extend this local solution to a global one, simply propagate the solution out of the base neighborhood along null geodesics. If the propagated field does not reach a point $q \in M$, set $\Phi(q) = 0$. If two null geodesics from the base neighborhood cross at q, obtain $\Phi(q)$ by adding the propagated fields.

Consider next the toroidal spacetime $T_{(1,r)}$ obtained from two-dimensional Minkowski spacetime by identifying the points (x, t) and (x', t') when $x = x'$ mod 1 and $t' = t$ mod r, where $r > 0$ is a real number. For r rational, $T_{(1,r)}$ is benign with respect to the scalar wave equation, as shown by Yurtsever.[37] Through any point on $T_{(1,r)}$ there is a time slice that lifts to many $t = $ constant surfaces in the Minkowski covering spacetime $\mathbb{R}^2, \eta_{ab}$. Consider one such surface, say, $t = 0$. Any solution $\Phi(x, t)$ on $T_{(1,r)}$ induces on $t = 0$ initial data $\Phi_0(x) \equiv \Phi(x, 0)$ and $\dot{\Phi}_0(x) \equiv d/dt\, \Phi(x, t)|_{t=0}$. By considering solutions on $\mathbb{R}^2, \eta_{ab}$ of the wave equation that develop from this initial data it follows that both Φ_0 and $\dot{\Phi}_0$ must be periodic with periods 1 and r respectively. Further, $\dot{\Phi}_0$ must satisfy the integral constraint

$$\int_{\{t=0\}} \dot{\Phi}_0(s)\, ds = 0.$$

When r is rational we can choose a small enough neighborhood of any point on $T_{(1,r)}$ such that arbitrary initial data Φ_0 and $\dot{\Phi}_0$ can be extended so as to meet the periodicity and integral constraints.

Friedman et al.[25] argue that the benignity property with respect to the scalar wave equation also holds for a class of chronology violating spacetimes that are asymptotically flat and globally Minkowskian except for a single wormhole that is threaded by CTCs. In some of these spacetimes there is a partial Cauchy surface S such that chronology violations lie entirely to the future of

S. It is argued that the formation of CTCs place no consistency constraints on initial data specified on S.

In all the examples considered so far the chronology violations are non-intrinsic in that they result from making identifications of points in a chronology preserving covering spacetime. Unfortunately, because of the non-trivial mathematics involved, almost nothing is known about the benignity properties of spacetimes with intrinsic chronology violations. I conjecture that Gödel spacetime is benign with respect to the scalar wave equation. If correct, this conjecture would cast new light on a puzzling feature of Gödel's own attitude towards the grandfather paradox in Gödel spacetime.[15] Basically his attitude was one of 'why worry' since the fuel requirements for a rocket needed to realize a time travel journey in Gödel spacetime are so demanding as to be impossible to meet by any practical scheme. But consistency constraints are constraints whether or not they can be tested by practical means. Thus, whatever puzzles arise with respect to the status of such constraints are unresolved by appeal to practical considerations. However, the above conjecture, if correct, suggests that for non-self-interacting systems the consistency constraints in Gödel spacetime are much milder than one might have thought. Similarly, Gödel's remarks can be interpreted as suggesting that the constraints will also be mild for self-interacting systems. Some further information on this matter is presented in section 8.[38]

It should be emphasized at this juncture that in spite of the connotations of the name, benignity does not necessarily imply physics as usual. For it does not imply that there are no non-trivial consistency constraints nor that the constraints cannot be detected locally. Benignity implies only that the constraints cannot be felt in sufficiently small neighborhoods, but this is compatible with their being felt in regions of a size that we typically observe.

To give an example of a non-benign spacetime we can return to $T_{(1,r)}$ and choose r to be an irrational number. Now the periodicity constraints on the initial data cannot be satisfied except for Φ_0 and $\dot{\Phi}_0$ constant. The integral constraint then requires that $\dot{\Phi}_0 = 0$, with the upshot that the only solutions allowed are $\Phi =$ constant everywhere. No local solution that allows the tiniest variation in Φ can be extended to a global solution.

I turn now to the question of the status of the consistency constraints for chronology violating spacetimes. By way of introduction, I note the assertion

---

[38] Gödel took the possibility of time travel to support the conclusion that time is 'ideal'. For discussions of Gödel's argument, see Savitt (1994) and Yourgrau (1991).

of the Consortium that a time traveller 'who went through a wormhole [and thus around a CTC] and tried to change the past would be *prevented by physical law* from making the change...' (Friedman et al.[25] p. 1928; italics added). One way of interpreting this assertion is in line with my working hypothesis that the consistency constraints entailed by the presence of CTCs in a nomologically accessible chronology violating world are laws of that world. This position is arguably endorsed by the M-R-L account of laws. For it is plausible that in each of the above examples the consistency constraints would appear as axioms or theorems of (each of) the best overall true theories of the world in question.

For the reasons discussed in sections 5 and 6, such a result is devoutly to be desired. But to play the devil's advocate for a moment, one might charge that the result is an artifact of the examples chosen. Each of these examples involves a spacetime with a very high degree of symmetry, and it is this symmetry one suspects of being responsible for the relative simplicity of the consistency constraints. If this suspicion is correct, then the consistency constraints that obtain in less symmetric spacetimes may be so complicated that they will not appear as axioms or theorems of any theory that achieves a good compromise between strength and simplicity. Due to the technical difficulties involved in solving for the consistency constraints in non-symmetric spacetimes, both the devil and his advocate may go blue in the face if they hold their breaths while waiting for a confirmation of their suspicions. If the devil's advocate should prove to be correct, the proponent of the naturalization could still find comfort if the cases where the naturalization fails could be deemed to be very remote possibilities.

An illustration of how time travel can be justly deemed to involve very remote possibilities occurs when the chronology violating world W is so far from actuality that, although the laws of the actual world are *true of* W, they are not *laws of* W except in a very attenuated sense. In the case of $T_{(1,r)}$ with r irrational we saw that $\Phi = $ constant are the only allowed solutions of the scalar wave equation. I take it that $\Box \Phi = 0$ will not appear as an axiom of a best theory of the $T_{(1,r)}$ world so that the scalar wave equation is demoted from fundamental law status. Presumably, however, $\Phi = $ constant will appear as an axiom in any best theory. Of course, this axiom entails that $\Box \Phi = 0$; but it also entails any number of other differential equations that are incompatible with one another and with the scalar wave equation when Φ is not constant. Thus, in the $T_{(1,r)}$ world the scalar wave equation and its rivals will have much the same status as 'All unicorns are red', 'All unicorns are blue', etc. in a world where it is a law that there are no unicorns. In this sense $\Box \Phi = 0$ has been supplanted as a law. This still is a case where physical possibility$_2$ is naturalized by reduction in the strong sense to physical possibility$_1$ since the consistency constraint is stated in purely local terms. But the more remote the possibility the less interesting the reduction. And in this case the possibility can be deemed to be very remote since the relation of nomological accessibility has become non-symmetric – by construction the toroidal world in question is nomologically accessible from the actual world, but the converse is not true because the toroidal law $\Phi =$ constant is violated here.

One might expect that such supplantation will take place in any world with a spacetime structure that is not benign. What I take to be a counterexample to this expectation is provided by the four-dimensional toroidal spacetime $T_{(1,1,1,1)}$ obtained from four-dimensional Minkowski spacetime by identifying the points (x, y, z, t) and (x', y', z', t') just in case the corresponding coordinates are equal mod 1. This spacetime is not benign with respect to the scalar wave equation. But solutions are not constrained to be constant; in fact, the allowed solutions form an infinite-dimensional subspace of the space of all solutions.[37] The consistency constraint imposes a high frequency cut off on plane wave solutions propagating in certain null directions. This constraint by my reckoning is simple and clean enough to count as an M-R-L law, but it supplements rather than supplants $\Box \Phi = 0$.

It is possible in principle to verify by means of local observations that we do not inhabit a non-causally regular region of some non-benign spacetime. And if we indulge in an admittedly dangerous inductive extrapolation from the above examples, we can conclude that we do not in fact inhabit a non-regular region. For it follows from the (source-free) Maxwell equations – which we may assume are laws of our world – that the components of the electromagnetic field obey the scalar wave equation. But the electromagnetic fields in our portion of the universe do not satisfy the restrictions which (if the induction is to be believed) are characteristic of non-benign spacetimes. Further experience with other examples of non-benign spacetimes may serve to strengthen or to refute this inference.

It is also possible to use local observations to rule out as models for our world certain benign chronology violating spacetimes, but only on the assumption that we have looked at large enough neighborhoods to reveal the consistency constraints indicative of these spacetimes. It is not easy to see how we would come by a justification for this enabling assumption.

8 Test fields on chronology violating spacetimes: self-interacting case

I begin with a reminder of two lessons from previous sections. First, the grandfather paradox is in the first instance a way of pointing to the presence of consistency constraints on local physics in a chronology violating spacetime. And second, while the usual discussion of the grandfather paradox assumes a self-interacting system, we have found that non-trivial and indeed very strong constraints can arise even for non-self-interacting systems. Of course, one would expect that the constraints for self-interacting systems will be even more severe. To test this expectation one could carry out an analysis that parallels that of section 7 by considering a test field obeying an equation such as $\Box\Phi = k\Phi^3$ ($k =$ constant), which implies that solutions do not superpose. Results are not available in the literature. What are available are results for the simpler and more artificial case of perfectly elastic billiard balls.

One assumes that between collisions the center of mass of such a ball traces out a timelike geodesic in the background spacetime and that in a collision the laws of elastic impact are obeyed. Under these assumptions the initial trajectory δ of the (two-dimensional) billiard ball in the spacetime of figure 3(b) leads to an inconsistent time development. The ball trips around the cylinder and participates in a grazing collision with its younger self, knocking its former self from the trajectory that brought about the collision (grandfather paradox for billiard balls). We saw in the previous section that the cylinder is benign for the scalar wave equation. But obviously the corresponding property fails for billiard balls: for any point x on the cylinder it is not the case that there is a sufficiently small neighborhood $N(x)$ such that any timelike geodesic segment on N, representing the initial trajectory of the ball, can be extended to a globally consistent trajectory. Nor are the forbidden initial conditions of measure zero in any natural measure.

For a single sufficiently small billiard ball, Gödel spacetime is benign, not just because all timelike geodesics are open but also because they are not almost closed. And it seems safe to conjecture that Gödel spacetime is benign with respect to any finite system of small billiard balls since it seems implausible that collisions among a finite collection can be arranged so as to achieve the sustained acceleration needed to instantiate a closed or almost closed timelike curve.

Echeverria et al.[39] have studied the behavior of billiard balls in two types of wormhole spacetimes that violate chronology. The first type is called the 'twin paradox spacetime' because the relative motion of the wormhole mouths gives rise to a differential aging effect which in turn leads to CTCs since the wormhole can be threaded by future directed timelike curves. This spacetime contains a partial Cauchy surface S. Chronology is violated only to the future of S, indeed to the future of $H^+(S)$. The other type of wormhole spacetime is called the 'eternal time machine spacetime' because CTCs that traverse the wormhole can reach arbitrarily far into the past. There are no partial Cauchy surfaces in this arena; but because of asymptotic flatness the notion of past null infinity $\mathscr{I}^-$ is well defined, and initial data can be posed on $\mathscr{I}^-$.[40]

For initial conditions (specified on S for the twin paradox spacetime or on $\mathscr{I}^-$ for the eternal time travel spacetime) that would send the billiard ball into the wormhole, one might expect to find that strong consistency constraints are needed to avoid the grandfather paradox. But when Echeverria et al. searched for forbidden initial conditions, they were unable to find any. Thus, it is plausible, but not proven, that for each initial state of the billiard ball, specified in the non-chronology violating region of the spacetime, there exists a globally consistent extension. Mikheeva and Novikov[41] have argued that a similar conclusion holds for an inelastic billiard ball.

Surprisingly, what Echeverria et al. did find was that each of many initial trajectories had a countably infinite number of consistent extensions. The consistency problem and the phenomenon of multiple extensions is illustrated in figure 4. In figure 4(a) we have yet another instance of the grandfather paradox. The initial trajectory ζ, if prolonged without interruption, takes the billiard ball into mouth 2 of the wormhole. The ball emerges from mouth 1 along η, collides with its younger self, converting ζ into ζ' and preventing itself from entering the wormhole. Figures 4(b) and 4(c) show how ζ admits of self-consistent extensions. In figure 4(b) the ball suffers a grazing collision which deflects it along trajectory ζ''. It then reemerges from mouth 1 along ψ and suffers a glancing blow from its younger self and is deflected along ψ'.

[39] Echeverria, Klinkhammer, & Thorne (1991).
[40] Intuitively, $\mathscr{I}^-$ is the origin of light rays that come in from spatial infinity. For a formal definition, see Hawking & Ellis (1973).
[41] Mikheeva & Novikov (1992).

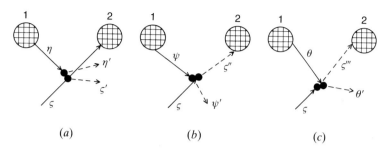

Fig. 4. Self-inconsistent and self-consistent billiard ball motions in a wormhole spacetime (after Echeverria et al.[39])

Readers can provide the interpretation of figure 4(c) for themselves. The demonstration of the existence of initial conditions that admit an infinite multiplicity of consistent extensions involves the consideration of trajectories that make multiple wormhole traversals; the details are too complicated to be considered here.

These fascinating findings on the multiplicity of extensions are relevant to the question of whether it is possible to operate a time machine; this matter will be taken up in section 9. Of more direct relevance to present concerns are the findings about consistency constraints for self-interacting systems. The results of Echeverria et al.[39] indicate that in the twin paradox spacetime, for instance, the *non*-chronology violating portion of the spacetime is benign with respect to all billiard ball trajectories, including those dangerous trajectories that take the ball into situations where the grandfather paradox might be expected. But the chronology violating region of this spacetime is most certainly *not* benign with respect to billiard ball motions. Perhaps it is a feature of non-benign spacetimes that the failure of benignity only shows up in the chronology violating region, but one example does not give much confidence.

The study of more complicated self-interacting systems quickly becomes intractable at the level of fundamental physics. What one has to deal with is a coupled set of equations describing the self- and cross-interactions of particles and fields. Deriving properties of solutions of such a set of equations for chronology violating spacetimes is beyond present capabilities. Instead one studies the behavior of 'devices' whose behavior is described on the macrolevel. The presumption is that if these devices were analyzed into fundamental constituents and if the field equations and equations of

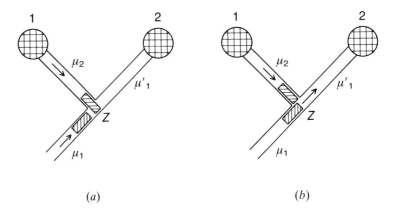

Fig. 5. Self-inconsistent and self-consistent motions for a piston device (after Novikov[42])

motion for these constituents were solved for some relevant range of initial and boundary conditions, then the solutions would display the behavior characteristic of the type of device in question. From this point of view, a perfectly elastic billiard ball might be considered to be one of the simplest devices.

A slightly more complicated device consists of a rigid Y shaped tube and a piston which moves frictionlessly in the tube. Imagine that the branches at the top of the Y are hooked to the mouths of the wormhole in the twin paradox spacetime. Because of the constraints the tube puts on the motion of the piston, it is not true that every initial motion of the piston up the bottom of the Y has a self-consistent extension. Figure 5(a) shows an initial motion that takes the piston up along the sections μ_1 and μ'_1, into the wormhole mouth 2, through the wormhole, out of mouth 1 at an earlier time, down the section μ_2 to the junction Z just in time to block its younger self from entering the μ'_1 section.

Novikov[42] argues that self-consistent solutions are possible if the device is made slightly more realistic by allowing the older and younger versions of the piston to experience friction as they rub against one another. In figure 5(b) the piston starts with the same initial velocity as in figure 5(a). But when the piston tries to pass through the junction Z it is slowed down by rubbing against its younger self. This slowing down means that when the piston

[42] Novikov (1992).

traverses the wormhole it will not emerge at an earlier enough time to block the junction but only to slow its younger self down. Novikov gives a semi-quantitative argument to show that, with self-friction present, for any initial motion of the piston that gives rise to an inconsistent/grandfather paradox evolution as in figure 5(a), there is a self-consistent extension as in figure 5(b). For initial trajectories that have the time travelling piston arriving at the junction well before its younger self reaches that point, there is also arguably a self-consistent continuation. Thus, it is plausible that there are no non-trivial consistency constraints on the initial motion of the piston up the μ_1 section of the tube.

The work of Wheeler and Feynman and Clarke[43] suggests that the absence of consistency constraints or at least the benignity of the constraints can be demonstrated generally for a class of devices for which the evolution is a continuous function of the parameters describing the initial conditions and the self-interaction, the idea being that fixed point theorems of topology can be invoked to yield the existence of a consistent evolution. However, Maudlin[44] showed that if the topology of the parameter space is complicated enough, a fixed point/self-consistent solution may not exist for some initial conditions. And one would suppose that in the general case the problem of deciding whether the relevant state space topology admits a fixed point theorem is as difficult as solving directly for the consistency constraints.

One might expect that with sufficiently complicated devices there may be no (or only rare) initial conditions that admit a self-consistent continuation in a chronology violating environment that allow the device to follow CTCs. Consider Novikov's device[42] consisting of a radio transmitter, which sends out a directed beam; a receiver, which listens for a signal; and a bomb. The device is programmed to detonate the bomb if and only if it detects a signal of a strength that would be experienced by being, say, 30 m from the device's transmitter. A self-consistent traverse of the wormhole of the twin paradox spacetime is possible if the device undergoes inelastic collisions; for then such a collision between the older and younger versions can produce a change in orientation of the transmitter such that the younger self does not receive the signal from its older self and, consequently, no explosion takes place. But one can think of any number of epicycles that do not admit of any obvious self-consistent solution. For example, as Novikov himself suggests,

[43] Wheeler & Feynman (1949) and Clarke (1977).
[44] Maudlin (1990).

the device could be equipped with gyro-stabilizers that maintain the direction of the radio beam.

It is all too easy to get caught up in the fascinating details of such devices and thereby to lose sight of the implications for what I take to be the important issues about time travel. As a way of stepping back, let me reiterate the point that came up in connection with the investigations of Echeverria et al.[39] of billiard ball motions in chronology violating spacetimes. The absence of consistency constraints or the benignity of these constraints with respect to initial conditions of a device, as specified in the *non*-chronology violating portion of the spacetime, does not establish the absence of consistency constraints or their benignity simpliciter. For example, assuming some self-friction of the piston of the Novikov piston device, there may be no non-trivial consistency constraints on the initial motion of the piston up the bottom of the Y. But there most certainly are constraints on the motion in the chronology violating portion of the spacetime, and these constraints are not benign. Consider any spacetime neighborhood that includes the junction Z at a time when the piston is passing Z. Passing from μ_1 to μ_1' with a speed v without rubbing against a piston coming down μ_2 is a physically possible local state for every v. But for some values of v there is no self-consistent extension. Thus, contrary to what some commentators have suggested, the recent work on the physics of time travel does not dissolve the paradoxes of time travel. Whatever exactly these paradoxes are, they rest on the existence of consistency constraints entailed by field equations/laws of motion in the presence of CTCs. Showing that those constraints are trivial would effectively dissolve the paradoxes. But all the recent work affirms the non-triviality of such constraints.

Suppose that there are devices that function normally in our world but which do not admit of any self-consistent evolution in a chronology violating background spacetime M, g_{ab}. What more would this show over and above that our world does not have the spatiotemporal structure of M, g_{ab}? Well suppose also that it followed from the first supposition that the laws which hold in the actual world and which govern the constituents of the devices do not have any globally consistent solutions in M, g_{ab}. Then these laws would presumably not be basic laws of a world with the spacetime M, g_{ab}. They could not even be true of such a world except in the trivial case where the constituent particles and fields were not present. Thus, we could conclude that the sort of spatiotemporal structure represented by M, g_{ab} is an extremely remote possibility. Alas, there is no good reason to think that the second supposition holds. All that follows from the non-existence of self-consistent histories for the devices is the non-existence of solutions of the relevant laws for the restricted ranges of fundamental variables that correspond to the normal operation of the device in our world. This is why it is important to get beyond 'devices' and tackle more directly the problems of consistency constraints for the fundamental laws of self-interacting systems. But as already noted there is little hope of rapid progress on this difficult project.

Does what we have learned about self-interacting systems give reason for optimism about my working hypothesis; namely, that insofar as a chronology violating world admits a set of M-R-L laws for test fields, those laws will subsume the consistency constraints forced out by the presence of CTCs? One might see a basis for pessimism deriving from the fact that in the wormhole spacetimes the constraints that obtain in different regions are different (e.g. no constraints on the initial conditions for the billiard ball in the non-chronology violating region of the twin paradox spacetime but non-trivial constraints in the chronology violating region). Since we want laws of nature to be 'universal' in the sense that they hold good for every region of spacetime, it might seem that the wormhole spacetimes dash the hope that the consistency constraints will have a lawlike status. But the hope is not to be extinguished so easily. To be 'universal', the constraint must be put in a general form; viz. for any region R, constraint $C(R)$ obtains iff _____, where the blank is filled in with conditions formulated in terms of suitably general (non-Goodmanized) predicates. The blank will need to be filled in not only with features of R but also with features of the relation of R to the rest of spacetime. So if the consistency constraints have law status, then the laws of a chronology violating wormhole spacetime cannot all be local. But that was only to be expected. The real concern is the one that surfaced in section 7; namely, that as the spacetime gets more and more complicated, the conditions that go into the blank may have to become so complicated that the consistency constraints will not qualify as M-R-L laws. Remember, however, that this concern is mitigated if the chronology violating worlds in question can be deemed to lie in the outer reaches of the space of worlds nomologically accessible from the actual world. Here I think that intuition pumping is useless until we have more concrete examples to serve as an anchor.

9 Is it possible for us to build and operate a time machine?

The question that forms the title of this section is not equivalent to the question of whether time travel is possible. If, for example, the Gödel universe is physically possible, then so is time travel. But in such cases it is the universe as a whole that is serving as a time machine and not something that we constructed and set into motion.

In trying to characterize a time machine we face something of a conundrum. To make sure that the CTCs are due to the operation of the time machine we could stipulate that there is a time slice S, corresponding to a time before the machine is turned on, such that there are no CTCs in $J^-(S)$. Furthermore, by going to a covering spacetime if necessary, we can guarantee that S is achronal and, thus, a partial Cauchy surface. But then by construction only time travel to the future of S is allowed, which immediately eliminates the kind of time travel envisioned in the typical time travel story of the science fiction literature. I do not see any easy way out of this conundrum, and for present purposes I will assume that such an S exists.

Next, one would like a condition which says that only local manipulations of matter and energy are involved in the operation of the machine. Requiring that the spacetime be asymptotically flat would be one approach – intuitively, the gravitational field of the machine falls off at large distances. But on one hand this requirement precludes many plausible cosmologies; and on the other hand it is not evident that a condition on the space structure guarantees that something funny with the causal structure was not already in progress before the machine was switched on.

It is even more delicate to pin down what it means to say that switching on the time machine produces CTCs. Programming the time machine corresponds to setting initial conditions on the partial Cauchy surface S. These conditions together with the coupled Einstein–matter equations determine a unique evolution for the portion of $J^+(S)$ contained in $D^+(S)$. But $D^+(S)$ is globally hyperbolic and therefore contains no CTCs.[45] Moreover, the future boundary $H^+(S)$ of $D^+(S)$ is always achronal. The notion that the initial conditions on S are responsible for the formation of CTCs can perhaps be captured by the requirement that some of the null geodesic generators of $H^+(S)$ are closed

---

[45] Global hyperbolicity is equivalent to the existence of a Cauchy surface. Thus, $D^+(S), g_{ab}|_{D^+(S)}$ considered as a spacetime in itself is at the top of the part of the causality hierarchy discussed in section 2.

or almost closed, indicating that CTCs are on the verge of forming.[46] Perhaps it should also be required that any appropriate extension of the spacetime across $H^+(S)$ contains CTCs.

One need not be too fussy about the sufficient conditions for the operation of a time machine if the goal is to prove negative results, for then one need only fix on some precise necessary conditions. An example of such a result was obtained recently by Hawking.[26] In concert with the above discussion he assumes the existence of a partial Cauchy surface S such that $H^+(S)$ separates the portion of spacetime with CTCs from the portion without CTCs. In Hawking's terminology, $H^+(S)$ is a *chronology horizon*. If all the past directed generators of $H^+(S)$ are contained in a compact set, then $H^+(S)$ is said to be *compactly generated*.

Theorem (Hawking). Let M, g_{ab}, T^{ab} be a cosmological model satisfying Einstein's field equations (with or without cosmological constant). Suppose that M, g_{ab} admits a partial Cauchy surface S and that T^{ab} satisfies the null energy condition. Then (a) if S is non-compact, $H^+(S)$ cannot be non-empty and compactly generated, and (b) if S is compact, $H^+(S)$ can be compactly generated but matter cannot cross $H^+(S)$.

How effective is this formal result as an argument against time machines? At best, part (b) of the Theorem shows that the operator of the time machine cannot himself sample the fruits of his labor. And both parts rely on $H^+(S)$ being compactly generated, which means physically that the generators of $H^+(S)$ do not emerge from a curvature singularity nor do they come in from 'infinity'. These prohibitions might be motivated by the idea that if the appearance of CTCs is to be attributed to the operation of a time machine, then the CTCs must result from the manipulation of matter–energy in a finite region of space. But it seems to me that this motivation is served best not by Hawking's requirement that $H^+(S)$ is compactly generated but by the combination of requirements that $H^+(S)$ contains closed or almost closed null geodesic generators and that $H^+(S)$ is *compactly causally generated* in the sense that $\overline{I^-(H^+(S))} \cap S$ is compact. If these requirements hold, then what happens if a finite region of space $Q \equiv \overline{I^-(H^+(S))} \cap S$ causally generates a development $D^+(Q)$ on whose future boundary $H^+(Q) \supset H^+(S)$ a causality violation occurs, indicating that CTCs are just about to form. $H^+(S)$ can be compactly causally generated without being compactly generated. But if the former condition is substituted for the latter, it remains to be seen whether a

[46] $H^+(S)$ is a null surface that is generated by null geodesics.

version of Hawking's chronology protection theorem will continue to be valid. (I conjecture that it will not.)

A different approach to showing that the laws of physics are unfriendly to the enterprise of building a time machine would be to try to show that the operation of the machine involves physical instabilities. More specifically, in terms of the setting suggested above, one would try to show that a chronology horizon is necessarily unstable. This approach links back to the problem of the behavior of test fields on chronology violating spacetimes; indeed, the stability property can be explicated in terms of the existence of extensions of solutions of the test field. To return to the example of a scalar field Φ obeying the wave equation $\Box \Phi = 0$, consider arbitrary initial data on the partial Cauchy surface S (the values on S of Φ and its normal derivative to S) of finite energy.[47] Each such data set determines a unique solution in the region $D^+(S)$. Then $H^+(S)$ can be said to be *completely stable* for Φ iff every such solution has a smooth extension across $H^+(S)$.[48] Further $H^+(S)$ can be said to be *generically stable* for Φ iff a generic solution has a smooth extension across $H^+(S)$.[49]

It has been argued by Morris et al.[20] and by Friedman and Morris[50] that some asymptotically flat wormhole spacetimes with CTCs have stable chronology horizons. However, these examples violate the weak energy condition of classical GTR.[51] Such violations are tolerated in quantum field theory. But recent investigations indicate that quantum field theory may require an averaged or integrated version of the energy conditions whose violation is entailed by the maintenance of traversable wormholes.[52]

An example of a vacuum solution to EFE that contains a partial Cauchy surface S with CTCs to the future of S and where the chronology horizon $H^+(S)$ is generically unstable for Φ is the two-dimensional version of Taub–NUT spacetime (see Hawking and Ellis[11] pp. 170–8). Here the chronology horizon is not only compactly generated but is itself compact; indeed, it is generated by a smoothly closed null geodesic λ. Each time the tangent vector of λ is transported parallel to itself around the loop it is expanded by a

---

[47] See Yurtsever (1991) for a definition of this notion and for a more precise specification of the stability property.
[48] See Yurtsever (1991) for a formulation of the relevant smoothness conditions.
[49] The space of initial conditions has a natural topology so that a generic set of solutions can be taken to be one that corresponds to an open set of initial conditions.
[50] Friedman & Morris (1991a,b).
[51] This follows from the results of Tipler (1976, 1977) which show that the weak energy condition prevents the kind of topology change that occurs with the development of wormholes.
[52] Wald & Yurtsever (1991) give a proof of the *averaged null energy condition* for a massless scalar field in a curved two-dimensional spacetime.

factor of $\exp(h)$, $h > 0$, indicating a blueshift. Now consider a generic high frequency wave packet solution to $\Box\Phi = 0$ 'propagating to the right'. As it nears $H^+(S)$ it experiences a blueshift each time it makes a circuit, and as an infinite number of circuits are needed to reach $H^+(S)$, the blueshift diverges. This is already indicative of an instability, but to demonstrate that the divergent blueshift involves an instability that prevents an extension across $H^+(S)$ it has to be checked that the local energy density of the wave packet diverges as $H^+(S)$ is approached.[53] That is in fact the case in this example.[26] One could then reason that when Φ is not treated merely as a test field but as a source for the gravitational field, spacetime singularities will develop on $H^+(S)$ thereby stopping the spacetime evolution and preventing the formation of CTCs that would otherwise have formed beyond the chronology horizon.[20]

However, the classical instability of chronology horizons does not seem to be an effective mechanism for ensuring chronology protection; for example, Hawking's results[26] indicate that among compactly generated chronology horizons, a non-negligible subset are classically stable. Even in cases where instability obtains, one can wonder how this instability undermines the feasibility of operating a time machine. Then insofar as the time machine involves non-zero values of Φ, it cannot succeed. The worst case of instability would be complete instability with respect to Φ, i.e. no solution of $\Box\Phi = 0$ other than $\Phi \equiv 0$ is extendible across $H^+(S)$. The next worst case would involve instability that is not complete but is so generic that only a set of solutions of measure zero admit extensions across $H^+(S)$. Here one could argue that if the time machine operator chose the parameter setting at random (with respect to the preferred measure on initial conditions), there would be a zero probability of hitting on a setting that would lead to successful operation of the machine. This would not be a proof of the impossibility of time travel using a time machine but only a demonstration that initiating the journey requires luck. Perhaps some stronger conclusion can be derived, but I do not see how. Measure zero arguments are commonly assumed to have a good deal of force, but it is hardly ever explained why.

The quantum instability of chronology horizons is currently under intensive investigation.[54] It seems that the expectation value of the (renormalized) stress

---

[53] The energy density depends on the behavior of the two-dimensional cross sectional area of a pencil of geodesics.

[54] See Hawking (1992), Kim & Thorne (1991), and Klinkhammer (1992).

energy tensor of a quantum field diverges as the chronology horizon is approached, with the divergence being stronger for a compactly generated horizon than for a non-compactly generated horizon. In the semi-classical approach to quantum gravity, the expectation value of the stress energy tensor is fed back into EFE to determine the effects on the spacetime geometry. Whether or not the divergence of the stress energy tensor on the chronology horizon produces an alteration of the spacetime sufficient to prevent the formation of CTCs is still controversial.

If a time machine can be constructed, the main puzzle about its operation is not the grandfather paradox but something quite different. The implicit assumption in the science fiction literature is that when the time machine is switched on, some definite scenario will unfold, as determined by the settings on the machine. But the still imperfectly understood physics of time travel hints at something quite at variance with these expectations. In the first place there may be different extensions of the spacetime across the chronology horizon $H^+(S)$.[55] In the second place, even when the spacetime extension is chosen and treated as a fixed background for test fields, billiard balls, and other devices, the equations which govern these systems may permit a multiplicity, perhaps even an infinite multiplicity, of extensions across $H^+(S)$ and into the time travel region. The point is not simply that one does not know the upshot of turning on the time machine but rather that the upshot is radically underdetermined on the ontological level. And thus the new puzzle: how does the universe choose among these ontologically distinct possibilities? Of course, it is unfair to demand a mechanism for making the choice if 'mechanism' implies determinism, for that is what is expressly ruled out in this situation. But it is equally unsatisfactory to respond with nothing more than the formula that it is just a matter of chance which option will unfold. If 'just a matter of chance' is to be more than an incantation or a recapitulation of the puzzle, then 'chance' must mean something like objective propensity. But the physics of classical GTR provides no basis for saying that there are objective probabilities of, say, 0.7 and 0.3 respectively for scenarios (b) and (c) in figure 4. Here quantum mechanics may come to the rescue of time travel by showing that for any initial quantum state describing the motion of the billiard ball before it enters the region of CTCs, there is a well defined probability for each of the subsequent classically consistent extensions.[56]

[55] One can conjecture that if $H^+(S)$ is stable, then suitable extensions of the spacetime across $H^+(S)$ will be unique.
[56] Results of this character have been announced by Klinkhammer and Thorne; see Echeverria, Klinkhammer, & Thorne (1991).

However, when CTCs are present it is to be expected that the time evolution will not be unitary.[57] The loss of unitarity is both a problem and an opportunity: a problem because it means that quantum mechanics in its standard form breaks down, and an opportunity because unitarity is a key ingredient in generating the measurement problem. Perhaps even more disturbing than the loss of unitarity is a seeming arbitrariness in the probabilities computed by the path integral approach. Apparently, probabilities for the outcomes of measurements made before $H^+(S)$ will depend on where the sum over histories is terminated to the future of $H^+(S)$.[58] Clearly how CTCs mesh or fail to mesh with quantum mechanics will be an exciting area of investigation for some time to come.

10 Conclusion

If nothing else, I hope that this chapter has made it clear that progress on understanding the problems and prospects of time travel is not going to come from the sorts of contemplations of the grandfather paradox typical of past philosophical writings. Using modal logic to symbolize the paradox, armchair reflections on the concept of causation, and the like are not going to yield new insights. The grandfather paradox is simply a way of pointing to the fact that if the usual laws of physics are supposed to hold true in a chronology violating spacetime, then consistency constraints emerge. The first step to understanding these constraints is to define their shape and content. This involves solving problems in physics, not armchair philosophical reflections.

But philosophy can help in understanding the status of the consistency constraints. Indeed, the existence of consistency constraints is a strong hint – but nevertheless a hint that most of the literature on time travel has managed to ignore – that it is naive to expect that the laws of a time travel world that is nomologically accessible from our world will be identical with the laws of our world. I explored this matter under the assumption that laws of nature are to be constructed following the analysis of Mill–Ramsey–Lewis. In some time travel worlds it is plausible that the M-R-L laws include the consistency constraints; in these cases the grandfather paradox has a satisfying resolution. In other cases the status of the consistency constraints remains obscure; in these cases the grandfather paradox leaves a residual itch. Those who wish to scratch the

[57] See Goldwirth, Perry, & Piran (1993).
[58] See Thorne (1991).

itch further may want to explore other analyses of laws. Indeed, time travel would seem to provide a good testing ground for competing analyses of laws.

I do not see any prospect for proving that time travel is impossible in any interesting sense. It may be, however, that it is not physically possible to operate a time machine that manufactures CTCs. But if so, no proof of this impossibility has emerged in classical GTR. If the operation of the time machine is feasible there emerges a new puzzle: a setting of the parameters on the time machine may correspond to many different scenarios in the time travel region. The problem here is not that the operation of the time machine is unpredictable or calls into play an element of indeterminacy; rather, the problem lies in providing an objective content to the notion of chance in this setting. Quantum mechanics is, of course, the place to look for such content. But standard quantum mechanics is hard to reconcile with CTCs. And it would be a little surprising and more than a little disturbing if Gödelian time machines, which seemed to be characterizable in purely classical relativistic terms, turn out to be inherently quantum mechanical. Is nothing safe from the clutches of the awful quantum?

Acknowledgments

Thanks are due to a number of people for helpful comments on an earlier draft of this paper, especially Aristides Arageorgis, Gordon Belot, Tim Maudlin, Laura Reutche, and Steven Savitt. I am also grateful to the ZiF, University of Bielefeld, for its support of this project.

References

Affleck, I. (1989) 'Quantum spin chains and the Haldane gap', *Journal of Physics: Condensed Matter*, **1**, 3047–72.

Aharonov, Y., Albert, D., Casher, A., & Vaidman, L. (1986) 'Novel properties of preselected and postselected ensembles'. In *New Techniques and Ideas in Quantum Measurement Theory*, ed. D. M. Greenberger, pp. 417–21. New York: New York Academy of Sciences.

Aharonov, Y., Albert, D., & Vaidman, L. (1988) 'How the result of a measurement of a component of the spin of a spin-1/2 particle can turn out to be 100', *Physical Review Letters*, **60**, 1351–4.

Aharonov, Y., Anandan, J., & Vaidman, L. (1993) 'Meaning of the wave function', *Physical Review A*, **47**, 4616–26.

Aharonov, Y., Bergmann, P. G., & Lebowitz, J. L. (1964) 'Time symmetry in the quantum process of measurement', *Physical Review*, **134B**, pp. 1410–16. Reprinted in Wheeler & Zurek (1983).

Aharonov, Y. & Bohm, D. (1959) 'Significance of electromagnetic potentials in the quantum theory', *Physical Review*, **115**, 485–91.

Aharonov, Y. & Vaidman, L. (1990) 'Properties of a quantum system during the interval between 2 measurements', *Physical Review A*, **41**, 11–20.

(1993) 'Measurement of the Schrödinger wave of a single particle', *Physics Letters A*, **178**, 38–42.

Albert, D. (1983) 'On quantum-mechanical automata', *Physics Letters A*, **98**, 249–52.

(1986) 'How to take a photograph of another Everett World'. In *New Techniques and Ideas in Quantum Measurement Theory*, ed. D. M. Greenberger, pp. 498–502. New York: New York Academy of Sciences.

(1992) *Quantum Mechanics and Experience*. Cambridge, MA: Harvard University Press.

Altshuler, B. L. & Lee, P.A. (1988) 'Disordered electronic systems', *Physics Today*, **41**, 36–44.

Ao, P. & Rammer, J. (1992) 'Influence of an environment on equilibrium properties of a charged quantum bead constrained to a ring', *Superlattices and Microstructures*, **11**, 265–8.

Arnold, V. & Avez, A. (1968) *Ergodic Problems of Classical Mechanics*. New York: Benjamin.

Arntzenius, F. (1990) 'Physics and common causes', *Synthese*, **82**, 77–96.

References

Aronov, A. G. & Sharvin, Y. G. (1987) 'Magnetic flux effects in disordered conductors', *Reviews of Modern Physics*, **59**, 755–79.

Atkins, P. W. (1986) 'Time and dispersal: the second law'. In *The Nature of Time*, ed. R. Flood & M. Lockwood, pp. 80–98. Oxford: Basil Blackwell Ltd.

Ballentine, L. E. & Jarrett, J. P. (1987) 'Bell's theorem: does quantum mechanics contradict relativity?', *American Journal of Physics*, **55**, 696–701.

Barrett, M. & Sober, E. (1992) 'Is entropy relevant to the asymmetry between prediction and retrodiction?', *British Journal for the Philosophy of Science*, **42**, 141–60.

Beardsley, T. M. ('T.M.B.') (1988) 'Cosmic quarrel: fearless philosopher finds fly in the ointment of time', *Scientific American*, **261**, 22–6.

Belinfante, F. J. (1975) *Measurement and Time Reversal in Objective Quantum Theory*. Oxford: Pergamon.

Bell, J. S. (1975) 'On wave packet reduction in the Coleman–Hepp model', *Helvetica Physica Acta*, **48**, 93–8. Reprinted in Bell (1987).

  (1982) 'On the impossible pilot wave', *Foundations of Physics*, **12**, 989–99. Reprinted in Bell (1987).

  (1987) *Speakable and Unspeakable in Quantum Mechanics*. Cambridge: Cambridge University Press.

  (1990) 'Against "Measurement"', *Physics World*, **3**, 33–40.

Belnap, N. (1991) 'Before refraining: concepts for agency', *Erkenntnis*, **34**, 137–69.

  (1992) 'Branching space-time', *Synthese*, **92**, 385–434.

Benioff, P. (1980) 'The computer as a physical system: a microscopic quantum-mechanical Hamiltonian model of computers as represented by Turing Machines', *Journal of Statistical Physics*, **22**, 563–91.

  (1982) 'Quantum mechanical models of Turing Machines that dissipate no energy', *Physical Review Letters*, **48**, 1581–5.

  (1986) 'Quantum mechanical Hamiltonian models of computers'. In *New Techniques and Ideas in Quantum Measurement Theory*, ed. D. M. Greenberger, pp. 475–86. New York: New York Academy of Sciences.

Berthiaume, A. & Brassard, G. (1994) 'Oracle Quantum Computing', *Journal of Modern Optics*, **41**, 2521–35.

Black, M. (1956) 'Why cannot an effect precede its cause?', *Analysis*, **16**, 49–58.

Blatt, J. (1959) 'An alternative approach to the ergodic problem', *Progress of Theoretical Physics*, **22**, 745–55.

Bohr, N. (1936) 'Can quantum-mechanical description of physical reality be considered complete?', *Physical Review*, **48**, 696–702. Reprinted in Wheeler & Zurek (1983).

  (1949) 'Discussion with Einstein on epistemological problems in atomic physics'. In *Albert Einstein: Philosopher-Scientist*, ed. P. Schilpp, pp. 199–241. La Salle, IL: Open Court. Reprinted in Wheeler & Zurek (1983).

Boltzmann, L. (1866) 'Uber die Mechanische Bedeutung des Zweiten Haupsatzes der Wärmetheorie', *Sitzungsberichte der Kaiserlichen Akademie der Wissenschaften in Wien, Klasse IIa*, **53**, 195–220.

  (1871) 'Uber das Warmegleichgewict Zwischen Mehratomigen Gasmolekülen', *Sitzungsberichte der Kaiserlichen Akademie der Wissenschaften in Wien, Klasse IIa*, **63**, 397–418.

References

(1872) 'Weitere Studien über das Wärmegleichgewicht unter Gasmolekülen', *Sitzungsberichte der Kaiserlichen Akademie der Wissenschaften in Wien, Klasse IIa*, **66**, 275–370.

(1896–8) *Vorlesungen über Gastheorie*. Leipzig: J. A. Barth. Translated 1964.

(1897) 'On Zermelo's paper "on the mechanical explanation of irreversible processes"', *Annalen der Physik*, **60**, 392–8. Translated and reprinted in Brush (1966), chapter 10.

(1964) *Lectures on Gas Theory*. English translation of Boltzmann (1896–8) by S. Brush. Berkeley and Los Angeles: University of California Press.

Bondi, H. (1962) 'Physics and cosmology', *The Observatory*, **82**, 133–43.

Broad, C. D. (1938) *Examination of McTaggart's Philosophy*, vol. 2, part I. Cambridge: Cambridge University Press. Reprinted, New York City: Octagon Books (1976).

Brooks, D. & Wiley, E. O. (1986) *Evolution as Entropy*. Chicago: University of Chicago Press.

Brillouin, L. (1962) *Science and Information Theory*. New York: Academic Press.

Brown, B. (1992) 'Defending backward causation', *Canadian Journal of Philosophy*, **22**, 429–44.

Brush, S. (1966) *Kinetic Theory*, vol. 2. Oxford: Pergamon Press.

Caldeira, A. O. & Leggett, A. J. (1981) 'Influence of dissipation on quantum tunneling in macroscopic systems', *Physical Review Letters*, **46**, 211–14.

(1983a) 'Quantum tunneling in a dissipative system', *Annals of Physics*, **149**, 374–456.

(1983b) 'Path integral approach to quantum Brownian motion', *Physica A*, **121**, 587–616.

(1985) 'Influence of damping on quantum interference: an exactly solvable model', *Physical Review A*, **31**, 1059–66.

Callan, C. G. & Freed, D. (1992) 'Phase diagram of the dissipative Hofstadter model', *Nuclear Physics B*, **374**, 543–66.

Carroll, J. (1994) *Laws of Nature*. New York: Cambridge University Press.

Carroll, S. M., Farhi, E., & Guth, A. H. (1992) 'An obstacle to building a time machine', *Physical Review Letters*, **68**, 263–6.

Carter, B. (1971) 'Causal structure in spacetime', *General Relativity and Gravitation*, **1**, 349–91.

Chakravarty, S. & Leggett, A. J. (1984) 'Dynamics of the 2-state system with ohmic dissipation', *Physical Review Letters*, **52**, 5–9.

Chapman, T. (1982) *Time: A Philosophical Analysis*. Dordrecht: D. Reidel.

Charlton, N. & Clarke, C. J. S. (1990) 'On the outcome of Kerr-like collapse', *Classical and Quantum Gravity*, **7**, 743–9.

Chen, Y. C. & Stamp, P. C. E. (1992) 'Quantum mobility of a dissipative Wannier–Azbel–Hofstadter particle', UBC preprint.

Christodoulou, D. (1984) 'Violation of cosmic censorship in the gravitational collapse of a dust cloud', *Communications in Mathematical Physics*, **93**, 171–95.

Clarke, C. J. S. (1977) 'Time in general relativity'. In *Foundations of Space-Time Theories*, ed. J. Earman, C. Glymour, & J. Stachel, pp. 94–108. Minneapolis: University of Minnesota Press.

References

Clarke, J., Cleland, A. N., DeVoret, M. H., Esteve, D., & Martinis, J. M. (1988) 'Quantum mechanics of a macroscopic variable: the phase difference of a Josephson junction', *Science*, **239**, 992–7.

Clauser, J. F. & Shimony, A. (1978) 'Bell's theorem: experimental tests and implications', *Reports on Progress in Physics*, **44**, 1881–1927.

Clifton, R. K., Redhead, M. L. G., & Butterfield, J. N. (1991) 'Generalization of the Greenberger–Horne–Zelinger algebraic proof of nonlocality', *Foundations of Physics*, **21**, 149–84.

Collier, J. (1986) 'Entropy in evolution', *Biology and Philosophy*, **1**, 5–24.

Courbage, M. & Prigogine, I. (1983) 'Intrinsic irreversibility in classical dynamical systems', *Proceedings of the National Academy of Sciences USA*, **80**, 2412–16.

Crow, J. F. & Kimura, M. (1970) *An Introduction to Population Genetics Theory*. Edina, MN: Burgess Publishing Company.

Cutler, C. (1992) 'Global structure of Gott's two-string spacetime', *Physical Review D*, **45**, 487–94.

Davies, P. C. W. (1974) *The Physics of Time Asymmetry*. Berkeley and Los Angeles: University of California Press. 2nd edn. (1977).

 (1977) *Space and Time in the Modern Universe*. Cambridge: Cambridge University Press.

 (1983) 'Inflation and time asymmetry in the Universe', *Nature*, **301**, 398–400.

Davies, P. C. W. & Twamley, J. (1993) 'Time-symmetric cosmology and the opacity of the future light cone', *Classical and Quantum Gravity*, **10**, 931–45.

De, U. K. (1969) 'Paths in universes having closed time-like lines', *Journal of Physics A (Ser. 2)*, **2**, 427–32.

de Broglie, L. (1939) *Matter and Light*. New York: W. W. Norton and Co.

de Bruyn Ouboter, R. & Bol, D. (1982) 'On the influence of dissipation on the transition mechanism of magnetic flux at low temperatures in a superconducting loop closed with a low-capacitance superconducting point contact', *Physica B*, **112**, 15–23.

Deser, S. (1993) 'Physical obstacles to time travel', *Classical and Quantum Gravity*, **10**, S67–S73.

Deser, S., Jakiw, R., & 't Hooft, G. (1992) 'Physical cosmic strings do not generate closed timelike curves', *Physical Review Letters*, **68**, 267–9.

d'Espagnat, B. (1979) 'The quantum theory and reality', *Scientific American*, **241**, 158–81.

Deutsch, D. (1985) 'Quantum theory, the Church–Turing principle, and the universal quantum computer', *Proceedings of the Royal Society A*, **400**, 97–117.

 (1989) 'Quantum computational networks', *Proceedings of the Royal Society A*, **425**, 73–90.

 (1991) 'Quantum mechanics near closed timelike curves', *Physical Review D*, **44**, 3197–217.

Deutsch, D. & Jozsa, R. (1992) 'Rapid solution of problems by quantum computation', *Proceedings of the Royal Society A*, **439**, 553–8.

De Witt, B. S. & Graham, N. eds. (1973) *The Many-Worlds Interpretation of Quantum Mechanics*. Princeton: Princeton University Press.

References

Dobson, A. (1905) *Collected Poems*. London: Kegan Paul, Trench, Trübner & Co. Ltd.

Dummett, M. (1960). 'A defence of McTaggart's proof of the unreality of time', *Philosophical Review*, **69**, 497–504. Reprinted in Dummett (1978).

(1964) 'Bringing about the past', *Philosophical Review*, **73**, 338–59. Reprinted in Dummett (1978).

(1978) *Trust and Other Enigmas*. Cambridge, MA: Harvard University Press.

(1986) 'Causal loops'. In *The Nature of Time*, ed. R. Flood & M. Lockwood, pp. 135–69. Oxford: Basil Blackwell Ltd.

Dwyer, L. (1975) 'Time travel and changing the past', *Philosophical Studies*, **27**, 341–50.

(1977) 'How to affect but not change the past', *Southern Journal of Philosophy*, **15**, 383–5.

(1978) 'Time travel and some alleged logical asymmetries between past and future', *Canadian Journal of Philosophy*, **8**, 15–38.

Dyson, F. J. (1979) 'Time without end: physics and biology in an open universe', *Reviews of Modern Physics*, **51**, 447–60.

Earman, J. (1972) 'Implications of causal propagation outside the null cone', *Australasian Journal of Philosophy*, **50**, 223–37.

(1974) 'An attempt to add a little direction to "the problem of the direction of time"', *Philosophy of Science*, **41**, 15–47.

(1987) 'The problem of irreversibility'. In *PSA: 1986*, vol. 2, ed. A. Fine & P. Machamer, pp. 226–33. East Lansing, MI: Philosophy of Science Association.

(1993a) 'In defense of laws: reflections on Bas van Fraassen's *Laws and Symmetry*', *Philosophy and Phenomenological Research*, **53**, 413–19.

(1993b) 'Cosmic censorship'. In *PSA: 1992*, vol. 2, ed. D. Hull, M. Forbes, & K. Okruhlik. East Lansing, MI: Philosophy of Science Association. In Press.

(1994) 'Observability, horizons, and common causes', preprint.

Echeverria, F., Klinkhammer, G., & Thorne, K. S. (1991) 'Billiard balls in wormhole spacetimes with closed timelike curves: classical theory', *Physical Review D*, **44**, 1077–99.

Eddington, A. S. (1928) *The Nature of the Physical World*. Cambridge: Cambridge University Press.

Ehrenfest, P. & Ehrenfest, T. (1959) *The Conceptual Foundations of the Statistical Approach in Mechanics*. Ithaca, NY: Cornell University Press.

Ehring, D. (1987) 'Personal identity and time travel', *Philosophical Studies*, **52**, 427–33.

Everett, H. (1957) '"Relative State" formulation of quantum mechanics', *Reviews of Modern Physics*, **29**, 454–62.

Feynman, R. P. (1949) 'The theory of positrons', *Physical Review*, **76**, 749–59.

(1982) 'Simulating physics with computers', *International Journal of Theoretical Physics*, **21**, 467–88.

(1986) 'Quantum mechanical computers', *Foundations of Physics*, **16**, 507–31.

Feynman, R. P., Leighton, R. B., & Sands, M. (1965) *The Feynman Lectures on Physics: Quantum Mechanics*. Reading, MA: Addison-Wesley.

Feynman, R. P. & Vernon, F. L. (1963) 'The theory of a general quantum system interacting with a linear dissipative system', *Annals of Physics*, **24**, 118–173.

References

Fine, A. (1989) 'Do correlations need to be explained?' In *Philosophical Consequences of Quantum Theory*, ed. J. T. Cushing & E. McMullin, pp. 175–94. Notre Dame, IN: University of Notre Dame Press.

Fisher, R. A. (1930) *The Genetical Theory of Natural Selection*. Oxford: Clarendon Press.

Flew, A. (1954) 'Can an effect precede its cause?', *Proceedings of the Aristotelian Society*, suppl. vol. 38, 45–62.

Friedman, J. L. & Morris, M. S. (1991a) 'The Cauchy problem for the scalar wave equation is well defined on a class of spacetimes with closed timelike curves', *Physical Review Letters*, **66**, 401–4.

(1991b) 'The Cauchy problem on spacetimes with closed timelike curves', *Annals of the New York Academy of Sciences*, **631**, 173–81.

Friedman, J. L., Morris, M. S., Novikov, I. D., Echeverria, F., Klinkhammer, G., Thorne, K. S., & Yurtsever, U. (1990) 'Cauchy problems in spacetimes with closed timelike curves', *Physical Review D*, **42**, 1915–30.

Frolov, V. P. (1991) 'Vacuum polarization in a locally static multiply connected spacetime and a time-machine problem', *Physical Review D*, **43**, 3878–94.

Frolov, V. P. & Novikov, I. D. (1990) 'Physical effects in wormholes and time machines', *Physical Review D*, **42**, 1057–65.

Gardner, M. (1988) 'Time travel'. In *Time Travel and Other Mathematical Bewilderments*, pp. 1–14. New York: W. H. Freeman.

Gell-Mann, M. & Hartle, J. B. (1991) 'Alternative decohering histories in Quantum Mechanics'. In *Proceedings of the 25th International Conference on High Energy Physics, Singapore, August 2–8, 1990*, ed. K. K. Phua & Y. Yamaguchi, pp. 1303–10. Singapore: World Scientific.

(1994) 'Time symmetry and asymmetry in quantum mechanics'. In *Proceedings of the NATO Workshop on the Physical Origin of Time Asymmetry, Mazagon, Spain, September 30–October 4, 1991*, ed. J. J. Halliwell, J. Perez-Mercador, & W. H. Zurek. Cambridge: Cambridge University Press. Also in *Proceedings of the First International A. D. Sakharov Conference on Physics, Moscow, USSR, May 27–31, 1991*.

Geroch, R. P. (1967) 'Topology in General Relativity', *Journal of Mathematical Physics*, **8**, 782–6.

Gibbons, G. W. & Hawking, S. W. (1992) 'Kinks and topology change', *Physical Review Letters*, **69**, 1719–21.

Gödel, K. (1949a) 'An example of a new type of cosmological solution to Einstein's field equations of gravitation', *Reviews of Modern Physics*, **21**, 447–50.

(1949b) 'A remark about the relationship between relativity theory and idealistic philosophy'. In *Albert Einstein: Philosopher-Scientist*, ed. P. Schilpp, pp. 557–62. La Salle, IL: Open Court. Reprinted in Yourgrau (1990) and Gödel (1990).

(1952) 'Rotating universes in General Relativity Theory'. In *Proceedings of the International Congress of Mathematicians, Cambridge, MA, USA, August 30–September 6, 1950*, vol. I, pp. 175–81. Providence: American Mathematical Society. Reprinted in Gödel (1990).

(1990) *Collected Works*, vol. 2, ed. S. Feferman, J. W. Dawson, Jr., S. C. Kleene, G. H. Moore, R. M. Solovay, & J. van Heijenoort. New York and Oxford: Oxford University Press.

References

Gold, T. (1962) 'The arrow of time', *American Journal of Physics*, **30**, 403–10.

Goldwirth, D. S., Perry, M. J., & Piran, T. (1993) 'The breakdown of Quantum Mechanics in the presence of time machines', *General Relativity and Gravitation*, **25**, 7–13.

Gott, J. R. (1991) 'Closed timelike curves produced by pairs of moving cosmic strings: exact solutions', *Physical Review Letters*, **66**, 1126–9.

Greenberger, D. M. ed. (1986) *New Techniques and Ideas in Quantum Measurement Theory*. New York: New York Academy of Sciences.

Greenberger, D. M., Horne, M. A., Shimony, A., & Zeilinger, A. (1990) 'Bell's theorem without inequalities', *American Journal of Physics*, **58**, 1131–43.

Griffiths, R. B. (1984) 'Consistent histories and the interpretation of quantum mechanics', *Journal of Statistical Physics*, **36**, 219–72.

Grünbaum, A. (1973) *Philosophical Problems of Space and Time*. Dordrecht: Reidel. 2nd edn.

Gutzwiller, M. (1990) *Chaos in Classical and Quantum Mechanics*. Berlin: Springer-Verlag.

Hahn, E. (1950) 'Spin echoes', *Physical Review*, **80**, 589–94.

  (1953) 'Free nuclear induction', *Physics Today*, **6**, 4–9.

Haldane, F. D. M. (1983) 'Non-linear field theory of large-S Heisenberg antiferromagnets: semiclassically quantized solutions of the 1-d easy axis Néel state', *Physical Review Letters*, **50**, 1153–6.

Harrison, J. (1971) 'Dr Who and the philosophers: or time-travel for beginners', *Proceedings of the Aristotelian Society*, suppl. vol. 45, 1–24.

Hartle, J. B. (1991) 'The quantum mechanics of cosmology'. In *Quantum Mechanics and Baby Universes: Proceedings of the 1989 Jerusalem Winter School for Theoretical Physics*, ed. S. Coleman, J. B. Hartle, T. Piran, & S. Weinberg, pp. 65–157. Singapore: World Scientific.

  (1993) 'The spacetime approach to quantum mechanics', *Vistas in Astronomy*, **37**, 569–83. Also in *Topics in Quantum Gravity and Beyond: Essays in Honor of Louis Witten on his Retirement*, ed. F. Mansouri & J. J. Scanio, pp. 17–32. Singapore: World Scientific.

  (1994) 'Spacetime quantum mechanics and the quantum mechanics of spacetime'. To appear in *Gravitation and Quantization: Proceedings of the 1992 Les Houches Summer School*, ed. B. Julia & J. Zinn-Justin. Amsterdam: North Holland. In Press.

Hawking, S. W. (1975) 'Particle creation by black holes', *Communications in Mathematical Physics*, **43**, 199–220.

  (1976) 'Black holes and thermodynamics', *Physical Review D*, **13**, 191–7.

  (1985) 'Arrow of time in cosmology', *Physical Review D*, **33**, 2489–95.

  (1988) *A Brief History of Time*. New York: Bantam.

  (1992) 'The chronology protection conjecture', *Physical Review D*, **46**, 603–11.

  (1994) 'The No Boundary Proposal and the arrow of time', *Vistas in Astronomy*, **37**, 559–68. Also in *Proceedings of the NATO Workshop on the Physical Origin of Time Asymmetry, Mazagon, Spain, September 30–October 4, 1991*, ed. J. J. Halliwell, J. Perez-Mercador, & W. H. Zurek. Cambridge: Cambridge University Press.

Hawking, S. W. & Ellis, G. F. R. (1973) *The Large Scale Structure of Space-time*. Cambridge: Cambridge University Press.

References

Hawking, S. W. & Halliwell, J. J. (1985) 'Origins of structure in the universe', *Physical Review D*, **31**, 1777–91.

Heller, E. J. & Tomsovic, S. (1993) 'Postmodern Quantum Mechanics', *Physics Today*, **46**(7), 38–46.

Horwich, P. (1987) *Asymmetries in Time*. Cambridge, MA: The MIT Press.

Hospers, J. (1967) *An Introduction to Philosophical Analysis*, Englewood Cliffs, NJ: Prentice-Hall. 2nd edn.

Howard, D. (1990) '"Nicht sein kann was nicht sein darf", or the prehistory of EPR, 1909–1935: Einstein's early worries about the Quantum Mechanics of composite systems'. In *Sixty-two Years of Uncertainty*, ed. A. I. Miller, pp. 61–111. New York: Plenum Press.

Hughes, R. I. G. (1989) *The Structure and Interpretation of Quantum Mechanics*. Cambridge, MA and London, England: Harvard University Press.

Jammer, M. (1974) *The Philosophy of Quantum Mechanics*. New York: Wiley.

Jarrett, J. P. (1984) 'On the physical significance of the locality conditions in the Bell arguments', *Noûs*, **18**, 569–89.

Jaynes, E. (1965) 'Gibbs vs. Boltzmann entropies', *American Journal of Physics*, **33**, 391–8.

Joshi, P. S. (1985) 'Topological properties of certain physically significant subsets of spacetime'. In *A Random Walk in Relativity and Cosmology*, ed. N. Dadhich, J. K. Rao, J. V. Narliker, & C. V. Vishceswara, pp. 128–36. New York: John Wiley.

Kagan, Y. A. (1992) 'Quantum diffusion in solids', *Journal of Low Temperature Physics*, **87**, 525–69.

Kagan, Y. A. & Prokof'ev, N. V. (1992) 'Quantum tunneling diffusion in solids'. In *Quantum Tunneling in Condensed Media*, ed. Y. A. Kagan & A. J. Leggett, pp. 37–143. Amsterdam: Elsevier.

Kelly, J. L. (1955) *General Topology*. Princeton: Van Nostrand.

Khinchin, A. I. (1949) *Mathematical Foundations of Statistical Mechanics*. New York: Dover Publications.

Kim, J. (1978) 'Supervenience and nomological incommensurables', *American Philosophical Quarterly*, **15**, 149–56.

Kim, S.-W. & Thorne, K. S. (1991) 'Do vacuum fluctuations prevent the creation of closed timelike curves?', *Physical Review D*, **43**, 3929–47.

Kittel, C. & Kroemer, H. (1980) *Thermal Physics*. New York: W. H. Freeman and Company. 2nd edn.

Klinkhammer, G. (1992) 'Vacuum polarization of scalar and spinor fields near closed null geodesics', *Physical Review D*, **46**, 3388–94.

Kriele, M. (1989) 'The structure of chronology violating sets with compact closure', *Classical and Quantum Gravity*, **6**, 1606–11.

  (1990) 'Causality violations and causality', *General Relativity and Gravitation*, **22**, 619–23.

Krylov, N. (1979) *Works on the Foundations of Statistical Physics*. Princeton: Princeton University Press.

Kuchar, K. (In Press) 'Canonical quantum gravity'. To appear in *1993 Proceedings of the 13th International Conference on General Relativity and Gravitation*, ed. C. Kozameh. Bristol: IOP Publishing.

References

Landau, L. D. & Lifshitz, E. M. (1965) *Quantum Mechanics. Course in Theoretical Physics*, vol. 3. Oxford: Pergamon.
  (1981) *Physical Kinetics. Course in Theoretical Physics*, vol. 9. Oxford: Pergamon.
Landsberg, P. T. (1970) 'Time in statistical physics and special relativity', *Studium Generale*, **23**, 1108-58.
  (1978) *Thermodynamics and Statistical Mechanics*. New York: Dover Publications Inc.
  (1982) *The Enigma of Time*. Bristol: Adam Hilger Ltd.
Lanford, O. (1976) 'On a derivation of the Boltzmann equation', *Asterique*, **40**, 117-137. Reprinted in Lebowitz & Montroll (1983).
Laplace, P. S. (1820) *Theorie Analytique des Probabilités*. Paris: V. Courcier.
Layzer, D. (1975) 'The arrow of time', *Scientific American*, **234**, 56-69.
  (1990) *Cosmogenesis: The Growth of Order in the Universe*. Oxford: Oxford University Press.
Lebowitz, J. (1983) 'Microscopic dynamics and macroscopic laws'. In *Long-Time Prediction in Dynamics*, ed. C. Horton, L. Reichl, & V. Szebehely, pp. 3-19. New York: Wiley.
Lebowitz, J. & Montroll, E. eds. (1983) *Nonequilibrium Phenomena I: The Boltzmann Equation*. Amsterdam: North-Holland.
Leggett, A. J. (1980) 'Macroscopic quantum systems and the quantum theory of measurement', *Supplement to Progress of Theoretical Physics*, **69**, 80-100.
  (1984) 'Quantum tunneling in the presence of an arbitrary linear dissipation mechanism', *Physical Review B*, **30**, 1208-18.
  (1986) 'Quantum measurement at the macroscopic level'. In *The Lessons of Quantum Theory*, ed. J. de Boer, E. Dal, & O. Ulfbeck, pp. 35-57. Amsterdam: Elsevier.
  (1987a) 'Reflections on the quantum measurement paradox'. In *Quantum Implications: Essays in Honor of David Bohm*, ed. B. J. Hiley & F. D. Peat, pp. 85-104. London: Routledge.
  (1987b) 'Quantum mechanics at the macroscopic level'. In *Matter and Chance: Proceedings of the 1986 les Houches Summer School*, ed. J. Souletie, J. Vannaimenus, & R. Stora, pp. 396-507. Amsterdam: North Holland.
  (1988) 'The quantum mechanics of a macroscopic variable: some recent results and current issues'. In *Frontiers and Borderlines in Many-Particle Physics*, ed. R. A. Broglia & J. R. Schrieffer. Italian Physical Society.
Leggett, A. J., Charkravarty, S., Dorsey, A. T., Fisher, M. P. A., Garg, A., & Zwerger, W. A. (1987) 'Dynamics of the dissipative 2-state system', *Reviews of Modern Physics*, **59**, 1-85.
Leggett, A. J. & Garg, A. (1985) 'Quantum mechanics versus macroscopic realism: is the flux there when nobody looks?', *Physical Review Letters*, **54**, 857-60.
Le Poidevin, R. & Macbeath, M. eds. (1993) *The Philosophy of Time*. Oxford: Oxford University Press.
Lewis, D. K. (1973) *Counterfactuals*. Cambridge: Cambridge University Press.
Lewis, D. K. (1976) 'The paradoxes of time travel', *American Philosophical Quarterly*, **13**, 145-52. Reprinted in Lewis (1986b).
  (1979) 'Counterfactual dependence and time's arrow', *Noûs*, **13**, 455-76. Reprinted in Lewis (1986b).

(1986a) *The Plurality of Worlds.* Oxford: Blackwell.

(1986b) *Philosophical Papers,* vol. 2. Oxford: Oxford University Press.

Lewis, G. N. (1926) *The Anatomy of Science.* New Haven: Yale University Press.

Linde, A. (1987) 'Inflation and quantum cosmology'. In *Three Hundred Years of Gravitation,* ed. S. W. Hawking & W. Israel, pp. 604–30. Cambridge: Cambridge University Press.

London, F. & Bauer, E. (1939) *La Theorie de L'Observation en Mécanique Quantique.* Herman et Cie: Paris. Translated and reprited in Wheeler & Zurek (1983).

Loss, D., di Vincenzo, D. P., & Grinstein, G. (1992) 'Suppression of tunneling by interference in 1/2-integer spin particles', *Physical Review Letters,* **69**, 3232–5.

Lossev, A. & Novikov, I. D. (1992) 'The Jinn of the time machine: non-trivial self-consistent solutions', *Classical and Quantum Gravity,* **9**, 2309–27.

MacBeath, M. (1982) 'Who was Dr Who's father?', *Synthese,* **51**, 397–430.

Malament, D. (1984) '"Time travel" in the Gödel Universe'. In *PSA: 1984,* vol. 2, ed. P. D. Asquith & P. Kitcher, pp. 91–100. East Lansing, MI: Philosophy of Science Association.

Margolus, N. (1986) 'Quantum computation'. In *New Techniques and Ideas in Quantum Measurement Theory,* ed. D. M. Greenberger, pp. 487–97. New York: New York Academy of Sciences.

Maudlin, T. (1990) 'Time travel and topology'. In *PSA: 1990,* vol. 1, ed. A. Fine, M. Forbes, & L. Wessels, pp. 303–15. East Lansing, MI: Philosophy of Science Association.

McCall, S. (1990) 'Choice trees'. In *Truth or Consequences: Essays in Honor of Nuel Belnap,* ed. J. M. Dunn & A. Gupta, pp. 231–44. Dordrecht: Kluwer.

(1994) *A Model of the Universe.* Oxford: Clarendon Press.

McTaggart, J. M. E. (1908) 'The unreality of time', *Mind,* **18**, 457–84.

Mellor, D. H. (1981) *Real Time.* Cambridge: Cambridge University Press.

Mermin, N. D. (1985) 'Is the Moon there when nobody looks? Reality and the quantum theory', *Physics Today,* **38**(4), 38–47.

(1990) 'What's wrong with these elements of reality?', *Physics Today,* **43**(6), 9–11.

Mikheeva, E. V. & Novikov, I. V. (1992) 'Inelastic billiard ball in a spacetime with a time machine', *Physical Review D.* In Press.

Mill, J. S. (1904) *A System of Logic.* New York: Harper and Row.

Morowitz, H. (1986) 'Entropy and nonsense', *Biology and Philosophy,* **1**, 473–6.

Morris, M. S., Thorne, K. S. & Yurtsever, U. (1988) 'Wormholes, time machines, and the weak energy condition', *Physical Review Letters,* **61**, 1446–9.

Nagel, E. (1961) *The Structure of Science.* New York & Burlingame: Harcourt, Brace & World, Inc.

Ne'eman, Y. (1970) 'CP and CPT violations, entropy and the expanding universe', *International Journal of Theoretical Physics,* **3**, 1–5.

Newton, I. (1686) *Mathematical Principles of Natural Philosophy,* 2 vols. Translated by Andrew Motte, revised by Florian Cajori. Berkeley: University of California Press (1934).

Novikov, I. D. (1989) 'An analysis of the operation of a time machine', *Soviet Journal of Experimental and Theoretical Physics,* **68**, 439–43.

(1992) 'Time machines and self-consistent evolution in problems with self-interaction', *Physical Review D,* **45**, 1989–94.

References

Olenick, R. P., Apostol, T. M., & Goodstein, D. L. (1985) *The Mechanical Universe: Introduction to Mechanics and Heat*. Cambridge: Cambridge University Press.

Ori, A. (1991) 'Rapidly moving cosmic strings and chronology protection', *Physical Review D*, **44**, 2214–5.

Ornstein, D. (1975) 'What does it mean for a mechanical system to be isomorphic to the Bernoulli flow?'. In *Dynamical Systems, Theory and Application*, ed. J. Moser, pp. 209–33. Berlin: Springer-Verlag.

Oszvath, I. (1967) 'Homogeneous Lichnerowicz Universes', *Journal of Mathematical Physics*, **8**, 326–44.

Page, D. N. (1983) 'Inflation does not explain time asymmetry', *Nature*, **304**, 39–41.

Park, D. (1972) 'The myth of the passage of time'. In *The Study of Time*, ed. J. T. Fraser, F. C. Haber, & G. H. Müller, pp. 110–21. Berlin: Heidelberg, and New York: Springer-Verlag.

Partridge, R. B. (1973) 'Absorber theory of radiation and the future of the universe', *Nature*, **244**, 263–5.

Pears, D. (1957) 'The priority of causes', *Analysis*, **17**, 54–63.

Peierls, R. E. (1979) *Surprises in Theoretical Physics*. Princeton: Princeton University Press.

Penrose, O. (1979) 'Foundations of statistical mechanics', *Reports on Progress in Physics*, **42**, 1937–2006.

Penrose, R. (1964) 'Conformal treatment of infinity'. In *Relativity, Groups, and Topology*, ed. C. De Witt & B. De Witt, pp. 565–73. New York: Gordon and Breach.

  (1969) 'Gravitational collapse: the role of general relativity', *Nuovo Cimento*, **1**, special number, 252–76.

  (1979) 'Singularities and time-asymmetry'. In *General Relativity: An Einstein Centenary Survey*, ed. S. W. Hawking & W. Israel, pp. 581–638. Cambridge: Cambridge University Press.

  (1989) *The Emperor's New Mind*. Oxford and New York: Oxford University Press.

  (1991) Personal correspondence, 28.1.91 & 21.2.91.

Pfarr, J. (1981) 'Time travel in Gödel's space', *General Relativity and Gravitation*, **13**, 1073–91.

Pohl, F. (1990) *The World at the End of Time*. New York: Ballentine.

Pound, R. V. & Rebka, G. A., Jr. (1960) 'Apparent weight of photons', *Physical Review Letters*, **4**, 337–41.

Price, H. (1989) 'A point on the arrow of time', *Nature*, **340**, 181–2.

  (1991) 'The asymmetry of radiation: reinterpreting the Wheeler–Feynman argument', *Foundations of Physics*, **21**, 959–75.

Prokof'ev, N. V. & Stamp, P. C. E. (1993) 'Giant spins and topological decoherence: a Hamiltonian approach', *Journal of Physics: Condensed Matter*, **5**, L663–L670.

Quine, W. V. O. (1951) 'Two dogmas of empiricism', *Philosophical Review*, **60**, 20–43. Reprinted in Quine (1961).

  (1961) *From a Logical Point of View*. Cambridge, MA: Harvard University Press. 2nd edn, revised.

References

Ramsey, F. P. (1978) 'Law and causality'. In *Foundations: Essays in Philosophy, Logic, Mathematics, and Economics*, ed. D. H. Mellor, pp. 128–51. Atlantic Highlands: Humanities Press.

Reichenbach, H. ed. (1956) *The Direction of Time*. Berkeley and Los Angeles: University of California Press. Reprinted (1971).

Rhim, W., Pines, A., & Waugh, J. (1971) 'Time reversal experiments in dipolar coupled spin systems', *Physical Review B*, **3**, 684–95.

Ritchie, N. W. M., Story, J. G., & Hulet, R. G. (1991) 'Realisation of a measurement of a "weak value"', *Physical Review Letters*, **66**, 1107–10.

Sachs, R. G. (1987) *The Physics of Time Reversal*. Chicago and London: University of Chicago Press.

Savitt, S. (1991) 'Critical notice of Paul Horwich's *Asymmetries in Time*', *Canadian Journal of Philosophy*, **21**, 399–417.

(1994) 'The replacement of time', *Australasian Journal of Philosophy*, **72**, 463–74.

Shapere, A. & Wilczek, F. (1989) *Geometric Phases in Physics*. Singapore: World Scientific.

Shimony, A. (1989) 'Conceptual foundations of quantum mechanics'. In *The New Physics*, ed. P. C. W. Davies, pp. 373–95. Cambridge: Cambridge University Press.

Sidles, J. A. (1992) 'Folded Stern–Gerlach experiment as a means for detecting nuclear magnetic resonance in individual nuclei', *Physical Review Letters*, **68**, 1124–7.

Sikkema, A. E. & Israel, W. (1991) 'Black-hole mergers and mass inflation in a bouncing universe', *Nature*, **349**, 45–7.

Sklar, L. (1974) *Space, Time, and Spacetime*. Berkeley and Los Angeles: University of California Press.

(1980) 'Semantic analogy', *Philosophical Studies*, **38**, 217–34. Reprinted in Sklar (1985).

(1981) 'Up and down, left and right, past and future', *Noûs*, **15**, 111–29. Reprinted in Sklar (1985) and Le Poidevin & Macbeath (1993).

(1985) *Philosophy and Spacetime Physics*. Berkeley: University of California Press.

(1993) *Physics and Chance: Philosophical issues in the foundations of statistical mechanics*. Cambridge: Cambridge University Press.

Smith, J. W. (1986) 'Time travel and backward causation'. In *Reason, Science, and Paradox*, pp. 49–58. London: Croom Helm.

Sneed, J. (1971) *The Logical Structure of Mathematical Physics*. Dordrecht: Reidel.

Sober, E. (1988) 'The principle of the common cause'. In *Explanation and Causation: Essays in Honor of Wesley Salmon*, ed. J. Fetzer, pp. 211–28. Dordrecht: Reidel.

Stamp, P. C. E. (1988) 'Influence of paramagnetic and Kando impurities on macroscopic quantum tunneling in SQUIDs', *Physical Review Letters*, **61**, 2905–8.

(1992) 'Magnets get their act together', *Nature*, **359**, 365–6.

(1993) 'Dissipation and decoherence in quantum magnetic systems', *Physica B*, **197**, 133–43.

Stamp, P. C. E., Chudnovsky, E. M., & Barbara, B. (1992) 'Quantum tunneling of magnetization in solids', *International Journal of Modern Physics B*, **6**, 1355–473.

References

Stapp, H. P. (1993) 'Significance of an experiment of the Greenberger–Horne–Zeilinger kind', *Physical Review A*, **47**, 847–53.

Stein, H. (1970) 'On the paradoxical time-structures of Gödel', *Philosophy of Science*, **37**, 589–601.

Swinburne, R. (1968) *Space and Time*. London: Macmillan.

Taylor, R. (1955) 'Spatial and temporal analogies and the concept of identity', *Journal of Philosophy*, **52**, 599–612. Reprinted in *Problems of Space and Time*, ed. J. J. C. Smart, pp. 381–96. New York: The Macmillan Company (1964).

Tesche, C. D. (1990) *Physical Review Letters*, **64**, 2358.

Thom, P. (1975) 'Time-travel and non-fatal suicide', *Philosophical Studies*, **27**, 211–16.

't Hooft, G. (1992) 'Causality in $(2+1)$-dimensional gravity', *Classical and Quantum Gravity*, **9**, 1335–48.

Thorne, K. S. (1991) 'Do the laws of physics permit closed timelike curves?', *Annals of the New York Academy of Sciences*, **631**, 182–93.

Tipler, F. J. (1974) 'Rotating cylinders and the possibility of global causality violation', *Physical Review D*, **9**, 2203–6.

(1976) 'Causality violation in asymptotically flat space-times', *Physical Review Letters*, **37**, 879–82.

(1977) 'Singularities and causality violation', *Annals of Physics*, **108**, 1–36.

Tolman, R. C. (1917) *The Theory of Relativity of Motion*. Berkeley: University of California Press.

(1934) *Relativity, Thermodynamics, and Cosmology*. Oxford: Oxford University Press.

Unruh, W. G. (1986) 'Quantum measurement'. In *New Techniques and Ideas in Quantum Measurement Theory*, ed. D. M. Greenberger, pp. 242–9. New York: New York Academy of Sciences.

(1994) 'The measurability of the wave function', *Physical Review A*.

Unruh, W. G. & Wald, R. M. (1989) 'Time and the interpretation of canonical quantum gravity', *Physical Review D*, **40**, 2598–614.

van Fraassen, B. C. (1989) *Laws and Symmetry*. Oxford: Oxford University Press.

Vessot, R. F. C., Levine, M. W., Mattison, E. M., Bloomberg, E. L., Hoffmann, T. E., Nystrom, G. U., et al. (1980) 'Test of relativistic gravitation with a space-borne hydrogen maser', *Physical Review Letters*, **45**, 2081–4.

von Delft, J. & Henley, C. (1992) 'Destructive interference in spin tunneling problems', *Physical Review Letters*, **69**, 3236–9.

von Neumann, J. (1935) *The Mathematical Foundations of Quantum Mechanics*. Princeton: Princeton University Press.

Wald, R. & Yurtsever, U. (1991) 'General proof of the averaged null energy condition for a massless scalar field in two-dimensional curved space-time', *Physical Review D*, **44**, 403–16.

Watanabe, S. (1955) 'Symmetry of physical laws part III. Prediction and retrodiction', *Reviews of Modern Physics*, **27**, 179–86.

Weir, S. (1988) 'Closed time and causal loops: a defense against Mellor', *Analysis*, **48**, 203–9.

Wells, H. G. (1895) *The Time Machine*. London: William Heinemann.

Wheeler, J. A. & Feynman, R. P. (1945) 'Interaction with the absorber as the mechanism of radiation', *Reviews of Modern Physics*, **17**, 157–81.

References

(1949) 'Classical electrodynamics in terms of direct inter-particle action', *Reviews of Modern Physics*, **21**, 425–34.

Wheeler, J. A. & Zurek, W. eds. (1983) *Quantum Theory and Measurement*. Princeton: Princeton University Press.

Wicken, J. (1987) *Entropy, Thermodynamics, and Evolution*. Oxford: Oxford University Press.

Wigner, E. P. (1963) 'The problem of measurement', *American Journal of Physics*, **31**, 6–15. Reprinted in Wigner (1979).

(1964) 'Two kinds of reality', *The Monist*, **48**, 248–64. Reprinted in Wigner (1979).

(1979) *Symmetries and Reflections*. Woodbridge, CT: Ox Bow Press.

Williams, D. C. (1951) 'The myth of passage'. Reprinted in *The Philosophy of Time*, ed. R. M. Gale, pp. 98–116. New York: Doubleday & Company (1967).

Yourgrau, P. (1990) *Demonstratives*. Oxford: Oxford University Press.

(1991) *The Disappearance of Time*. Cambridge: Cambridge University Press.

Yurtsever, U. (1990) 'Test fields on compact space-times', *Journal of Mathematical Physics*, **31**, 3064–78.

(1991) 'Classical and quantum instability of compact Cauchy horizons in two dimensions', *Classical and Quantum Gravity*, **8**, 1127–39.

Zeh, H.-D. (1989) *The Physical Basis of the Direction of Time*. New York, Berlin, and Heidelberg: Springer-Verlag.

Zeldovich, Ya. B. & Novikov, I. D. (1971) *Relativistic Astrophysics*; volume 1, *Stars and Relativity*; volume 2, *The Structure and Evolution of the Universe*. Chicago: University of Chicago Press.

Zemach, E. M. (1968) 'Many times', *Analysis*, **8**, 145–51.

Zurek, W. H. (1984) 'Reversibility and stability of information processing systems', *Physical Review Letters*, **53**, 391–5.

Index

'afterness', entropy and 224–5, 229
Aharonov, Y. 49–53, 97–8, 130

backward causation, time travel 270–7
becoming 7–12, 217–29
Belinfante, F. J. 97–8
Bell, J. S. 121–2, 169–70
benignity, time travel 294–9, 302
Benioff, P., quantum computer 134–8, 151–4
Bernoulli systems 197
Big Bang 55–6, 66, 67–94, 289
Big Crunch 67–8, 69, 70, 72, 73–8
bilking argument, closed causal chains 261–7
biological populations, second law of
 thermodynamics and 4–5, 230–3, 241–52
black holes 5–6, 56–7, 83–5, 86, 117–18
Boltzmann, L. 9, 193–4, 195, 200, 219–20
branched models of spacetime 11–12, 155–72
 definition of measurement 169–72
 non-locality 162–9
 objective time flow 155–9
 quantum probabilities 159–62
 stochastically branching spacetime
  topology 11–12, 173–88
 time asymmetry and thermodynamic
  laws 205–9

Caldeira–Leggett model 122–7, 144–5
canonical ensemble 239, 253–5
Carroll, J. 290
Cauchy surfaces 275–6
causation
 experienced succession and entropy 221–9
 time travel 5, 259–67, 268–310
Chapman, T., two times 273

chronology horizons 305–9
Clarke, C. J. S. 301
classical mechanics, emergence of from
 quantum mechanics 108, 109, 139–51
closed causal chains, time travel 12, 259–67
closed timelike curves (CTCs) 11, 12, 33,
 269–70
consistency constraints 269, 284–87
 non-self-interacting case 297, 294, 295
 self-interacting case 298, 301, 302, 303
Gödelian time travel 272–3, 277–80
grandfather paradox 281–4, 287
perceptual and physical time 225
time machine to produce 304–9, 310
coherence see decoherence
Coleman–Hepp model, quantum computer
 and 135, 151, 153
computations, measurements
 differentiated 134
consciousness see minds and consciousness
consistency constraints, time travel 269–70,
 284–7, 309–10
 general theory of relativity and 288
 laws and occurrent facts 290–2
 non-self-interacting case 292–7
 particle horizons 289
 self-consistency 285–6, 288, 290
 self-interacting case 297–303
consistent histories approach, decoherence
 63–4
cosmology 6
 double standard in reasoning about time 5,
  66–94
 case against symmetric time-reversing
  universe 81–6

325

Index

Hawking and the Big Crunch 73-8
inflationary model 71-2
observability and coherence of symmetric time-reversing universe 86-91, 92
Penrose explanation 78-81
formalism to describe quantum mechanics 49, 61-5
Gödelian time travel 277-80
gravity and time 23-34
quantum gravity 54-61
quantum mechanics and time 34-53
time asymmetry and thermodynamic laws 205-9
coupling process, entropy 232-3
biological populations 241-52
statistical mechanics of 234-8
curvature of spacetime 31(fn)
Weyl hypothesis 78-9, 206

Davies, P.
consciousness and time flow 156
cosmology and time asymmetry 12-13, 68-9, 71-2, 82-3, 86(fn), 206
De, U. K., CTCs 278
decoherence
emergence of classical properties 108, 109, 139-51
formalism for condition setting at arbitrary times 61-5
mental processes and free will 128-9
quantum measurement paradox and 101, 132-9
wave-functions and the quantum arrow 122-7
determinism, time travel 282-3
dissipation
quantum computer 133-9
wave-functions and the quantum arrow 122-7
Dobson, A., moving Now 8

Echeverria, F., wormholes 298-9, 302
Eddington, A. S. 1, 2, 9, 222-3, 226, 228, 229
eigenvectors and eigenvalues 39-40, 42, 43, 49, 61-5
Einstein's Field Equations (EFE), Gödelian time travel 277-80, 281

entropy *see* second law of thermodynamics
EPR experiments, spacetime branched models 162-7
ergodic theory 195-6, 197, 199-200, 209, 212-15, 234
eternal time machine spacetime 298
Everett, H., many-worlds 120, 130
experienced succession, becoming and entropy 9-10, 217-29

Feynman, R. P. 260, 264-7, 301
quantum computer 134-8, 151-4
fields, fluctuations in 56-7
Fisher, R. A. 230-1, 252
free will 129, 282, 291-2
Friedman, J. L., time travel 285-6, 293-5, 306
future directedness
becoming 7-12
of causation, time travel 5, 259-67, 268-310

garbling sequences 98-9
general theory of relativity (GTR)
black holes 5-6
causal structure of relativistic spacetimes 274-7
consistency constraint, time travel 288
Gödelian time travel 277-80, 281
gravity and time 24-34
quantization problems 54-61
genetic models, entropy 231-2, 241-52
GHZ experiments, spacetime branched models 162-3, 167-9
Gödel, K.
perceptual and physical time 225
simultaneity 10-11, 12
time travel 269, 271-3, 277-80
causal structure of relativistic spacetimes 274-7, 288
closed causal chains 259-60, 263, 264
grandfather paradox 280-4, 294
Gold, T. 67-8, 69, 70, 80, 81-91, 92
Gott, J. R., CTCs 279-80
grandfather paradox, time travel 280-4, 287, 294, 299, 309

Index

gravity
  gravitational argument for symmetry (GAS) 70, 77–8, 79–80, 82, 93
  quantum gravity 54–61, 118–21
  sensed downness, experienced succession and entropy 9, 218–21, 224
  thermodynamic and radiation arrows and 110–18
  as time 23–34

Hamiltonians
  quantum computer 135–8, 151, 153–4
  quantum fractal system 150–1
  spin-boson problem 140
  two-slit systems 144–5
Hartle, J. B., quantum gravity 60–1
Hausdorff spaces 177–8, 180, 182
Hawking, S. W. 59–60, 73–8, 305–6, 307
Heisenberg equations of evolution 37–8
Heisenberg representation 37
Hermitian operators 36, 38–40, 136–8
Hilbert space 35–8, 39, 42, 43, 55–8
Hofstadter (WAH) problem 150–1
H-theorem 193–4

inflationary model, cosmology, time asymmetry 71–2
initial conditions
  quantum mechanics and role of time 41–53
  statistical mechanics and thermodynamic laws 191–216
interventionism, thermodynamic laws 203–5, 211–12
irreversibility see quantum mechanics, measurement paradox

Jaynes, E. 202–3

KAM theorem 197–8
Kerr model, CTCs 279
Khinchin, A. I. 238–41, 245, 246, 247, 249
kinetic equations, time asymmetry and
  thermodynamic laws 191–216
  cosmology and branch systems 205–9
  formal approaches to irreversibility 195–9
  history of problem 193–5
  interventionism 203–5, 211–12

Krylov's solution 209–12
'parity of reasoning' problem 199–203
Prigogine's solution 212–15
Kramer's theorem, example 146
Krylov's solution 209–12
K-systems 197

Landsberg, P. T., The Now 7, 8
Lanford, O. 195, 200
Leggett, A. J. 122–7, 144–5
Lewis, D., M-R-L laws 290–2, 295, 303, 309
Lewis, G. N., two-way time 2–4

Mackie, J. L., time symmetry 218
macrorealism (MR) 15, 97, 101–6
macroscopic quantum processes
  decoherence and dissipation 126–7, 140–3
  quantum interference 127–30
Malament, D., CTCs 278
many-worlds theory 120, 130, 173–88
Maudlin, T., consistency 301
Maxwell law and equations, time travel 283–4, 296
McTaggart, J. M. E. 8–9
measurement paradox see quantum mechanics, measurement paradox
microcanonical ensemble, probability distribution 234, 239, 253–5
Mill, J. S., M-R-L laws 290–2, 295, 303, 309
minds and consciousness
  branched models of time flow 156–7, 187
  decoherence and 128–9
  experienced succession and entropy 9–10, 217–29
  quantum computer 138–9
  in quantum measurement paradox 120
mixing property, equilibrium 196–8, 209
Morris, M. S. 279, 306
MR (macrorealism) 15, 97, 101–6
M-R-L laws, time travel 290–2, 295, 303, 309

natural selection, entropy and 230–1, 241–52
negentropy, thermodynamic and radiation arrows and gravity 111–12
Newton, I., time and gravity 23–4

Index

no boundary condition (NBC), time asymmetry 73–8
Novikov, I. D. 285, 300–2
Now, The, movement of 7–8, 11

one-way time 3
Oszvath, I., CTCs 278, 279

parity of reasoning problem 199–203
Penrose, O., causality 201
Penrose, R. 4–6
  cosmology and time asymmetry 69, 78–81
    time reversal invariance 15–18, 46–9, 83–6, 92
    Weyl curvature 78–9, 206
perceptual time, physical time and 224–9
pilot wave theory 186–7
positron theory, closed causal chains 260, 264–7
precognition, closed causal chain 260, 263–4
Prigogine's solution 212–15
psychological time 5
  see also minds and consciousness

quantum arrow
  measurement calculus 130–3
  orthodox standpoint 110, 119, 127–8
  see also quantum mechanics, measurement paradox
quantum diffusion 149–50
quantum fractal system 150–1
quantum gravity 4, 54–61, 118–21
quantum mechanics
  branched models of spacetime 11, 155–72
  closed timelike curves and time machines 306–9, 310
  cosmology and time 4, 34–53, 110
  formalism for condition setting at arbitrary times 49, 61–5
  measurement paradox 97–106, 107–55
    branched interpretation 169–72
    decoherence and emergence of classical properties 108, 109, 139–51
    measurement calculus and quantum arrow 130–3
    quantum computer 109, 133–9, 151–4

thermodynamic and radiation arrow 5, 109–18
  wave-functions and the quantum arrow 121–30
  stochastically-branching spacetime topology 11–12, 173–88
  quantum probabilities, branched models of spacetime 159–62

radiation arrow 5, 109–18
radiation theories, symmetric time-reversing universe 85, 86(fn), 87–91
Ramsey, F. P.
  meaning of terms 223, 226, 227
  M-R-L laws 290–2, 295, 303, 309
realism, experienced succession 227–8
reductionism, succession and entropy 9, 218–29
relativity see general theory of relativity; special theory of relativity
representationalism, experienced succession 227–8
reversible measurements, quantum mechanics see quantum mechanics, measurement paradox

Schrödinger evolution 170, 171–2, 187
Schrödinger representation 37, 44
second law of thermodynamics
  Eddington's view 1, 2
  entropy
    biological populations 4–5, 230–3, 241–52
    coupling process 232–8, 241–52
    description and simple properties of 231–3, 252–3
    experienced succession and becoming 9–10, 217–29
    statistical mechanics 233–41, 253–5
  Lewis's two-way time 2–4
  statistical mechanics and time asymmetry 191–216
  see also thermodynamic arrow
self-consistency (PSC), time travel 285–6, 288, 289, 290, 300, 301–2
special theory of relativity (STR)
  becoming 10–11

328

Index

branched models of spacetime 156–7, 162–9
gravity and time 24–6
time travel, twin paradox 270–2, 281
spin systems
   quantum computer 135–8, 151–3
   time asymmetry and thermodynamic laws 203–4
   time symmetry of quantum mechanics 44–6, 49–53
   two-slit systems 145–7
spin-boson problem, decoherence 140–3
SQUID ring, quantum measurement 101–2
statistical mechanics 4–5
   biological populations 4–5, 230–3, 241–52
   time asymmetry 191–216
stochastically branching spacetime topology 11–12, 173–88
succession, becoming and entropy 9–10, 217–29
symmetry of time 4, 6
   cosmological theories
      quantum mechanics 4, 15–18, 34–53, 99–106
      statistical mechanics and thermodynamic laws 205–9
      thermodynamic arrow 66–94
   experienced succession and entropy 9–10, 217–29
   measurement calculus and quantum arrow 131–3
   statistical mechanics and thermodynamic laws 191–216
      cosmology and branch systems 205–9
      formal approaches to irreversibility 195–9
      history of problem 193–5
      interventionism 203–5, 211–12
      Krylov's solution 209–12
      'parity of reasoning' problem 199–203
      Prigogine's solution 212–15

Taub–NUT model, CTCs 278, 279, 307
Taylor, R., temporal flow 7
temporal asymmetry and symmetry *see* symmetry of time

thermodynamic arrow 5, 6
   cosmological theories 5, 66–94
      gravitational argument for symmetry (GAS) 70, 77–8, 79–80, 82, 92
      Hawking and the Big Crunch 73–8
      inflationary model 71–2
      symmetric time-reversing universe 81–91, 92
   determines radiation arrow 109, 110–18
   explanations for 6, 191–216
   quantum measurement paradox and 99–106, 120–1
   see also second law of thermodynamics
Thorne, K. S., CTCs 279
time travel 5, 12, 33, 268–310
   backward causation 270–7
   causal structure of relativistic spacetimes 274–7
   closed causal chains 259–67
   consistency constraints 269–70, 284–7, 309–10
   general theory of relativity and 288
   laws and occurrent facts 290–2
   non-self-interacting case 292–7
   particle horizons 289
   self-consistency 285–6, 288, 290
   self-interacting case 297–303
   paradoxes of 280–4, 287, 294, 299, 309
   time machines 304–9, 310
time-reversal invariance 12–19, 46–9, 67–8, 69, 70, 80
   case against 81–6
   observability and coherence of 86–91
   see also quantum mechanics, measurement paradox
Tipler, F. J., CTCs 279
topological decoherence 147
topological spacetime, stochastically branching 11–12, 173–88
twin paradox spacetime 298, 299, 300
two-state systems, decoherence 140–7
two-way time 2–4

Vessot, R. F. C. 25–6
von Neumann, J. 119–20

WAH (Hofstadter) problem 150–1

Index

Wahlund coupling 243–4, 248, 249, 250, 252
wave function 59–60
    collapse of 44, 119–20
    no boundary condition 73–8
    quantum arrow and 108, 121–30, 132–3
Wellsian time travel 269, 270–3
Weyl curvature hypothesis 78–9, 206
Wheeler, J. A. 301

Wheeler–DeWitt equation 59–60
white holes 5, 86, 117–18, 206
wormholes 279, 295, 298–9, 300–2, 303, 306

Yurtsever, U., CTCs 279, 292–3

Zemach, E. M., two times 273